Analysis and Optimal Control of Phase-Field Transition System: Fractional Steps Methods

Authored by

Costică Moroşanu

Faculty of Mathematics
"Al.I. Cuza" University of Iasi
Romania

CONTENTS

Foreword

This book is concerned with the analysis and optimal control of the phase-field transition mathematical model described by Caginalp phase-field parabolic system, which was proposed in 1986 to replace the classical two-phase Stephan problem. With respect to the Stephan free boundary model of phase transition, the Caginalp model describes better and more accurately the physical processes of melting and solidification and is now widely accepted in mathematical physics. It is also a more convenient way to represent the moving boundary of melting and solidification processes.

From mathematical point of view, the great advantage is that the new model is represented by a parabolic semilinear system with a nonlinearity of Ginzburg-Landau type, while the classical Stephan model is described by a parabolic variational inequality, which is a parabolic multivalued equation hardly to solve or approximate.

In this book the author, who has several important contributions in theory of otimal control problems of phase-field systems, has rigorously presented the existence and approximation theory of this phase-field system as well as the theory of otimal control problems governed by this system. Applications to control of solidification region in specific industrial processes and numerical tests are also included. Most of the results presented in this book are new or taken from author's previous works. It is in our opinion a good and useful work for all pure and applied mathematicians interested in phase transition mathematical models.

Viorel BARBU
University "Al. I. Cuza" and
Romanian Academy

Preface

Last years have shown a great interest in studying applied mathematics problems. One of the most important classes of such problems is the class of systems with phase change (systems with free boundary) encountered in numerous problems of physics, such as: melting ice, crystal formation, diffusion of oxygen in an absorbent tissue, solidification in continuous casting, etc.

From mathematical point of view, a *free boundary problem* can be considered as a nonlinear parabolic equation with limit values, for which is unknown the solution and its field.

In mathematics literature, for the phenomenon of solidification are known more mathematical models designed to describe the free boundary. Remember in this connection the Stefan problem as well as the phase-field transition system (Caginalp's model). Among the papers and monographs devoted to the study of free boundary problems of Stefan type, we recall those signed by V. Barbu [1], V. Barbu & N. Barron [2], J.L. Lions [3], C. Saguez [4], V. Arnăutu & P. Neittaanmäki [5] and D. Tiba [6].

The phase-field transition system, known also as a *phase-field system*, was introduced in literature by G. Caginalp [7]. This model has been established as an extension (a refinement) of the classical two phase Stefan problem to capture the effects of *surface tension*, *supercooling*, *superheating*, etc. Phase-field model (often called as *phase-field transition system*) in the form in which it was introduced by Caginalp [7], consists in the following two nonlinear differential equations of parabolic type ($Q = [0, T] \times \Omega$):

$$u_t + \frac{\ell}{2}\varphi_t = k\Delta u + f \qquad \text{in } Q, \tag{1}$$

$$\tau\varphi_t = \xi^2\Delta\varphi + \frac{1}{2a}(\varphi - \varphi^3) + 2u + g \qquad \text{in } Q, \tag{2}$$

subject to the boundary conditions

$$u = u_\partial(x), \qquad \varphi = \varphi_\partial(x) \qquad\qquad x \in \partial\Omega, \tag{3}$$

and initial conditions

$$u(0,x) = u_0(x), \qquad \varphi(0,x) = \varphi_0(x) \qquad x \in \Omega, \tag{4}$$

where u, φ are the unknown functions, $f = g = 0$, and $\Omega, T, \ell, k, \tau, \xi, a$, are described in detail in Chapter 2.

On the one hand, the content of this book is dedicated to the approximation (from theoretical and numerical point of view) of equations (1)-(2) in the presence of different types of boundary conditions and, on the other hand, analysis of some boundary optimal control problems, governed by this system.

The present work is conceived on the basis of the results and methods used by the author in notes [8]-[37] and it consists in a preface, five chapters, two annexes and bibliography.

Several types of boundary conditions were considered in the context of this work, namely ($\Sigma = [0,T] \times \partial\Omega$):

$$\frac{\partial}{\partial\nu}u + hu = w(t,x), \qquad \frac{\partial}{\partial\nu}\varphi = 0 \qquad \text{on} \qquad \Sigma, \tag{3$'$}$$

$$\frac{\partial}{\partial\nu}u + hu = w(t)g(x), \qquad \varphi = 1 \qquad \text{on} \qquad \Sigma, \tag{3$''$}$$

$$\frac{\partial}{\partial\nu}u + hu = 0, \qquad \varphi = 0 \qquad \text{on} \qquad \Sigma, \tag{3$'''$}$$

$$\frac{\partial}{\partial\nu}u = 0, \qquad \frac{\partial}{\partial\nu}\varphi = 0 \qquad \text{on} \qquad \Sigma, \tag{3^{iv}}$$

$$u = 0, \qquad \varphi = 0 \qquad \text{on} \qquad \Sigma. \tag{3^{v}}$$

In the first sections of Chapter 1 we recall notations, definitions and the main spaces of functions, some results about approximation of the nonlinear equations in Banach spaces and basic notions from numerical analysis, etc, frequently used in the next chapters. In the last two sections of this chapter we present a more detailed descriptions about the phase-field system and about the continuous-casting process of steel - the main industrial technology involved in our numerical experiments.

In Chapter 2 we will deal, on one hand, with the study of the existence, uniqueness, regularity, and estimates of the solution of phase-field transition system and, on the other hand, we will analyze the convergence of some approximating schemes of fractional steps type associated with this nonlinear system. Section 2.1 studies is the existence, uniqueness and regularity of the solution of the phase-field transition system subject to the non-homogeneous Cauchy-Neumann boundary conditions (**Theorem 2.1**). Such kind of conditions allow to the phase-field system (Caginalp's model) to be considered as a model of heat transfer from the surface of product to the environment, when we assume that the heat is extracted by convection and conduction in the continuous casting process. The purpose of Section 2.2 is to treat the phase-field system with a general nonlinearity in φ. The existence, uniqueness, regularity and estimates of the solution it is proved (**Theorem 2.3**) for this relevant case. Basic tools in this approach are the Leray-Schauder degree theory, the L_p-theory of linear parabolic equations, properties of the Nemytskij operator, and a priori estimates in $L^p(Q)$. Section 2.3 reviews a fractional steps scheme corresponding to (1), (2), (3^{v}) and (4), with $f = g = 0$. The obtained result is included in **Proposition 2.1** and the idea of the proof is inspired from the work of V. Barbu & M. Iannelli [38]. The next Section is devoted to the extension of some results known for m-accretive operators (see, e.g., Barbu [39] and Brézis [40]) to ω-m-accretive operators (**Theorem 2.4**), with a fixed real number ω. As an application of the Theorem 2.4, it is shown how the nonlinear phase-field transition system (with homogeneous Neumann-Neumann boundary conditions - (3^{iv})) can be decoupled in two simples systems. In the last Sections of Chapter 2 we consider the phase-field transition system (1)-(2) and (4) with two different boundary conditions:

- homogeneous Cauchy-Dirichlet boundary conditions - ($3^{'''}$);
- non-homogeneous Cauchy-Neumann boundary conditions - ($3'$).

The main results are included in **Theorem 2.5** and **Theorem 2.6**, and assert that (in certain assumptions on data u_0, φ_0 and w) for $\varepsilon > 0$, the solution $(u^\varepsilon, \varphi^\varepsilon)$ of approximating scheme converge to the weak solution (u^*, φ^*) of problem (1), (2), (4) and ($3'''$) or ($3'$), respectively. It is also proved that the weak solution of approximating scheme is a strong solution; we are thus in front of a constructive way to demonstrate the existence solution in phase-field transition system. The methods used in demonstration of this convergence results are those of compactness (in particular Helly-Foias theorem). The weak stability of approximating scheme corresponding to boundary conditions ($3^{'''}$), is also proved (**Corollary 2.1**).

Some types of boundary optimal control problems, governed by the nonlinear phase-field transition system, are introduced and analyzed in

Chapter 3. The aim of the Section 3.1 is to prove (for later use) a priori estimates in $L^2([0,T];H^2(\Omega))$ for unknown u, φ in phase-field system, in the presence of following boundary conditions: $\frac{\partial}{\partial\nu}u + hu = w(t,x)$, $\frac{\partial}{\partial\nu}\varphi = 0$ on Σ and, $\frac{\partial}{\partial\nu}u + hu = w(t)g(x)$, $\varphi = 1$ on Σ, where $w : \Sigma \to I\!R$ (or $w : [0,T] \to I\!R$) represents *the boundary control* (the temperature of the surrounding at $\partial\Omega$). Both cases (w depending explicitly on t and x or only on t) allow the model of phase-field to be involved in realistic numerical simulations in the metallurgic industry, and not only.

In Section 3.2 we will prove the existence of an optimal control (**Proposition 3.3**) for the problem stated there. The proof of this result is based on estimates established in the previous Section. The distributed optimal control problem governed by phase-field system has been analyzed in works done by: Z. Chen & K.-H. Hoffmann [41], M. Heinkenschloss & E.W. Sachs [42], M. Heinkenschloss & F. Troltzsch [43], as well as K.-H. Hoffmann & L. Jiang [44]. Boundary optimal control problem governed by the classical Stefan problem in two phases was studied by V. Barbu [1], A. Friedman [45], C. Saguez [4], D. Tiba [6].

We associate to the optimal control problem introduced in 3.2 an approximating optimal control problem for which we prove, first, the existence of an optimal control. Besides the existence of an optimal control, the convergence of the optimal solution of approximating problem to the optimal solution of the original problem is proved in the Section 3.3. The result is included in (**Theorem 3.1**). For the approximating problem, necessary optimality conditions (**Theorem 3.2**) are established in the next Section. Such a problem was studied, for an optimal control problem governed by nonlinear and parabolic variational inequalities by V. Barbu in [46].

In Section 3.5 is defined another type of boundary optimal control problem (an inverse problem). Necessary optimality conditions (maximum principle) for such sort of problem are given by **Theorem 3.4**. A likewise problem was studied in V. Arnăutu [47] but governed by the Stefan problem.

An optimal control problem, with the distributed control acting on a subset $\omega \subset \Omega$ and with the state constraint in time variable (P_S), is analyzed in Section 3.6. The necessary optimality conditions for (P_S), **Theorem 3.5**, were obtained by passing to the limit for $\varepsilon \to 0$ in the approximating control process considered (the adequate penalty problem P_S^ε, $\varepsilon > 0$).

The last Section of this Chapter is dedicated to a non-homogeneous boundary optimal control problem (P). The main result of this Section (**Theorem 3.6**) amounts to saying that problem (P) can be approximated for $\varepsilon \to 0$ by the sequence of problems (P^{ε}). The convergence of the approximating process leads to an idea of numerical approximation of the optimal control of problem (P), namely (see Chapter 4, algorithm `cpht-2D`, step **P2**), at every iteration *iter*, the computation of the approximate solution corresponding to the nonlinear phase-field transition system is substituted by computation of the approximate solution for an ordinary equation and a linear system. Hence a large amount of time is saved concerning computations.

Chapter 4 is oriented to the numerical analysis of the problems stated in Chapters 2 and 3.[1]

The first Section is dedicated to the approximation of solution to the phase-field system in 1D. The discrete state equations were constructed using a *First-order Implicit Backward Difference Formula* (**1-IMBDF**), (see and S.J. Ruuth [48]). Three numerical methods to compute the approximate solution (*Newton method*, the *fractional steps method* and *cubic spline method*) are developed and, corresponding, conceptual algorithms are presented (`algnewton1D`, `algfrac1D`, `al-gspline1D`). The use of the *fractional steps method* simplifies the numerical computation due to its decoupling feature (compares the algorithms `algnewton1D` and `algfrac1D`). Stability conditions (**Proposition 4.1**, **Proposition 4.2**) for fractional and cubic spline approximation are established too.

Section 4.2 deals with the approximation of solution to the phase-field system in 2D via fractional steps method. An implicit (backward) finite difference scheme in time and a **f**inite **e**lement **m**ethod (**fem**) in space are used to construct the discrete equations. An conceptual algorithm (`algfracfem2D`) have been introduced in order to compute the approximate solution.

In the next Section we will deal with the approximation of the boundary optimal control in $(\mathrm{P}^{\varepsilon}_{inv})$ stated by Theorem 3.4 in Section 3.5. The main novelty brought by the conceptual algorithm `InvPHT1D` is that the computation of the approximate solution corresponding to the nonlinear system (4.1) is replaced by the calculation of the approximate solution for an ordinary equation and a linear system (compare step **P1** in C. Moroşanu [14] with the steps **P1-P2** in the present algorithm). Numerical experiments join the abstract treatment.

---

[1] parts of this material have been partial elaborated under the support of Contract CEx 05-D11-**84**/28.10.2005, financed by Romanian Ministry of Education and Research.

Two conceptual algorithms of gradient type (`scpth-2D`, `cpth-2D`) to compute the optimal control (see Theorem 3.5) and the suboptimal control (see Theorem 3.7) of problems (P_S) and (P^{ε}), introduced in Chapter 3, are presented at the end of this Chapter. Each algorithm is implemented and accompanied by numerical experiments.

Aspects of the implementation of conceptual algorithms developed in Chapter 4 and numerical results obtained through their use, are given in the last Chapter. Thus the reader may be convinced on the accuracy of programs written and can easily understand the importance of methods developed in this material. All programs were written in MATLAB language and are endowed with sufficient comments for facilitate tracking them.

We complete the work with two annexes. The existence and regularity for the solution of one linear system (*auxiliary linear system*), having a structure similar with those in the associated approximating schemes of nonlinear phase-field transition system, it is proved in ANNEXE A (**Theorem A.1**). Concerning the methods used in the proof, an essential difference with respect to Chapter 2 consists in that here we make a priori estimates in $L^2(Q)$ instead of $L^p(Q)$. The estimation technique used is also different from those used for the rest of the book. In ANNEXE B we have presented details about the implementation of the **fem** (**f**inite **e**lement **m**ethod).

Optimal control problems treated in this work reflect the continuous casting process of steel. To emphasize more deeply the practical nature of the issue studied in this work, we will make a brief presentation of the general principle of operation for a continuous casting equipment:

> *The steel is brought over the installation in a pot from which will flow in the distributor and then in the crystallizer (cast-iron mould from copper - cooled by water circulation). On leaving the crystallizer, solidified front (which surrounds steel that is still liquid) pass in secondary cooling zones, intended to ensure the end of the solidification process.*

The leadership of a continuous casting equipment is based first on a good knowledge of the overall thermal evolution (the thickness of the front solidified at the exit of the crystallizer, the temperature distribution along the body in casting, the amount of heat extracted by each cooling zone, etc.) To develop a continuous casting equipment is motivated mainly by improving the quality of cast steel and by reducing the cost price. Development of numerical methods for calculating the effective solution of some equations that make up the theoretical model is thus essential. The problems of control study may serve to define the optimal strategy for the development of automation on this issue.

A good part of this material was exposed by the author to a group of Ph.Ds students of Huazhong Normal University in Wuhan, China, constituting at the same time the support-course for "Special chapters of informatics" thought to the IVth year students of "Al.I. Cuza" University in Iasi, Faculty of Mathematics, Applied Mathematics Department.

Finally, to Mr Acad.Prof.dr. V. Barbu I wish to express my gratitude for his encouragement and valuable suggestions, made during the drafting of this book. I am deeply grateful to Mr Prof.dr. Calin Ignat to whom I owe the first steps towards research. Also, warm thanks to members of the scientific seminar *Nonlinear analysis and optimal control* of the Faculty of Mathematics of the University "Al.I. Cuza University in Iasi for interesting discussions and collegiality.

Dr. Costică MOROŞANU,

"Al.I. Cuza" University of Iasi,

Faculty of Mathematics.

E-Mail: costica.morosanu@uaic.ro

List of Symbols

$\mathbb{N}$	the set of all positive integers (natural numbers),
Z	the set of all integers,
$\mathbb{R}$	the set of all real numbers,
$\mathbb{R}^n$	the n dimensional Euclidean space,
$\{x \in X : \ P(x)\}$	the subset of X containing all elements that have the property $P(x)$,
$[a, b]$	the closed interval $\{x \in \mathbb{R} : \ a \leq x \leq b\}$,
(a, b)	the open interval $\{x \in \mathbb{R} : \ a < x < b\}$,
$[a, b), \ (a, b]$	semiclosed intervals,
$u^+ = \max\{u, 0\}$	positive part of u,
$u^- = -\inf\{u, 0\}$	negative part of u,
supp u	the support of the function u,
$\lVert x \rVert$	$= \sqrt{x_1^2 + \cdots + x_n^2}$ - the Euclidean norm of $x = (x_1, \cdots, x_n)$,
$B_\rho(x)$	$= \{y \in \mathbb{R}^n : \lVert x - y \rVert < \rho\}$ - the open ball of radius ρ and center x,
∂u	the subdifferential of the convex function u,
∇u	$= \left(\frac{\partial}{\partial x_1} u, \cdots, \frac{\partial}{\partial x_n} u\right)$ - the gradient of the differentiable function u on $\mathbb{R}^n$,
$u', u_t, u_x, \frac{\partial}{\partial t} u, \frac{\partial}{\partial x} u$	different notations of derivatives of $u(t, x)$,
$\frac{\partial}{\partial \nu} u, \ \nu \cdot \nabla u$	directional derivative of u in the direction of ν (the outward normal derivative),
$x \cdot y = \sum_{i=1}^{n} x_i y_i$	the dot product of the vectors $x, y \in \mathbb{R}^n$,
(x, y)	the scalar product of the vectors

	$x, y \in H$ (a Hilbert space),
$\text{div } u = \sum_{i=1}^{n} \frac{\partial}{\partial x_i} u_i$	the divergence of $u = (u_1, \cdots, u_n)$,
$\Delta = \sum_{i=1}^{n} \frac{\partial^2}{\partial x_i^2}$	the Laplace operator,
$L(X, Y)$	the space of linear continuous operators between normed spaces X and Y,
$\\|x\\|_{L(X,Y)}$	the norm of a linear normed space $L(X, Y)$,
X^*	the dual of the space X,
dom(A)	the domain of the mapping A,
$R(A)$	the range of the mapping A,
A^*	the adjoint of the linear operator A,
$\Omega \subset I\!R^n$	a bounded domain in $I\!R^n$,
$\Gamma = \partial\Omega$	the boundary of Ω,
Q, Σ	$[0, T] \times \Omega$, $[0, T] \times \partial\Omega$, respectively,
$\int_{\Omega} u(x, y)\, dxdy$	integral over a domain $\Omega \subset I\!R^2$,
$\int_{0}^{T} \int_{\Omega} u(t, x)\, dtdx$	integral over Q,
$C(\Omega)$ $(C(\bar{\Omega}))$	the space of all continuous real valued functions on Ω (respectively on $\bar{\Omega}$),
$C^k(\Omega)$	the space of all continuously differentiable functions on Ω of order up to and including k,
$\mathcal{D}(\Omega), C_0^\infty(\Omega)$	the space of all indefinitely differentiable functions with compact support in Ω,
$\mathcal{D}'(\Omega)$	the space of all distributions on Ω (the dual of $C_0^\infty(\Omega)$),
$L^p(\Omega)$, $1 \leq p < \infty$	Lebesgue space of measurable functions u defined on Ω for which $\int_{\Omega} \|u\|^p dx < \infty$,
$L^\infty(\Omega)$	Lebesgue space of measurable essentially bounded functions defined on Ω,

$D_i u \overset{\text{not}}{=} \frac{\partial}{\partial x_i} u$	the first generalized derivative of u,
$D^\alpha u = D_1^{\alpha_1} \cdots D_n^{\alpha_n} u$	αth generalized derivative of u ($\alpha = (\alpha_1, \cdots, \alpha_n)$ multi-index),
$W^{k,p}(\Omega)$	the Sobolev space: $\{u \in L^p(\Omega) : D^\alpha u \in L^p(\Omega), \lvert\alpha\rvert \le k\}$ where D^α is in the sense of distributions,
$W_0^{k,p}(\Omega)$	the closure of $C_0^\infty(\Omega)$ in the $W^{k,p}(\Omega)$ norm,
$H^k(\Omega)$	the Sobolev spaces $W^{k,2}(\Omega)$,
$H_0^k(\Omega)$	the Sobolev spaces $W_0^{k,2}(\Omega)$,
$H_0^1(\Omega)$	the closure of $C_0^\infty(\Omega)$ in $H^1(\Omega)$,
$u'(t)$, $\frac{d}{dt}u(t)$	the derivative of the function $u : [0,T] \to X$,
$C^k([0,T]; X)$	the space of all continuously differentiable functions $u : [0,T] \to X$, X a Banach space, of order up to and including k,
$L^p([0,T]; X)$	the space of p-integrable functions $u : [0,T] \to X$, X a Banach space, u measurable and $\int_0^T \lVert u(t)\rVert_X^p dt < \infty$,
$\mathcal{D}'([0,T]; X)$	the space of all continuous functions from $\mathcal{D}([0,T])$ to X,
$W_p^{2k,k}(Q)$	the Sobolev space: $\{y \in L^p(Q) : \frac{\partial^r}{\partial t^r} \frac{\partial^s}{\partial x^s} y \in L^p(Q)$, for $2r+s \le 2k\}$,
$W_p^{2,1}(Q)$	$= L^p([0,T]; W^{2,p}(\Omega)) \cup W^{1,p}([0,T]; L^p(\Omega))$,
$H^{2,1}(Q)$	$= W_2^{2,1}(Q)$,
$AC([0,T]; X)$	the space of absolutely continuous functions from $[0,T]$ to X,
$BV([0,T]; X)$	the space of bounded variation functions on $[0,T]$ with values in X,
$T_K(v)$	the tangent cone to K at v,
$N_K(v)$	the normal cone to K at v,

$P_k(\Omega)$	the space of polynomials of degree at most k defined on Ω,
u^k	k-th vector in a sequence,
$\|u\|_p$	p-norm of u,
$A = (a_{ij})$	matrix with a_{ij} in i-th row and j-th column,
A^T	the transpose of A,
A^{-1}	the inverse of A,
I	identity matrix,
$\|A\|_p$	p-norm of A,
$\mathcal{T}_h$	triangulation (partition),
h	mesh parameter,
$\tilde{e}$	the reference element of $\mathcal{T}_h$,
V_n	finite element space,
φ_k	Courant basis function,
a.e.	almost everywhere,
q.e.d.	quod erat demonstrandum,
δ_{ij}	Kronecker's symbol: $\delta_{ij} = 1$ for $i = j$, $\delta_{ij} = 0$ otherwise,
$\mathcal{O}(x)$	asymptotic order symbol (big $\mathcal{O}$); (any function obeying $\|\mathcal{O}(x)\| \leq c$, $c > 0$),
$o(x)$	asymptotic order symbol (small o); (any function obeying $\lim\limits_{x\to 0} \|o(x)\|/\|x\| = 0$),
u_j^i, $u_{i,j}$	numerical approximation of $u(t,x)$ at grid point x_j and time level t_i,

Note. Other notations which are used in the text are explained when they are used.

CHAPTER 1

PRELIMINARIES

Costică MOROŞANU

Faculty of Mathematics, "Al.I. Cuza" University of Iasi, Romania

Abstract: The aim of this Chapter is to provide some basic results that we will use frequently in the chapters that follow. Most results presented here are given without proof. We refer the reader to the appropriate monographs or textbooks where details can be found.

Keywords: real functions, functions of bounded variation, integral formulas, abstract approximation theory, nonlinear functional analysis, distributions, operator theory, semigroups of nonlinear operators, optimal control, numerical analysis, finite difference methods, finite element method, nonlinear systems, fractional steps method, phase changes, mathematical modeling, applications in industry.

1.1 Fundamental inequalities

First, we recall some classical inequalities that we will use in the sequel. Let Ω be a bounded measurable subset in $\mathbb{R}^n$ $(n \geq 1)$ with sufficiently smooth boundary $\partial\Omega$.

Cauchy's inequality (with ε):

$$ab \leq \frac{\varepsilon}{2}a^2 + \frac{1}{2\varepsilon}b^2, \quad \forall\, a, b \in \mathbb{R} \text{ and } \varepsilon > 0. \tag{1.1}$$

Buniakowsky–Schwarz inequality. Let f and g be real and integrable functions on $\Omega \subset \mathbb{R}^n$, $n \geq 1$. Then

$$\int_\Omega f(x)g(x)\, dx \leq \left(\int_\Omega f(x)^2\, dx\right)^{\frac{1}{2}} \left(\int_\Omega g(x)^2\, dx\right)^{\frac{1}{2}}. \tag{1.2}$$

The Buniakowsky–Schwarz inequality is an integral analogue of the inequality (1.1).

Hölder's inequality. Let $p, q > 1$ such that $\frac{1}{p} + \frac{1}{q} = 1$. If f and g are real functions defined on Ω and if $|f|^p$ and $|g|^q$ are integrable functions on Ω, then

$$\int_\Omega f(x)g(x)\,dx \le \left(\int_\Omega |f(x)|^p\,dx\right)^{\frac{1}{p}} \left(\int_\Omega |g(x)|^q\,dx\right)^{\frac{1}{q}}. \tag{1.3}$$

Corresponding to $p = q = 2$ one obtain the inequality (1.2). More generally, we have

$$\left|\int_\Omega \prod_{i=1}^{n} f_i(x)\,dx\right| \le \prod_{i=1}^{n} \left(\int_\Omega |f_i(x)|^{p_i}\,dx\right)^{\frac{1}{p_i}} \tag{1.4}$$

where $f_i \in L^{p_i}(\Omega)$, $i = \overline{1,n}$, and $\sum_{i=1}^{n} \frac{1}{p_i} = 1$.

Young's inequality. Let f be a real-valued, continuous and strictly increasing function on $[0, c]$, with $c > 0$. If $f(0) = 0$, $x_1 \in [0, c]$ and $x_2 \in [0, f(c)]$, then

$$x_1 x_2 \le \int_0^{x_1} f(x)\,dx + \int_0^{x_2} f^{-1}(x)\,dx \tag{1.5}$$

where f^{-1} is the inverse function of f. Applying the inequality (1.5) to the function $x \mapsto f(x) = x^{p-1}$, $p > 1$ (all the conditions assumed for f are satisfied), we get

$$ab \le \frac{1}{p}a^p + \frac{1}{q}b^q \quad \text{for all } a, b \ge 0, \tag{1.6}$$

where $\frac{1}{p} + \frac{1}{q} = 1$.

Gronwall's inequality. Let $x, \varphi, \psi \in C(0,T)$ - the space of continuous functions from $[0,T]$ to $I\!R$, $\psi(t) \geq 0$, a.e.[1] $[0,T]$, where $T \leq +\infty$. If x satisfies the inequality

$$x(t) \leq \varphi(t) + \int_0^t \psi(s)x(s)\, ds \qquad t \in [0,T], \tag{1.7}$$

then the following estimate is true

$$x(t) \leq \varphi(t) + \int_0^t \varphi(s)\psi(s)\, e^{\int_s^t \psi(\tau)\, d\tau}\, ds \qquad t \in [0,T]. \tag{1.8}$$

Gronwall-Bellman's inequality. Let $m \in L^1(0,T)$ - the space of Lebesgue integrable functions from $[0,T]$ to $I\!R$, $m \geq 0$, a.e. on $[0,T]$, and x_0 be a positive constant. If $x \in C(0,T)$ satisfies the inequality

$$x(t) \leq x_0 + \int_0^t m(s)x(s)\, ds \qquad t \in [0,T], \tag{1.9}$$

then

$$x(t) \leq x_0\, e^{\int_0^t m(s)\, ds} \qquad t \in [0,T]. \tag{1.10}$$

1.2 Basic concepts of real analysis

Definition 2.1 (support). Let $f : X \to Y$ be a function from a *topological space*[2] (X, τ_X) to a *vector space*[3] Y. Then we define the *support* of f by

$$supp\ (f) = \overline{\{x \in X : \ f(x) \neq 0\}},$$

a closed subset of X.

---

[1] a.e. - almost everywhere

[2] see: **http://en.wikipedia.org/wiki/topological space**

[3] see: **http://en.wikipedia.org/wiki/Vector space**

Definition 2.2 (completeness). A sequence $\{x_n\}_{n=1}^{\infty}$ in a *metric space*[4] (X, d) is *Cauchy* provided that

$$\lim_{m,n\to\infty} d(x_n, x_m) = 0.$$

Definition 2.3. If $(X, \|\cdot\|)$ is a normed vector space, then we say $\{x_n\}_{n=1}^{\infty} \subset X$ is a *Cauchy sequence* if $\lim\limits_{m,n\to\infty} \|x_n - x_m\| = 0$.

Definition 2.4. A metric space (X, d) is *complete* if all Cauchy sequences are convergent sequences.

Definition 2.5. A *Banach space* is a complete normed vector space, i.e. if every $\{x_n\}_{n=1}^{\infty} \subset X$ which is Cauchy is convergent ($\{x_n\}_{n=1}^{\infty} \subset X$ is convergent iff there exists $x \in X$ such that $\lim\limits_{n\to\infty} \|x_n - x\| = 0$, or, every Cauchy sequence in X has a limit in X with respect to the metric $d(x, y) = \|x - y\|$).

If X and Y are two topological spaces, let

$\mathbf{C(X, Y)}$ - denote the continuous functions from $X \to Y$.

If Y is a Banach space, let

$$\mathbf{BC(X, Y)} = \{f \in C(X, Y) : \sup_{x\in X} \|f(x)\|_Y < \infty\}$$

and

$$\mathbf{C_c(X, Y)} = \{f \in C(X, Y) : \; supp\ (f) \text{ is compact}\}.$$

Definition 2.6. A *connected space* is a topological space which cannot be represented as the disjoint union of two or more nonempty subsets.

Definition 2.7. A *domain* is an open connected subset of a real vector space.

The spaces $C^k(\Omega)$. The differential operator $\dfrac{\partial}{\partial x_i}$, $1 \le i \le n$, will be written D_i and if $\alpha = (\alpha_1, \cdots, \alpha_n)$ is a multi-index, by D^α will be denoted the differential operator

$$D^\alpha = D_1^{\alpha_1} \cdots D_n^{\alpha_n} = \frac{\partial^{[\alpha]}}{\partial x_1^{\alpha_1} \cdots \partial x_n^{\alpha_n}} \tag{2.1}$$

[4] see: http://en.wikipedia.org/wiki/metric space

where $[\alpha] = \alpha_1 + \cdots + \alpha_n$. If $\alpha_i = 0$ for some i, $D_i^{\alpha_i}$ is the identity operator; if $[\alpha] = 0$ then D^α is the identity.

i. For $k \in I\!N$, we denote by the symbol $\mathbf{C^k(\Omega)}$ the set of all functions $f = f(x)$ defined on Ω, derivatives $D^\alpha f$ of which up to the order $|\alpha| \leq k$ are continuous in Ω.

ii. By the symbol

$$\mathbf{C(\bar{\Omega})} \equiv C^0(\bar{\Omega})$$

we denote the set of all functions $f = f(x)$ defined on $\bar{\Omega}$ which are continuous in $\bar{\Omega}$. Obviously, we have $C^0(\bar{\Omega}) \subset C^0(\Omega)$.

iii. By the symbol $\mathbf{C^k(\bar{\Omega})}$ we denote the set of all functions $f \in C^k(\Omega)$, whose derivatives $D^\alpha f$ have the property: for every derivative $D^\alpha f$ ($|\alpha| \leq k$) there exists a function $f_\alpha \in C^0(\bar{\Omega})$ for which

$$D^\alpha f(x) = f_\alpha(x) \quad \text{holds for} \quad x \in \Omega.$$

iv. Let $f = f(x)$ be a function defined on Ω and let S be the set of those points $x \in \Omega$ for which $f(x) \neq 0$. By the symbol $\mathbf{C_0^k(\Omega)}$, we denote the set of all function $f \in C^k(\Omega)$ for which *supp* f (see Definition 2.1) is compact and contained in Ω.

Green's formula. Let $\Omega \subset I\!R^n$ be an open, bounded subset with a C^1 boundary $\partial\Omega$.

a. If $u \in C^1(\overline{\Omega})$, $v \in C^1(\overline{\Omega})$ and $\Delta v \in C(\overline{\Omega})$ then

$$\int_\Omega u(x)\Delta v(x)\; dx = -\int_\Omega \nabla u(x)\nabla v(x)\; dx + \int_{\partial\Omega} u(x)\frac{\partial}{\partial\nu}v(x)\; dx. \quad (2.2)$$

b. If $u, v \in C^2(\Omega) \cap C^1(\overline{\Omega})$ and $\Delta u, \Delta v \in C(\overline{\Omega})$ then

$$\int_\Omega [u(x)\Delta v(x) - v(x)\Delta u(x)]\; dx = \int_{\partial\Omega} \left[u(x)\frac{\partial}{\partial\nu}v(x) - v(x)\frac{\partial}{\partial\nu}u(x)\right] dx. \quad (2.3)$$

In the above relations $\frac{\partial}{\partial\nu}\varphi$ is the directional derivative of φ in the direction of the outward pointing unit normal vector ν to the surface element dx, and

$$\frac{\partial}{\partial\nu}\varphi = \nabla\varphi \bullet \nu$$

where $\nabla\varphi = \left(\frac{\partial}{\partial x_1}\varphi, \cdots, \frac{\partial}{\partial x_n}\varphi\right)$, $\nu = (\nu_1, \cdots, \nu_n)$, and $\bullet$ represents the *inner (dot) product*[5] in $I\!R^n$ (also known as *scalar product*).

5 see: **http://en.wikipedia.org/wiki/inner product**

Bounded Variation (BV) functions. The *total variation* of a real-valued function f, defined from an interval $[a,b] \subset I\!R$ to $I\!R$ is the quantity

$$\bigvee_a^b (f) = \sup_{P \in \mathcal{P}} \sum_{i=0}^{n_P - 1} |f(x_{i+1}) - f(x_i)|,$$

where the supremum is taken over the set of all partitions:

$$\mathcal{P} = \{P = \{x_0, \cdots, x_{n_P}\} : \ P \text{ is a partition of } [a,b]\}.$$

Definition 2.8. A real-valued function $f : [a,b] \to I\!R$ is said to be of *bounded variation* ($BV\ function$) on a chosen interval $[a,b]$ if its *total variation* $V_0^T\ (f)$ is finite, i.e.

$$f \in BV([a,b]) \iff \bigvee_a^b (f) < +\infty.$$

Let

$\mathbf{BV([a,b])} = \{f : [a,b] \to I\!R : \ \bigvee_a^b (f) < \infty\}$ - the space of all real-valued functions with bounded variation on $[a,b]$.

1.3 The space of distributions

We will denote be

$\mathcal{D}(\boldsymbol{\Omega})$ - the space of *test functions* on Ω,

defined as follows:

i. a function $\varphi : \Omega \to I\!R$ is said to have compact support if there exists a compact subset ω of Ω such that $\varphi(x) = 0$ for all $x \in \Omega \backslash \omega$.
ii. the elements of $\mathcal{D}(\boldsymbol{\Omega})$ are the infinitely differentiable functions $\varphi : \Omega \to I\!R$ with compact support.

Definition 3.1. A sequence $\varphi_k \in \mathcal{D}(\Omega)$ is said to *converge* to $\varphi \in \mathcal{D}(\Omega)$ if the following two conditions hold[6]

[6] I.M. Gelfand and G.E. Shilov (1966-1968), Generalized functions, 1-5, Academic Press.

i. there is a compact set $\omega \subset \Omega$ containing the supports of all φ_k:

$$\bigcup_k \, supp\ (\varphi_k) \subset \omega;$$

ii. for each multi-index α, the sequence of partial derivatives $D^\alpha \varphi_k$ (see (2.1)) tends uniformly to $D^\alpha \varphi$.

With this definition of the limit of a sequence of elements of $\mathcal{D}(\Omega)$, $\mathcal{D}(\mathbf{\Omega})$ becomes a complete locally convex topological vector space.

Definition 3.2. A *distribution on* Ω is a linear functional $D : \mathcal{D}(\mathbf{\Omega}) \rightarrow I\!R$ such that

$$\lim_{n\to\infty} D(\varphi_n) = D(\lim_{n\to\infty} \varphi_n)$$

for any convergent sequence φ_n in $\mathcal{D}(\Omega)$.

We will denote be

$\mathcal{D}'(\mathbf{\Omega})$ - the space of all distributions on Ω.

$\mathcal{D}'(\Omega)$ is, equivalently, the continuous dual space of $\mathcal{D}(\Omega)$:

$$\mathcal{D}'(\Omega) \times \mathcal{D}(\Omega) \ni (D, \varphi) \longmapsto < D, \varphi > \in I\!R, \tag{3.1}$$

where by $< \cdot, \cdot >$ we have denoted the duality. For example, for each $x \in \Omega$ we can define the Dirac distribution $\delta_x \in \mathcal{D}'(\Omega)$ by $\varphi(x) = < \delta_x, \varphi >$, $\forall \varphi \in \mathcal{D}(\Omega)$.

Equipped with the *weak** - topology,[7] the space $\mathcal{D}'(\Omega)$ is a locally convex topological vector space.

Definition 3.3. A sequence $D_k \in \mathcal{D}'(\Omega)$ *converges* to a distribution D if and only if

$$< D_k, \varphi > \ \longrightarrow \ < D, \varphi > \qquad \text{for all test functions } \varphi.$$

Functions as distributions. The function $f : \Omega \rightarrow I\!R$ is called *locally integrable* if it is Lebesgue integrable (see Section I.4) over every compact subset ω of Ω. The *topology on* $\mathcal{D}(\Omega)$ *is defined as follows*: any locally integrable function f yields a continuous linear function on $\mathcal{D}(\Omega)$ (that is, an element of $\mathcal{D}'(\Omega)$ - denoted by D_f), whose value on the test function φ is given by the Lebesgue integral:

$$< D_f, \varphi > = \int_\Omega f\varphi \, dx.$$

[7] see: **http://en.wikipedia.org/wiki/weak topology**

The *test functions* are themselves locally integrable, and so define distributions. As such they are dense in $\mathcal{D}'(\Omega)$ with respect to the topology on $\mathcal{D}'(\Omega)$, there is a sequence $\varphi_n \in \mathcal{D}(\Omega)$ such that

$$< \varphi_n, \psi > \longrightarrow < D, \psi > \quad \text{for all } \psi \in \mathcal{D}(\Omega).$$

For any distribution $D \in \mathcal{D}'(\Omega)$ we can define $\frac{\partial}{\partial x_i} D \in \mathcal{D}'(\Omega)$ by the equality

$$\left\langle \frac{\partial}{\partial x_i} D, \varphi \right\rangle = -\left\langle D, \frac{\partial}{\partial x_i} \varphi \right\rangle. \tag{3.2}$$

When f is a differentiable function, the *distributional derivative* defined by (3.2) coincides with the usual derivative in the terms of the above identification. In fact, on the one side, by (3.2) we have

$$\left\langle \frac{\partial}{\partial x_i} D_f, \varphi \right\rangle = -\left\langle D_f, \frac{\partial}{\partial x_i} \varphi \right\rangle = -\int_\Omega f(x) \frac{\partial}{\partial x_i} \varphi(x)\, dx \quad \text{for any } \varphi \in \mathcal{D}(\Omega)$$

where $D_f \in \mathcal{D}'(\Omega)$. On the other side, by integrating by parts we get

$$\left\langle D_{\frac{\partial}{\partial x_i} f}, \varphi \right\rangle = \int_\Omega \frac{\partial}{\partial x_i} f(x) \varphi(x)\, dx = -\int_\Omega f(x) \frac{\partial}{\partial x_i} \varphi(x)\, dx, \tag{3.3}$$

since φ is vanishing on the frontier of Ω.

Definition 3.4. If $\alpha = (\alpha_1, \cdots, \alpha_n)$ is an arbitrary multi-index and ∂^α denotes the associated mixed partial derivative operator from $\mathcal{D}(\Omega)$ to $\mathcal{D}(\Omega)$, the mixed partial derivative $\partial^\alpha D$ of the distribution $D \in \mathcal{D}'(\Omega)$ (the *generalized derivative of* D) is defined by:

$$< \partial^\alpha D, \varphi >= (-1)^{[\alpha]} < D, \partial^\alpha \varphi > \quad \text{for all } \varphi \in \mathcal{D}(\Omega). \tag{3.4}$$

∂^α (see (2.1)) is a continuous operator on $\mathcal{D}'(\Omega)$.

We are now able to define the Sobolev spaces.

1.4 L^p and Sobolev spaces

The *Lebesgue measure* in mathematics represents a standard way of assigning a length, area or volume to subsets of Euclidean space; in particular, it is used to define *Lebesgue integration*. The measure of the Lebesgue measurable set A is denoted by $\mu(A)$. For example, if $A = [a,b] \times [c,d] \subset I\!R^2$ then $\mu(A) = (b-a)(d-c)$.

Consider (X, Σ, μ) a *measure space*[8] where X is the Euclidean space $I\!R^n$ or some Lebesgue measurable subset of it, Σ will be the σ-*algebra*[9] of all Lebesgue measurable subsets of X, and μ will be the Lebesgue measure. A function $f : X \to I\!K$ ($I\!R$ or $\mathbb{C}$) is *Lebesgue measurable* on X if the preimage of every closed interval is in Σ. If $\int_X |f| d\mu < \infty$ then f is called *Lebesgue integrable.*

Fatou's lemma. *If $\{f_k\}_{k \in I\!N}$ is a sequence of non-negative measurable functions, then*

$$\int \liminf_k f_k \, d\mu \leq \liminf_k \int f_k \, d\mu.$$

Dominated convergence theorem. *If $\{f_k\}_{k \in I\!N}$ is a sequence of measurable functions with pointwise limit f, and if there is a Lebesgue integrable function g such that $|f_k| \leq g$ for all k, then f is Lebesgue integrable and $\lim_k \int f_k \, d\mu = \int f \, d\mu$.*

L^p spaces. Let $1 \leq p < \infty$ and (X, Σ, μ) be a measure space, and consider the set of all measurable functions $f : X \to I\!K$ (C or $I\!R$) whose absolute value raised to the pth power has a finite Lebesgue integral, i.e.:

$$\|f\|_p \overset{\text{not}}{=} \left(\int_X |f|^p \, d\mu \right)^{\frac{1}{p}} < \infty.$$

By $\mathcal{L}^p(X, \mu)$ will be denoted the seminormed vector space of the set of pth power integrable functions together with the function $\| \cdot \|_p$. Since for any measurable function f we have that $\|f\|_p = 0$ if and only if $f = 0$ **a.e.** (**al**most **e**verywhere), the *kernel* of $\| \cdot \|_p$ does not depend on p, so

$$N \overset{\text{not}}{=} ker(\| \cdot \|_p) = \{f : \; f = 0 \;\; \mu \text{ a.e.}\}.$$

---

8 see: **http://en.wikipedia.org/wiki/measure space**

9 see: **http://en.wikipedia.org/wiki/σ-algebra**

Taking the *quotient space* of $\mathcal{L}^p(X,\mu)$ with respect to the kernel of $\|\cdot\|_p$, we get a normed vector space, denoted by $\mathbf{L^p(X},\mu)$. In the quotient space, two functions f and g are identified if $f = g$ a.e.

$\mathbf{L^0(X},\mathbf{\Sigma},\mu)$ - denote the space of (equivalent: "classes of") measurable functions on (X,Σ,μ).

Definition 4.1. We call *essential supremum* (bounded) of $f: X \to I\!\!K$ (denoted by *ess sup*):

$$ess\ sup\ f = \inf\{C \geq 0:\ \mu(\{x:\ f(x) > C\}) = 0\},$$

i.e. the smallest $C \geq 0$ for which $f(x) \leq C$ a.e. x.

By $\mathcal{L}^\infty(X,\mu)$ we will denote the set of all measurable functions from $X \to I\!\!K$ which are essentially bounded. For $f \in \mathcal{L}^\infty(X,\mu)$ denote by

$$\|f\|_\infty = ess\ sup\ |f|$$

and then

$$\mathcal{L}^\infty(X,\mu) = \{f:\ \|f\|_\infty < \infty\}.$$

As above, $\mathbf{L}^\infty(\mathbf{X},\mu)$ will be denoted the *quotient space* of $\mathcal{L}^\infty(X,\mu)$ with respect to the kernel of $\|\cdot\|_\infty$.

If the domain X has finite measure, than, for $1 \leq p < q < \infty$,

$$\|f\|_p \leq \mu(X)^{\frac{1}{p}-\frac{1}{q}}\|f\|_q \tag{4.1}$$

means the space $L^q(X,\Sigma,\mu)$ is *continuously embedded* in $L^p(X,\Sigma,\mu)$.

For $X = \Omega$ - an open set in $I\!\!R^n$, $n \geq 1$, and $\mu = dx = dx_1 \cdots dx_n$, we denote by $\mathbf{L^p}(\mathbf{\Omega})$, $1 \leq p < \infty$ (or $\mathbf{L}^\infty(\mathbf{\Omega})$) *the space of real functions defined on* Ω with the pth power absolutely integrable (or essentially bounded real functions), endweded with the norm:

$$\|f\|_{L^p(\Omega)} = \left(\int_\Omega |f(x)|^p\ dx\right)^{\frac{1}{p}}$$

or

$$\|f\|_{L^\infty(\Omega)} = ess\ sup\ \{|f(x)|,\ \ x \in \Omega\}.$$

$\mathbf{L^p}(\mathbf{\Omega})$, $1 \leq p < \infty$ and $\mathbf{L}^\infty(\mathbf{\Omega})$ are, respectively, Banach spaces with respect to the norms defined above.

Corollary 4.1. *Let $p \in (1, \infty]$ and $q = \dfrac{p}{p-1}$. Then to every bounded sequence $\{u_n\}_{n=1}^{\infty} \subset L^p(\Omega)$ there is a subsequence $\{\tilde{u}_n\}_{n=1}^{\infty}$ and an element $u \in L^p(\Omega)$ such that*

$$\lim_{n\to\infty} \langle \tilde{u}_n, g\rangle = \langle u, g\rangle \qquad \text{for all } \ g \in L^q(\Omega).$$

Definition 4.2. For any function $f \in L^p(\Omega)$, $1 \leq p \leq \infty$, we can define the distribution $D_f \in \mathcal{D}'(\Omega)$

$$f(\varphi) =< D_f, \varphi >= \int_{\Omega} f(x)\varphi(x)\, dx \quad \forall \varphi \in \mathcal{D}(\Omega), \tag{4.2}$$

where by $< \cdot, \cdot >$ we denote the duality (see (3.1)). Since $\mathcal{D}(\Omega)$ in dense in $L^p(\Omega)$, the application $f \mapsto D_f$ is injective and thus injection is continuous. Then we can identify $L^p(\Omega)$ with a subspace of $\mathcal{D}'(\Omega)$.

Sobolev spaces and inequalities. Sobolev space is a vector space of functions equipped with a norm that is a combination of L^p-norms of the function itself as well as its derivatives up to a given order; the derivatives are understood in a suitable weak sense to make the space complete (a Banach space).

Let Ω be any open set in $I\!R^n$, let k be a natural number and let $1 \leq p \leq \infty$.

The Sobolev space $\mathbf{W^{k,p}(\Omega)}$ $(k \geq 1)$ is the subspace of $L^p(\Omega)$ of functions y for which there exist $g_\alpha \in L^p(\Omega)$, $\alpha = (\alpha_1, \cdots, \alpha_n) \in I\!N^n$, with $0 \leq [\alpha] \leq k$, $[\alpha] = \alpha_1 + \cdots + \alpha_n$, such that

$$\int_{\Omega} yD^\alpha \varphi\, dx = (-1)^{[\alpha]} \int_{\Omega} g_\alpha \varphi\, dx \quad \forall \varphi \in \mathcal{D}(\Omega).$$

The function g_α will be called α–th *generalized derivative* of y (see and relation (3.4)) and will be denoted by $D^\alpha y$.

For any natural number $k \geq 1$, the Sobolev space $\mathbf{W^{k,p}(\Omega)}$ is organized as a Banach space with the norm

$$\|y\|_{W^{k,p}(\Omega)} = \left(\sum_{0\leq[\alpha]\leq k} \int_{\Omega} |D^\alpha y|^p dx \right)^{\frac{1}{p}} \qquad 1 \leq p < \infty,$$

and for $p = \infty$,

$$\|y\|_{W^{k,\infty}(\Omega)} = \max_{0 \le [\alpha] \le k} (\underset{x \in \Omega}{ess\ sup} |D^\alpha y(x)|).$$

When $p = 2$, $W^{k,2}(\Omega) \overset{\text{not}}{=} \mathbf{H^k(\Omega)}$ is a *Hilbert space* with the inner product[10]

$$\langle y, g \rangle_{H^k(\Omega)} = \sum_{[\alpha] \le k} \langle D^\alpha y, D^\alpha g \rangle .$$

The Sobolev space $\mathbf{W^{1,2}(\Omega)}$ is also denoted by $\mathbf{H^1(\Omega)}$ and the norm defined above reduces here to:

$$\|y\|_{H^1(\Omega)} = \left(\int_\Omega \left(|y|^2 + |Dy|^2 \right) dx \right)^{\frac{1}{2}} .$$

$H^1(\Omega)$ is a Hilbert space with an important subspace $\mathbf{H_0^1(\Omega)}$ - the space of functions in $H^1(\Omega)$ that vanish at the boundary $\partial\Omega$.

The **Poincaré inequality** states that there is a constant $C = C(\Omega)$ such that

$$\int_\Omega |y|^2 dx \le C^2 \int_\Omega |Dy|^2 dx \quad \forall y \in H_0^1(\Omega). \tag{4.3}$$

When Ω is bounded, the injection from $H_0^1(\Omega)$ to $L^2(\Omega)$ is compact.

Gagliardo-Nirenberg-Sobolev inequality. *If* $p \in [1, d)$ *then* $W^{1,p}(I\!R^n) \hookrightarrow L^{p^*}(I\!R^n)$

$$\|u\|_{L^{p^*}(I\!R^n)} \le \frac{1}{\sqrt{d}} \frac{p(d-1)}{d-p} \|\nabla u\|_{L^p(I\!R^n)} \quad \text{for all} \quad u \in W^{1,p}(I\!R^n) \tag{4.4}$$

where $p^* = \frac{dp}{d-p}$ *or equivalently* $\frac{1}{p^*} = \frac{1}{p} - \frac{1}{q}$.

Suppose $\Omega \subset I\!R^n$ *is bounded open set with* C^1 *boundary, then for all* $p \in [1, d)$ *and* $1 \le q \le p^*$ *there exists* $C = C(\Omega, p, q)$ *such that*

$$\|u\|_{L^q(\Omega)} \le C \|u\|_{W^{1,p}(\Omega)}. \tag{4.5}$$

[10] see: **http://en.wikipedia.org/wiki/inner product:**

Sobolev Embedding Theorem. *Suppose $\Omega = I\!R^n$ or $\Omega \subset I\!R^n$ is bounded open set with C^1 boundary, $p \in [1,\infty)$, $k, \ell \in I\!N$ with $\ell \leq k$.*

1. *If $p < \dfrac{d}{\ell}$, then $W^{k,p}(\Omega) \hookrightarrow W^{k-\ell,q}(\Omega)$ provided $q = \dfrac{dp}{d-p\ell}$, i.e. q solves $\frac{1}{q} = \frac{1}{p} - \frac{\ell}{d} > 0$ and there is a constant $C < \infty$ such that*

$$\|u\|_{W^{k-\ell,q}(\Omega)} \leq C\|u\|_{W^{k,p}(\Omega)} \quad \text{for all} \quad u \in W^{k,p}(\Omega). \tag{4.6}$$

2. *If $p > \dfrac{d}{k}$, then $W^{k,p}(\Omega) \hookrightarrow C^{k-(d/p)+}(\Omega)$ and there is a constant $C < \infty$ such that*

$$\|u\|_{C^{k-(d/p)+}(\Omega)} \leq C\|u\|_{W^{k,p}(\Omega)} \quad \text{for all} \quad u \in W^{k,p}(\Omega). \tag{4.7}$$

Rellich-Kondrachov Compactness Theorem. *Suppose $\Omega \subset I\!R^n$ is a precompact open subset with C^1 boundary, $p \in [1,d)$ and $1 \leq q \leq p^*$ then $W^{1,p}(\Omega)$ is compactly embedded in $L^q(\Omega)$.*

Corollary 4.2. *The inclusion of $W^{k,p}(\Omega)$ into $W^{k-\ell,q}(\Omega)$ is compact provided $\ell \geq 1$ and $\frac{1}{q} > \frac{1}{p} - \frac{\ell}{d} > 0$, i.e. $q < \frac{dp}{d-p\ell}$.*

Trace theorem. *There exist a constant $C > 0$ such that*

$$\|y\|_{L^p(\partial\Omega)} \leq C\|y\|_{W^{1,p}(\Omega)} \quad \forall y \in C^\infty(\bar{\Omega}), \tag{4.8}$$

i.e., the mapping $\gamma : C^\infty(\bar{\Omega}) \longrightarrow L^p(\partial\Omega)$, $\gamma y = y$ on $\partial\Omega$ is linear and continuous on $C^\infty(\bar{\Omega})$. Instead of γy we shall simply write y and this function (belonging to $L^p(\partial\Omega)$) will be called the *trace* of y on $\partial\Omega$.

If $y \in W^{k,p}(\Omega)$ one can define the trace of all derivatives up to the order $k-1$. This justifies the following definition

$$\mathbf{W_0^{k,p}}(\Omega) = \{v \in W^{k,p}(\Omega) : \ D^\alpha v = 0 \text{ a.e. on } \partial\Omega, \ \ \forall\ [\alpha] \leq k-1\}.$$

By $\mathbf{W^{k,p}}(\partial\boldsymbol{\Omega})$ we shall denote the corresponding Sobolev space on $\partial\Omega$ (see [49] for the definition of this space for noninteger k). We set $\mathbf{H^2}(\partial\boldsymbol{\Omega}) = W^{2,2}(\partial\Omega)$.

1.5 Vector-valued functions of a real variable

Let X be a real Banach space with the norm $\|\cdot\|_X$ and let $[0,T]$ be a fixed interval on the real axis.

By $\mathbf{L^p([0,T];X)}$ we will denote the space of all measurable functions $y : [0,T] \longrightarrow X$ such that

$$\|y\|_{L^p([0,T];X)} = \left(\int_0^T \|y(t)\|_X^p dt \right)^{\frac{1}{p}} < \infty$$

for $1 \leq p < \infty$, and

$$\|y\|_{L^\infty([0,T];X)} = \operatorname*{ess\ sup}_{t\in[0,T]} \|y(t)\|_X.$$

This is a Banach space.

Let $\mathcal{D}([0,T])$ be the space of all infinitely differentiable real valued functions on $[0,T]$ with compact support in $[0,T]$ (see Section I.3). We shall denote by $\mathcal{D}'(\mathbf{[0,T];X})$ - the space of all continuous functions from $\mathcal{D}([0,T])$ to X and we will called an element $y \in \mathcal{D}'([0,T];X)$ a *X–valued distribution on* $[0,T]$. If $y \in \mathcal{D}'([0,T];X)$ then his derivative $\frac{d^k}{dt^k}y \in \mathcal{D}'([0,T];X)$ is defined by the formula

$$D^k y(\varphi) = \frac{d^k}{dt^k}y = (-1)^k y \frac{d^k}{dt^k}\varphi \quad \forall \varphi \in \mathcal{D}([0,T]). \tag{5.1}$$

We shall denote by $\mathbf{W^{k,p}([0,T];X)}$ the space of all X–valued distributions $y \in \mathcal{D}'([0,T];X)$ with the derivatives (in the sense of distributions - see (5.1)) $D^j y \in L^p([0,T];X)$, for $j = \overline{0,k}$. Also, we denote

- $\mathbf{C([0,T];X)}$ – the space of continuous functions from $[0,T]$ to X;
- $\mathbf{AC([0,T];X)}$ – the space of absolutely continuous functions from $[0,T]$ to X;
- $\mathbf{W^{1,p}([0,T];X)} = \{y \in L^p([0,T];X) : \ Dy \in L^p([0,T];X)\}$.

On $Q = [0,T] \times \Omega$, $\Omega \subset I\!R^n$, $n \geq 1$, we will denote by $\mathbf{L^{r,p}(Q)}$ the Banach space consisting of all measurable functions on Q with the finite norm

$$\|y\|_{L^{r,p}(Q)} = \left(\int_0^T \left[\int_\Omega |y(t,x)|^p dx \right]^{\frac{r}{p}} \right)^{\frac{1}{r}}$$

where $r \geq 1$ and $p \geq 1$.

For $r = p$, $L^{p,p}(Q)$ will be denoted by $L^p(Q)$ and the norm $\|\cdot\|_{L^{p,p}(Q)}$ by $\|\cdot\|_{L^p(Q)}$.

For a given positive integer k and $1 \leq p \leq \infty$ we shall denote by $\mathbf{W_p^{2k,k}(Q)}$ the Sobolev space on Q:

$$\mathbf{W_p^{2k,k}(Q)} = \left\{ y \in L^p(Q) : \frac{\partial^r}{\partial t^r} \frac{\partial^s}{\partial x^s} y \in L^p(Q), \quad \text{for } 2r + s \leq 2k \right\}.$$

In particular we have

$$\mathbf{W_p^{2,1}(Q)} = \left\{ y : \; y, \frac{\partial}{\partial x_i} y, \frac{\partial^2}{\partial x_i \partial x_j} y, \frac{\partial}{\partial t} y \in L^p(Q) \right\}$$

equipped with the norm

$$\|y\|_{W_p^{2,1}(Q)} = \left(\int_Q \left[|y|^p + \sum_{i=1}^{n} \left| \frac{\partial}{\partial x_i} y \right|^p + \sum_{i,j=1}^{n} \left| \frac{\partial^2}{\partial x_i \partial x_j} y \right|^p + \left| \frac{\partial}{\partial t} y \right|^p \right] dxdt \right)^{\frac{1}{p}}.$$

$\mathbf{W_2^{1,0}(Q)}$ is the Hilbert space with scalar product

$$(u, v)_{W_2^{1,0}(Q)} = \int_Q (u \cdot v + u_{x_i} \cdot v_{x_i}) \; dtdx$$

and $\mathbf{W_2^{1,1}(Q)}$ is the Hilbert space with scalar product

$$(u, v)_{W_2^{1,1}(Q)} = \int_Q (u \cdot v + u_{x_i} \cdot v_{x_i} + u_t \cdot v_t) \; dtdx.$$

A zero over $W_2^{1,0}(Q), W_2^{1,1}(Q)$ means that only those elements of these spaces are taken which vanish on $\Sigma = [0, T] \times \partial\Omega$.

$\mathbf{C^{2,1}(Q)}$ is the set of all continuous functions in Q having continuous derivatives u_x, u_{xx}, u_t in Q.

Interpolation inequality: (see Lions [3, Lemma 5.1]) *Let B_0, B_1, B_2 be Banach spaces such that*

$$B_0 \subset B_1 \subset B_2, \qquad B_i \; \text{ reflexive for } i = 0, 1,$$

and the injection of B_0 into B_1 is compact. For every $\varepsilon > 0$ there exists $C_\varepsilon > 0$ such that

$$\|y\|_{B_1} \leq \varepsilon \|y\|_{B_0} + C_\varepsilon \|y\|_{B_2} \quad \forall y \in B_0. \tag{5.2}$$

Lions-Peetre embedding theorem: (see Lions [3, p.21]) The embedding theorem of Lions and Peetre ensures the existence of a continuous embedding $W_q^{2,1}(Q) \subset L^\mu(Q)$, where the number μ is defined as follows

$$\mu = \begin{cases} \infty, & \text{if } \dfrac{1}{q} - \dfrac{2}{N+2} < 0, \\ \text{any positive number} & \text{if } \dfrac{1}{q} - \dfrac{2}{N+2} = 0, \\ \left(\dfrac{1}{q} - \dfrac{2}{N+2}\right)^{-1} & \text{if } \dfrac{1}{q} - \dfrac{2}{N+2} > 0. \end{cases} \tag{5.3}$$

Basic result on the solvability of the boundary value problem with directional derivative in classes $W_p^{2,1}(Q)$. We denote by $L(t, x, \frac{\partial}{\partial t}, \frac{\partial}{\partial x})$ the linear parabolic differential operator with real coefficients (see [49])

$$L(t, x, \frac{\partial}{\partial t}, \frac{\partial}{\partial x}) = \frac{\partial}{\partial t} u - \sum_{i,j=1}^{n} a_{i,j}(t, x) \frac{\partial^2}{\partial x_i \partial x_j} u$$

$$+ \sum_{i=1}^{n} a_i(t, x) \frac{\partial}{\partial x_i} u + a(t, x) u.$$

We assume that the coefficients of the operator $L(t, x, \frac{\partial}{\partial t}, \frac{\partial}{\partial x})$ are defined in $(0, T) \times I\!R^n$.

In the cylindrical domain Q with lateral surface Σ we will consider the problem with directional derivative:

$$\begin{cases} L(t, x, \frac{\partial}{\partial t}, \frac{\partial}{\partial x}) u(t, x) = f(t, x) & \text{in } Q, \\ B(t, x, \frac{\partial}{\partial x}) u = \Phi(t, x) & \text{on } \Sigma, \\ u(0, x) = u_0(x) & x \in \Omega, \end{cases} \tag{5.4}$$

where $B(t, x, \frac{\partial}{\partial x}) u = \sum_{i=1}^{n} b_i(t, x) \frac{\partial}{\partial x_i} u + b(t, x) u$ on Σ.

We assume that the functions $b_i(t,x)$ satisfy everywhere on Σ the condition

$$\left|\sum_{i=1}^{n} b_i(t,x)\nu_i(x)\right| \geq \delta > 0,$$

or, equivalently:

$$b \cdot \nu \geq \delta > 0$$

where $b = (b_1, \cdots, b_n)$, $\nu = (\nu_1, \cdots, \nu_n)$ and it means that the vector b does not at any point lie in the tangent plane to Σ. If we take $b = \nu$, the boundary conditions of (5.4) can be written in the form

$$\sum_{i=1}^{n} \nu_i \cdot \frac{\partial}{\partial x_i} u + b(t,x)u = \frac{\partial}{\partial \nu} u + b(t,x)u \qquad \text{on } \Sigma.$$

We will consider the problem (5.4) in the classes $W_p^{2k,k}(Q)$ the elements of which tend to zero in a certain sense for $|x| \to \infty$. We will also assume that the functions f, u_0, Φ in (5.4) satisfy the compatibility conditions[11] for $x \in \partial\Omega$, $t = 0$.

We introduce the norm

$$\|y\|_{L^r(Q)}^{(loc)} = \sup_q \|y\|_{L^r(Q)}$$

where the supremum is taken over the whole cylinder $q = (0,T) \times \omega$, the bases ω of which are the intersection of a domain Ω with same domain of unit measure (a cube, for example).

Remark 5.1. If the domain Ω is bounded, then the set of functions with finite norm $\|y\|_{L^r(Q)}^{(loc)}$ coincides with $L^r(Q)$.

Theorem 5.1. *Let $p > 1$. Suppose that the coefficients a_{ij} of the operator L are bounded continuous functions in Q, while the coefficients a_i and a have finite norms $\|a_i\|_{L^r(Q)}^{(loc)}$ and $\|a\|_{L^s(Q)}^{(loc)}$ respectively, with*

$$r = \begin{cases} \max\{p, n+2\} & \text{for } p \neq n+2, \\ n+2+\varepsilon & \text{for } p = n+2, \end{cases}$$

[11] we will say that the compatibility conditions of order $k \geq 0$ are fulfilled for problem (5.4) if $u^{(j)}(x) = \frac{\partial^j}{\partial t^j} u(t,x)|_{t=0, x\in\partial\Omega} = \frac{\partial^j}{\partial t^j}\Phi|_{t=0} = \Phi^{(j)}(x)$, $j = 0, \cdots, k$.

$$s = \begin{cases} \max\{p, \frac{n+2}{2}\} & \textit{for } p \neq \frac{n+2}{2}, \\ \\ \frac{n+2}{2} + \varepsilon & \textit{for } p = \frac{n+2}{2} \end{cases}$$

and ε being an arbitrarily small positive number. Suppose, further, that the quantities $\|a_i\|^{(loc)}_{L^r(Q_{t,t+\tau})}$ and $\|a\|^{(loc)}_{L^s(Q_{t,t+\tau})}$ tend to zero for $\tau \to 0$. Let $\partial\Omega \in \mathcal{C}^2$. Then, for any $f \in L^p(Q)$, $u_0 \in W_p^{2-\frac{2}{p}}(\Omega)$ and $\Phi \in W_p^{2-\frac{1}{p},1-\frac{1}{2p}}(\Sigma)$, with $p \neq 3$, satisfying in the case $p > 3$ the compatibility condition of zero order

$$u_0|_{\partial\Omega} = \Phi|_{t=0},$$

problem (5.4) has a unique solution $u \in W_p^{2,1}(Q)$. It satisfies the estimate

$$\|u\|_{W_p^{2,1}(Q)} \leq C\left(\|f\|_{L^p(Q)} + \|u_0\|_{W_p^{2-\frac{2}{p}}(\Omega)} + \|\Phi\|_{W_p^{2-\frac{1}{p},1-\frac{1}{2p}}(\Sigma)}\right). \quad (5.5)$$

1.6 Convex sets and functions

Let X be a real Banach space with the norm $\|\cdot\|_X$ and let X^* be denoted his dual space. $K \subseteq X$ is a *convex set* if $\lambda x_1 + (1-\lambda)x_2 \in K$ for any $x_1, x_2 \in K$ and any $\lambda \in [0,1]$.

Let K be a convex set from X. $y : K \to I\!R$ is a *convex function* if $y(\lambda x_1 + (1-\lambda)x_2) \leq \lambda y(x_1) + (1-\lambda)y(x_2)$, for any $x_1, x_2 \in K$ and any $\lambda \in [0,1]$.

The convex function $y : K \to I\!R$ is *proper* if $y(x) > -\infty$ for any $x \in K$ and y is not identically equal to $+\infty$.

The function $y : X \to I\!R$ is *lower semicontinuous* (in short, **l.s.c**) at x if for any sequence $\{x_n\}$ which is convergent to x, we have

$$y(x) \leq \liminf_{n\to\infty} y(x_n). \quad (6.1)$$

A function which is lower semicontinuous at each point of X is called *lower semicontinuous on X*.

Given a function y from a Banach space X to $I\!R$, the mapping $y' : X \times X \to I\!R$ defined by

$$y'(x,z) = \lim_{\lambda \downarrow 0} \frac{y(x+\lambda z) - y(x)}{\lambda}, \qquad x, z \in X,$$

(if the limit exists) is called the *directional derivative* of y at x in direction z.

Definition 6.1. The *directional derivative* for a scalar function $y(\vec{x}) = y(x_1, x_2, \cdots, x_n)$ along a vector $\vec{v} = (v_1, v_2, \cdots, v_n)$ is the function $\nabla_{\vec{v}}$ (or $D_{\vec{v}}$) defined by the limit

$$D_{\vec{v}}\, y(\vec{x}) = \lim_{h \to 0} \frac{y(\vec{x} + h\vec{v}) - y(\vec{x})}{h}.$$

If y is differentiable at $\vec{x}$, then $D_{\vec{v}}\, y(\vec{x})$ exists along vector $\vec{v}$, and one has

$$D_{\vec{v}}\, y(\vec{x}) = \nabla y(\vec{x}) \bullet \vec{v}$$

where ∇ denotes the gradient and $\bullet$ is the Euclidean inner (dot) product.

Definition 6.2. A *normal derivative* is a directional derivative taken in the direction normal (that is, orthogonal) to some surface in space. If the normal direction is denoted by $\vec{n}$, then $D_{\vec{n}}$ is sometimes denoted as $\frac{\partial}{\partial n}$, or $\frac{\partial}{\partial \nu}$.

Given a l.s.c, convex, proper function $y : X \to \bar{I\!R} = [-\infty, +\infty]$, the mapping $\partial y : X \to X^*$ defined by

$$\partial y(x) = \{x^* \in X^*;\ \ y(x) - y(z) \le (x - z, x^*),\ \ \forall z \in X\}$$

is called the *subdifferential* of y.

Let K be a closed convex subset of X. The function $I_K : X \to \bar{I\!R}$ defined by

$$I_K(x) = \begin{cases} 0 & \text{if } x \in K, \\ +\infty & \text{if } x \notin K, \end{cases}$$

is called the *indicator function* of K and

$$\partial I_K(x) = N_K(x) = \{x^* \in X^*,\ \ (x - u, x^*) \ge 0,\ \ \forall u \in K\}$$

where $N_K(x)$ is the *normal cone* at K in x.

Remark 6.1. The minimizing problem for a function $y(x)$ on a subset $K \subset X$ can be reformulated as a problem of minimization for the function $y(x) + I_K(x)$ over the space X where no constraints explicitly appear (for detailes, see Barbu & Precupanu [50], pp.184).

Projected gradient methods. Let K be a nonempty convex and closed subset of X and consider the problem of minimizing the function $y : X \to I\!R$ over K, i.e.,

$$(P) \qquad \begin{cases} \text{Find } u \in K \text{ such that} \\ y(u) \leq y(v) \text{ for any } v \in K. \end{cases}$$

The existence result for (P) is given by

Theorem 6.1. *If y is convex, proper and l.s.c and if K is convex, closed and bounded, then* (P) *has a solution. Moreover, if y is strictly convex then the solution of problem* (P) *is unique.*

The *projection operator* $P_K : X \to K$ is defined as follows

$$\|P_K v - v\|_X \leq \|w - v\|_X \quad \text{for any } w \in K. \tag{6.2}$$

A class of iterative minimization methods is given by

$$u^{(k+1)} = u^{(k)} + \rho_k w^{(k)} \tag{6.3}$$

where $w^{(k)} \in X$ is a search direction and $\rho \in I\!R$ is an advancing step. One element $w \in X$ is called *descent direction* for y at $u \in X$ if there exists $\bar{\rho} > 0$ such that

$$y(u + \rho w) < y(u), \quad \text{for any } \rho \in (0, \bar{\rho}].$$

If $w \in X$ satisfies

$$(\nabla y(u), w) < 0,$$

then w is a *descent direction* for y at x.

Generally, the projected gradient methods consists in: *given a descent direction $w^{(k)}$, we search the step ρ_k such that the cost value for the projected vector $u^{(k)} + \rho w^{(k)}$ be minimal, i.e.,*

$$y(P_K(u^{(k)} + \rho_k w^{(k)})) \leq y(P_K(u^{(k)} + \rho w^{(k)})), \quad \textit{for any } \rho > 0.$$

A projection method is described by

Rosen's Algorithm

P0: Choose $u_0 \in K$ and $\rho > 0$;
Set k:=0;

P1: $\bar{u}_k := P_K(u_k - \rho \nabla y(u_k))$;
$v_k := \bar{u}_k - u_k$;
P2: Compute $\lambda_k \in (0, 1]$ such that
$y(u_k + \lambda_k v_k)) = \min\{y(u_k + \lambda v_k), \quad 0 < \lambda \leq 1\}$;
P3: $u_{k+1} := u_k + \lambda_k v_k$;
P4: the stopping criterion
IF $\|u_{k+1} - u_k\| < \varepsilon$
THEN STOP (u_{k+1} is the solution)
ELSE $k := k + 1$; go to P1.

1.7 m-ω-Accretive operators in Hilbert spaces

Let H be a real Hilbert space with the scalar product $(\cdot, \cdot)$ and the induced norm $\|\cdot\|$.

A multivalued operator $A : D(A) \subset H \to 2^H$ is regarded as a subset of $H \times H$ which can be identified with its graph, where $D(A) = \{x \in H : \ Ax \neq \emptyset\}$ (the domain of A) and $Ax = \{y \in H : \ [x, y] \in A\}$, $\forall x \in H$. The elements of A will be written as $[x, y] \in A$ where $x \in D(A)$ and $y \in Ax$. For a given $A : D(A) \subset H \to 2^H$, we define the range and the inverse operator of A by

$$\begin{cases} R(A) = \bigcup\limits_{x \in D(A)} Ax, \\ A^{-1} = \{[y, x] : \ [x, y] \in A\}. \end{cases} \tag{7.1}$$

Definition 7.1.

a. A subset $A \subset H \times H$ is said to be *accretive* if for all $[x_1, y_1]$, $[x_2, y_2] \in A$, one has

$$(x_1 - x_2, y_1 - y_2) \geq 0. \tag{7.2}$$

b. A is said to be *m-accretive* (or *maximal monotone*) if A is accretive and $R(I + A) = H$.

c. A is said to be *ω-accretive* for some $\omega \in \mathbb{R}$, if $A + \omega I$ is accretive, i.e.,

$$(x_1 - x_2, y_1 - y_2) \geq -\omega \|x_1 - x_2\|^2, \qquad \forall y_i \in Ax_i, \ i = 1, 2. \tag{7.3}$$

d. A is said to be *m-ω-accretive* for some $\omega \in \mathbb{R}$, if $A + \omega I$ is m–accretive.

For a later use, we give a characterization of ω-accretiveness.

Proposition 7.1. *The operator* $A : D(A) \subset H \longrightarrow H$ (single-valued) *is* ω*-accretive for an* $\omega \in I\!R$, *if and only if* A *is in the class* $\mathcal{A}(\omega)$ (see Brézis & Pazy [51, p. 63]), i.e.,

$$(1-\lambda\omega)\|x_1-x_2\| \leq \|x_1+\lambda y_1-(x_2+\lambda y_2)\|, \tag{7.4}$$

$$\forall\,\lambda \geq 0,\ [x_i, y_i] \in A,\ i=1,2.$$

For the proof see Motreanu & Pavel [52, Proposition 3.2, p. 352].

Definitions of J_λ and A_λ. Proposition 7.1 allows to define the resolvent J_λ and the Yosida approximation A_λ associated to an ω-accretive operator A.

Let $u \in R(I+\lambda A)$ with $0<\lambda<\omega^{-1}$. It follows from (7.4) that there exist a unique $x_u \in D(A)$ and a unique $y_u \in Ax_u$ such that $u = x_u + \lambda y_u$. To verify the uniqueness of x_u and y_u we suppose that we have also $u = \tilde{x} + \lambda\tilde{y}$ for some $[\tilde{x}, \tilde{y}] \in A$. Then, by (7.4) it results $(1-\lambda\omega)\|x_u - \tilde{x}_u\| \leq 0$, so, $x_u = \tilde{x}_u$. Consequently, $y_u = \tilde{y}_u$.

Let us denote $J_\lambda u := x_u = (I+\lambda A)^{-1}u$ and $A_\lambda u := y_u = \dfrac{1}{\lambda}(u - x_u) = \dfrac{1}{\lambda}(I-J_\lambda)u$. Then, for each $0<\lambda<\omega^{-1}$, we have the operators

$$J_\lambda : R(I+\lambda A) \to D(A) \subset H,\quad J_\lambda(x+\lambda y) = x, \tag{7.5}$$

$$A_\lambda : R(I+\lambda A) \to R(A) \subset H,\quad A_\lambda(x+\lambda y) = y. \tag{7.6}$$

In addition, the equality below holds

$$u = J_\lambda u + \lambda A_\lambda u,\quad \forall\, u \in R(I+\lambda A). \tag{7.7}$$

The next property of an m-ω-accretive operator is useful in the sequel.

Lemma 7.1. *Let* A *be an* m*-*ω*-accretive operator with* $\omega > 0$. *If* $0<\lambda<\omega^{-1}$ *then* $R(I+\lambda A) = H$.

Proof. By Definition 7.1 b-d, it follows $R(I+\mu(A+\omega I)) = H$, $\forall \mu > 0$. Then, with a fixed $\mu > 0$, for each $z \in H$ one finds uniquely $x \in D(A)$ and $y \in Ax$ such that $(1+\mu\omega)x + \mu y = (1+\mu\omega)z$, i.e.,

$$x + \frac{\mu}{1+\mu\omega}y = z. \tag{7.8}$$

Denote $\lambda = \frac{\mu}{1+\mu\omega}$, for any $\mu > 0$. It is clear that λ is an arbitrary in $(0, \frac{1}{\omega})$. Then from (7.8) we derive that the assertion in Lemma 7.1 is true. **q.e.d.**

Thus for an m–ω–accretive operator, by Lemma 7.1, we have

$$J_\lambda : H \to H, \quad A_\lambda : H \to H \quad \forall\ 0 < \lambda < \omega^{-1}. \tag{7.9}$$

We give now the main properties of the operators J_λ and A_λ.

Lemma 7.2. *Let A be a m–ω–accretive operator with $\omega > 0$. Then, for each $0 < \lambda < \omega^{-1}$, we have:*
a. $\|J_\lambda u - J_\lambda v\| \le \frac{1}{1-\lambda\omega}\|u - v\|, \quad \forall\, u, v \in H$;
b. $\|J_\lambda u - u\| \le \frac{\lambda}{1-\lambda\omega}|Au|$, where $|Au| = \inf\{\|y\| : y \in Ax\}$, $\forall\, u \in D(A)$;
c. $(1 - \lambda\omega)\|A_\lambda u\| \le (1 - \mu\omega)\|A_\mu u\|$, for all $0 < \mu \le \lambda$ and $u \in H$;
d. $\|A_\lambda u - A_\lambda v\| \le \frac{1}{\lambda}\left(1 + \frac{1}{1-\lambda\omega}\right)\|u - v\|, \quad \forall u, v \in H$;
e. $A_\lambda u \in AJ_\lambda u$ and $\|A_\lambda u\| \le \dfrac{1}{1 - \lambda\omega}|Au|$, $\forall\, u \in H$;
f. $(A_\lambda u - A_\lambda v, u - v) \ge -\frac{\omega}{1-\lambda\omega}\|u - v\|^2, \quad \forall u, v \in H$.

Proof. a. For $u, v \in H$, let $[x_i, y_i] \in A$, $i = 1, 2$ be such that $u = x_1 + \lambda y_1$ and $v = x_2 + \lambda y_2$. Then, by (7.5) we have $J_\lambda u = x_1$, $J_\lambda v = x_2$ and so a. is a consequence of (7.4).

b. To prove b. let us consider $u \in D(A)$. Then $u = J_\lambda(u + \lambda y)$, $\forall y \in Au$, and making use of a., we get

$$\|J_\lambda u - u\| = \|J_\lambda u - J_\lambda(u + \lambda y)\| \le \frac{\lambda}{1 - \lambda\omega}\|y\|,$$

which implies b.

c. In order to prove c., we firstly establish following preliminary property

$$J_\mu\left(\frac{\mu}{\lambda}u + \frac{\lambda - \mu}{\lambda}J_\lambda u\right) = J_\lambda u, \quad \forall u \in H, \quad 0 < \mu \le \lambda < \omega^{-1}. \tag{7.10}$$

Indeed, for every $u \in H$, by (7.5) - (7.7) we can write

$$\frac{\mu}{\lambda}u + \frac{\lambda - \mu}{\lambda}J_\lambda u = J_\lambda u + \mu\frac{1}{\lambda(u - J_\lambda u)} = J_\lambda u + \mu AJ_\lambda u.$$

Using again (7.5) we get (7.10).

For an arbitrary $u \in H = R(I+\lambda A)$, by (7.7), (7.10) and assertion a. of Lemma 7.2, we obtain

$$\|A_\lambda u\| = \lambda^{-1}\|u - J_\lambda u\| \le \lambda^{-1}\|J_\mu u - u\| + \lambda^{-1}\|J_\mu u - J_\lambda u\|$$

$$= \mu\lambda^{-1}\|A_\mu u\| + \lambda^{-1}\|J_\mu u - J_\mu\Big(\frac{\mu}{\lambda}u + \frac{\lambda-\mu}{\lambda}J_\lambda u\Big)\|$$

$$\le \mu\lambda^{-1}\|A_\mu u\| + \lambda^{-1}\frac{1}{1-\mu\omega}\|u - \frac{\mu}{\lambda}u + \frac{\lambda-\mu}{\lambda}J_\lambda u\|$$

$$= \mu\lambda^{-1}\|A_\mu u\| + \lambda^{-2}\frac{1}{1-\mu\omega}(\lambda-\mu)\|u - J_\lambda u\|$$

$$= \mu\lambda^{-1}\|A_\mu u\| + \lambda^{-1}\frac{1}{1-\mu\omega}(\lambda-\mu)\|A_\lambda u\|.$$

We infer that property c. is true.

d. This follows from (7.7) and assertion a. of Lemma 7.2.

e. This property follows from (7.5) and (7.6).

f. By (7.7) and assertion a. it results

$$(A_\lambda u - A_\lambda v, u-v) = \lambda^{-1}[\|u-v\|^2 - (J_\lambda u - J_\lambda v, u-v)] = -\tfrac{\omega}{1-\lambda\omega}\|u-v\|^2.$$

The proof of Lemma 7.2 is complete. **q.e.d.**

The next result provides important properties of m–ω–accretive operators.

Proposition 7.2. *Let A be a m-ω-accretive operator with some $\omega > 0$. Then one has:*

i. A is demiclosed. In particular, A is closed.

ii. For every $x \in \overline{D(A)}$ one has $J_\lambda x \to x$ in H as $\lambda \downarrow 0$.

iii. Ax is a closed and convex subset of H, $\forall x \in H$.

iv. Given $\{x_\lambda\} \subset D(A)$ such that $x_\lambda \to x$ as $\lambda \downarrow 0$ and $\{A_\lambda x_\lambda\}$ is bounded, then $x \in D(A)$ and any weak limit point y of $\{A_\lambda x_\lambda\}$ belongs to Ax.

v. $\|A_\lambda x\| \to |Ax|$, as $\lambda \downarrow 0$ $\forall x \in D(A)$.

vi. for each $x \in D(A)$ one has $A_\lambda x \to A_0 x$ in H as $\lambda \downarrow 0$ where $A_0 x \in Ax$ is the unique element (called the minimal section) satisfying $\|A_0\| = |Ax|$.

Proof. *i.* Let $[x_n, y_n] \in A$ with $x_n \to x$ (strongly) and $y_n \rightharpoonup y$ (weakly) as $n \to \infty$. Since A is ω–accretive we have

$$(y_n - \tilde{y}, x_n - \tilde{x}) \geq -\omega \|x_n - \tilde{x}\|^2 \quad \forall [\tilde{x}, \tilde{y}] \in A.$$

Letting $n \to \infty$ in the inequality above, we get

$$(y - \tilde{y}, x - \tilde{x}) \geq -\omega \|x - \tilde{x}\|^2 \quad \forall [\tilde{x}, \tilde{y}] \in A.$$

The assumption on A to be m–ω–accretive ensures $[x, y] \in A$.

ii. Given $x \in \overline{D(A)}$, for every $\varepsilon > 0$ there exists $x_\varepsilon \in D(A)$ with $\|x - x_\varepsilon\| < \varepsilon$. The properties a. and b. in Lemma 7.2 imply

$$\|J_\lambda x - x\| \leq \|J_\lambda x - J_\lambda x_\varepsilon\| + \|J_\lambda x_\varepsilon - x_\varepsilon\| + \|x_\varepsilon - x\|$$

$$\leq \left(\tfrac{1}{1-\lambda\omega} + 1\right) \varepsilon + \tfrac{\lambda}{1-\lambda\omega} |Ax_\varepsilon|.$$

Since $\varepsilon > 0$ is fixed, we may chose $\delta = \delta(\varepsilon) > 0$ such that the right-hand side in the inequality above is less than $C\varepsilon$, with a constant $C > 0$ independent of ε, whenever $0 < \lambda < \delta$. So, assertion *ii.* is proved.

iii. In view of *i.* the set Ax is closed in H for each $x \in H$. In order to show the convexity of Ax, let $y_1, y_2 \in Ax$ and $t \in [0, 1]$. Using the property of A to be ω–accretive, it results

$$(\tilde{y} - (ty_1 + (1-t)y_2), \tilde{x} - x) = t(\tilde{y} - y_1, \tilde{x} - x) + (1-t)(\tilde{y} - y_2, \tilde{x} - x)$$

$$\geq -t\omega \|\tilde{x} - x\|^2 - (1-t)\omega \|\tilde{x} - x\|^2 = -\omega \|\tilde{x} - x\|^2 \quad \forall [\tilde{x}, \tilde{y}] \in A.$$

This implies $ty_1 + (1-t)y_2 \in Ax$ because A is m–ω–accretive.

iv. By Lemma 7.2 a., Proposition 7.2 *ii.* and the fact that $x_\lambda \to x$ in H, we deduce that $J_\lambda x_\lambda \to x$ as $\lambda \downarrow 0$. This property, combined with the assertion $A_\lambda x_\lambda \rightharpoonup y$ in H together with $A_\lambda x_\lambda \in AJ_\lambda x_\lambda$ (see Lemma 7.2 e.) and using Proposition 7.2 *i.*, leads to the desired conclusion.

v. In view of Lemma 7.2, properties c. and e., we see that

$$|Ax| \geq (1-\lambda\omega)\|A_\lambda x\| \geq (1-\mu\omega)\|A_\mu x\|, \;\; 0 < \lambda \leq \mu \;\; x \in D(A). \quad (7.11)$$

Hence the limit

$$\lim_{\lambda \downarrow 0} ((1 - \lambda\omega)\|A_\lambda x\|) = t \quad (7.12)$$

exists and

$$t \leq |Ax|. \tag{7.13}$$

So, by (7.12) we can conclude that

$$\lim_{\lambda \downarrow 0} \|A_\lambda x\| = t. \tag{7.14}$$

Along a subsequence we then have $A_\lambda x \rightharpoonup y$ for some $y \in H$, as $\lambda \downarrow 0$ (since H is reflexive). It is known that $A_\lambda x \in AJ_\lambda x$ and $J_\lambda x \to x$ for $x \in D(A)$ (see Lemma 7.2 e. and Proposition 7.2 *ii.*). Using Proposition 7.2 *i.*, we derive that $y \in Ax$, consequently

$$|Ax| \leq \|y\|. \tag{7.15}$$

If $y = 0$, from (7.15) we have that $|Ax| = 0$. From (7.11) we get that $A_\lambda x = 0$ (since $1 - \lambda\omega > 0$). So in this case *v.* is trivially verified. We assume now that $y \neq 0$. Since $A_\lambda x \rightharpoonup y$ as $\lambda \downarrow 0$, by (7.14) we obtain that

$$\|y\|^2 = \lim_{\lambda \downarrow 0} (A_\lambda x, y) \leq \lim_{\lambda \downarrow 0} \|A_\lambda x\| \|y\| = t\|y\|,$$

or equivalently,

$$\|y\| \leq t. \tag{7.16}$$

Combining (7.13)-(7.16) we arrive at the conclusion in *v.*

vi. Since for any $x \in D(A)$, Ax is convex and closed (see Proposition 7.2 *iii.*), there exists a unique element $A_0 \in Ax$ with $\|A_0 x\| = |Ax|$. Let $y \in H$ by any point fulfilling $A_\lambda x \rightharpoonup y$ in H as $\lambda \downarrow 0$ for some subsequence of $\{A_\lambda x\}$ (such an y exists because the set $\{A_\lambda x\}$ is bounded according to Proposition 7.2 *v.*). Using Proposition 7.2 *v.* it follows that

$$\|y\| \leq \liminf_{\lambda \downarrow 0} \|A_\lambda x\| = |Ax|$$

which implies $y = A_0 x$.

Therefore we established that $A_\lambda x \rightharpoonup A_0 x$ in H as $\lambda \downarrow 0$. In view of Proposition 7.2 *v.* we achieve the conclusion. **q.e.d.**

Proposition 7.3. *Let $\omega > 0$ and let A be a m–ω–accretive operator. Let C be a closed convex subset of H such that $C \subset \overline{D(A)}$. Then, for every $0 < \lambda < \omega^{-1}$ the following properties are equivalent:*

j. $J_\lambda(C) \subset C$;

jj. $(A_\lambda x, y) \geq 0, \ \forall\, [x, y] \in \partial I_C$;

jjj. $(A_\lambda x, (\partial I_C)_\mu(x)) \geq -\frac{\omega}{\mu(1-\lambda\omega)}\|x - P_C x\|^2$, $\forall\, x \in H$, $\forall\, \mu > 0$;

jv. $(A_0 x, (\partial I_C)_\mu(x)) \geq -\frac{\omega}{\mu}\|x - P_C x\|^2$, $\forall\, x \in D(A)$, $\forall\, \mu > 0$;

v. $(I_C)_\mu(S_A(t)x) \leq (I_C)_\mu(x) + \frac{\omega}{\mu}\int_0^t \|S_A(s)x - P_C S_A(s)x\|^2 ds$, $\forall\, x \in \overline{D(A)}$, $\forall\, \mu > 0$, $\forall\, t > 0$;

vj. $\|S_A(t)x - P_C S_A(t)x\|^2 \leq \|x - P_C x\|^2 e^{2\omega t}$, $\forall\, x \in \overline{D(A)}$, $\forall\, t > 0$.

Here $S_A(t)$, $t \geq 0$ stands for the semigroup generated on H by $-A$.

Proof. *j.⇒jj.* Let $y \in \partial I_C(x)$, i.e., $x \in C$ and $(y, z - x) \leq 0$, $\forall\, z \in C$. Then, by hypothesis *j.*, $J_\lambda x \in C$. Thus, taking into account (7.7), we have

$$(A_\lambda x, y) = -\frac{1}{\lambda}(y, J_\lambda x - x) \geq 0$$

which implies *jj.*

jj.⇒jjj. It is known that $(\partial I_C)_\mu(x) = \frac{1}{\mu}(x - P_C x)$ (see Brézis [53, p. 46]), $\forall\, \mu > 0$, where $P_C x$ denotes the projection of $x \in H$ on the set C in H. If $x \in C$, it is seen that the implication *jj.⇒jjj.* is trivially verified. We assume now $x \notin C$. Using Lemma 7.2 f., we can write

$$(A_\lambda x, (\partial I_C)_\mu(x)) = \frac{1}{\mu}(A_\lambda x, x - P_C x)$$

$$= \frac{1}{\mu}\left[(A_\lambda x - A_\lambda P_C x, x - P_C x) + (A_\lambda P_C x, x - P_C x)\right]$$

$$\geq \frac{1}{\mu}\left[-\frac{\omega}{1-\lambda\omega}\|x - P_C x\|^2 + (A_\lambda P_C x, x - P_C x)\right].$$

The characterization of the projection $P_C x$ means $(x - P_C x, z - P_C x) \leq 0$ which is equivalent to $[P_C x, x - P_C x] \in \partial I_C$. Then we may apply Proposition 7.3 *jj.* with $P_C x$ in place of x and $y = x - P_C x$. Thus *jjj.* follows.

jjj.⇒jv. Making use of Proposition 7.2 *vi.* we observe that the statement *jv.* is a direct consequence of *jjj.* letting $\lambda \downarrow 0$.

jv.⇒v. Let us firstly show that it is sufficient to verify (v) for $x \in D(A)$. For this, admitting that the result is true for $D(A)$, we consider $x \in \overline{D(A)}$. Let $\{x_n\} \subset D(A)$ with $x_n \longrightarrow x$ in H as $n \to \infty$. Then one has

$$(I_C)_\mu(S_A(t)x_n) \leq (I_C)_\mu(x_n) + \frac{\omega}{\mu}\int_0^t \|S_A(s)x_n - P_C S_A(s)x_n\|^2 ds, \ \ \forall\, t > 0.$$

We note that

$$\lim_{n\to\infty}\int_0^t \|S_A(s)x_n - P_C S_A(s)x_n\|^2 ds = \int_0^t \|S_A(s)x - P_C S_A(s)x\|^2 ds$$

which derives essentially from the estimates

$$\|S_A(s)x_n - S_A(s)x\| \le e^{\omega s}\|x_n - x\|, \quad \forall\, s \in [0,t]$$

(see relation (1.5) in Brézis and Pazy [54]) and the continuity of P_C. Passing to the limit as $n \to \infty$ the claim is established.

If $x \in D(A)$ than the map $t \mapsto (I_C)_\mu(S_A(t)x)$ is Lipschitz on bounded intervals so there exists

$$\frac{d}{dt}(I_C)_\mu(S_A(t)x) = \partial(I_C)_\mu(S_A(t)x)\frac{d}{dt}(S_A(t)x)$$

$$= -\partial(I_C)_\mu(S_A(t)x)A_0 S_A(t)x \le \frac{\omega}{\mu}\|S_A(t)x - P_C S_A(t)x\|^2.$$

Integrating on $[0,t]$ we obtain $v.$

$v.\Rightarrow vj.$ We have $(I_C)_\mu(S_A(t)x) = \frac{1}{2\mu}\|S_A(t)x - P_C S_A(t)x\|^2$ and by $v.$ we get

$$\tfrac{1}{2\mu}\|S_A(t)x\text{-}P_C S_A(t)x\|^2 \le \tfrac{1}{2\mu}\|x\text{-}P_C x\|^2 + \tfrac{\omega}{\mu}\int_0^t \|S_A(t)x\text{-}P_C S_A(t)x\|^2 ds.$$

Using Gronwall inequality, the above relation leads to $vj.$ **q.e.d.**

Remark 7.1. $S_A(t)(C) \subset C$, $\forall\, t \ge 0$.

Indeed, if $x \in C$ than $P_C x = x$ and than from $vj.$ we have $\|S_A(t)x - P_C S_A(t)x\|^2 \le 0$, i.e., $S_A(t)x = P_C S_A(t)x \in C$. So $S_A(t)x \in C$, for all $t \ge 0$.

1.8 Approximation of nonlinear equations in Banach spaces

Consider the Cauchy problem

$$\begin{cases} v'(t) + Av(t) \ni f(t) & t \in [0,T], \\ v(0) = v_0. \end{cases} \tag{8.1}$$

Definition 8.1. A *strong solution* of problem (8.1) is a function $v \in W^{1,1}((0,T];X) \cap C([0,T];X)$ such that

$$f(t) - v'(t) \in Av(t) \quad \text{a.e. } t \in (0,T], \quad v(0) = v_0.$$

Here

$$\mathbf{W^{1,1}((0,T];X)}=\{y \in L^1([0,T];X) : y' \in L^1((\delta,T];X),\ \forall \delta \in (0,T)\}.$$

Definition 8.2. Let $f \in L^1([0,T];X)$ and $\varepsilon > 0$ be given. We called an *ε–discretization* on $[0,T]$ of the equation $v' + Av \ni f$, a partition $0 = t_0 \le t_1 \le t_2 \le \cdots \le t_N$ of the interval $[0,t_N]$ and a finite sequence $\{f_i\}_{i=1}^N \subset X$ such that $t_i - t_{i-1} < \varepsilon$ for $i = 1,2,...,N$, $T - \varepsilon < t_N \le T$,

$$\sum_{i=1}^{N} \int_{t_{i-1}}^{t_i} \|f(s) - f_i\|_X \, ds < \varepsilon.$$

We shall denote by $D_A^\varepsilon(0 = t_0, t_1, ..., t_N; f_1, ..., f_N)$ this ε-discretization.

Definition 8.3. A *solution* to the ε-discretization $D_A^\varepsilon(0 = t_0, t_1, ..., t_N; f_1, ..., f_N)$ is a piecewise constant function $z : [0,t_N] \longrightarrow X$ whose value z_i on $(t_{i-1}, t_i]$ satisfy the equation

$$\frac{z_i - z_{i-1}}{t_i - t_{i-1}} + Az_i \ni f_i, \quad i = 1,2,...,N.$$

Definition 8.4. The function $z = \{z_i\}_{i=1}^N$ is called *ε–approximative solution* to the Cauchy problem (8.1) if it satisfies in addition

$$\|z(0) - v_0\|_X \le \varepsilon.$$

Definition 8.5. A *mild solution* of the Cauchy problem (8.1) is a function $v \in C([0,T];X)$ with the property that for each $\varepsilon > 0$ there is an ε–approximate solution z of $v' + Av \ni f$ on $[0,T]$ such that $\|v(t) - z(t)\|_X \le \varepsilon$ for all $t \in [0,T]$ and $v(0) = v_0$.

Crandall & Liggett theorem. *Let A be ω-accretive, satisfying the range condition $R(I + \lambda A) \supset D(A)$ for small $\lambda > 0$ and $v_0 \in \overline{D(A)}$. Then the Cauchy problem*

$$\begin{cases} \dfrac{dv}{dt} + Av \ni 0 \quad t > 0, \\ \\ v(0) = v_0, \end{cases} \tag{8.2}$$

has a unique mild solution v. Moreover

$$v(t) = \lim_{n\to\infty} \left(I + \frac{t}{n}A\right)^{-n} v_0 \tag{8.3}$$

uniformly in t on compact intervals.

Given two m–accretive operators $A, B \subset X \times X$ such that $A + B$ is m–accretive (see Definition 7.1 b.), one might expect that

$$S_{A+B}(t)x = \lim_{n\to\infty} \left(S_A\left(\frac{t}{n}\right) S_B\left(\frac{t}{n}\right)\right)^n x \quad \forall\, t \geq 0, \tag{8.4}$$

for all $x \in \overline{D(A) \cap D(B)}$, where by S_{A+B}, S_A, S_B we have denoted the semigruoups generated by $A + B$, A and B, respectively. Relation (8.4) is the *Lie-Trotter product* formula and one knows that it is true for C_0–semigruoups of contractions (see A. Pazy [55, pp. 92]). It is readily seen that (8.4) is equivalent with the convergence of *the fractional steps method* for the Cauchy problem

$$\begin{cases} \dfrac{dv}{dt} + Av + Bv \ni 0 \quad \text{in } [0, T], \\ v(0) = v_0, \end{cases} \tag{8.5}$$

i.e.,

$$\begin{cases} \dfrac{dv}{dt} + Av \ni 0 \quad \text{in } [i\varepsilon, (i+1)\varepsilon],\ i = 0, 1, ..., N-1,\ T = N\varepsilon, \\ v^+(i\varepsilon) = z(\varepsilon) \quad v^+(0) = v_0, \end{cases} \tag{8.6}$$

$$\begin{cases} \dfrac{dz}{dt} + Bz \ni 0 \quad \text{in } [0, \varepsilon], \\ z(0) = v^-(i\varepsilon), \end{cases} \tag{8.7}$$

where $v^+(i\varepsilon) = \lim\limits_{t\downarrow i\varepsilon} v(t)$, $v^-(i\varepsilon) = \lim\limits_{t\uparrow i\varepsilon} v(t)$.

In a general Banach space, the Lie-Trotter formula (8.4) is not convergent always. There are known several situations in which this formula is true, namely.

Brézis & Pazy theorem. *Let X and X^* be uniformly convex and let A, B be m–accretive single-valued operators on X such that $A+B$ is m-accretive and $S_A(t)$, $S_B(t)$ map $K=\overline{D(A)\cap D(B)}$ into itself. Then*

$$S_{A+B}(t)x = \lim_{n\to\infty}\left(S_A\left(\frac{t}{n}\right)S_B\left(\frac{t}{n}\right)\right)^n x \quad \forall x\in K, \tag{8.8}$$

and the limit is uniform in t on compact intervals.

The Brézis & Pazy theorem was extended by Y. Kobayashi [56] to multivalued operators A and B in a Hilbert space. More precisely, if A, B and $A+B$ are maximal monotone and if there is a nonempty closed convex set $C\subset\overline{D(A)\cap D(B)}$ such that $(I+\lambda A)^{-1}C\subset C$ and $(I+\lambda B)^{-1}C\subset C$ $\forall\lambda>0$, then

$$S_{A+B}(t)x = \lim_{n\to\infty}\left(S_A\left(\frac{t}{n}\right)S_B\left(\frac{t}{n}\right)\right)^n x \quad \forall x\in C, \tag{8.9}$$

uniformly in t on compact intervals.

The special case in which A and B are nonliniar operators (more exactlt A is ω-accretive (see Definition 7.1 c.) and B is continuous on a closed convex subset) is treated in V. Barbu & M. Iannelli [38]. Further on we recall this result

Barbu & Iannelli theorem. *Let Y be a real Banach space, let C be a closed subset of Y and $K=\overline{C\cap D(A)}$ be a convex subset of Y such that*

(H1) *A is ω–accretive and $R(I+\lambda A)\supset\overline{D(A)}$, $\forall\lambda\in(0,\lambda_0)$;*

(H2) *B is a continuous ω–accretive operator on C such that*

$$R(I+\lambda B)\supset C,\ \forall\,\lambda\in(0,\lambda_0);$$

(H3) *$R(I+\lambda(A+B))\supset K$, $(I+\lambda A)^{-1}K\subset K$, $(I+\lambda B)^{-1}K\subset K$;*

(H4) *For every $[x,y]\in A$, there exists $\{x_h\}\subset Y$ such that*

$$\lim_{h\longrightarrow 0} x_h = x,\quad \lim_{h\longrightarrow 0}\left\|\frac{1}{h}(x_h - e^{-Ah}x_h) - y\right\| = 0.$$

Then, for every $y_0\in K$, we have

$$\lim_{\varepsilon\longrightarrow 0} y_\varepsilon(t) = y(t),\ \forall\, t\geq 0, \tag{8.10}$$

and the limit is uniform on bounded t intervals.

Here, by $y(t)$, we have noted the generalized solution to the Cauchy problem (8.5) and by $y_\varepsilon(t)$ the solution of the corresponding approximative scheme (8.6)-(8.7).

1.9 Selected topics from numerical analysis

Finite-difference approximations to derivatives. Let $y(x)$ be a function defined on the interval $[a, b]$. Assume that the function y has n derivatives on this interval, and that all of them are finite and continuous functions of x. If $x \in [a, b]$ and for a given h, $x - h$ and $x + h$ are also in $[a, b]$, then by Taylor's theorem,

$$y(x+h) = y(x) + hy'(x) + \frac{1}{2}h^2 y''(x) + \frac{1}{6}h^3 y'''(x) + \cdots \tag{9.1}$$

$$y(x-h) = y(x) - hy'(x) + \frac{1}{2}h^2 y''(x) - \frac{1}{6}h^3 y'''(x) + \cdots. \tag{9.2}$$

Addition of this expansions gives

$$y(x+h) + y(x-h) = 2y(x) + h^2 y''(x) + O(h^4), \tag{9.3}$$

where $O(h^4)$ denotes terms containing fourth and higher powers of h. Neglect of terms of order h^4, relation (9.3) leads to

$$y''(x) = \frac{d^2 y}{dx^2}(x) \equiv \frac{1}{h^2}[y(x+h) - 2y(x) + y(x-h)] \tag{9.4}$$

with an error of order h^2. Thus we can approximate $y''(x)$ by the right-hand side in relation (9.4).

Substraction of equation (9.2) from equation (9.1) and neglect of terms of order h^3 leads to

$$y'(x) = \frac{dy}{dx}(x) \equiv \frac{1}{2h}[y(x+h) - y(x-h)] \tag{9.5}$$

with an error of order h. Equation (9.5) is called a *central-difference* approximation of $y'(x)$. We can also approximate $y'(x)$ by either the *forward-difference* formula

$$y'(x) = \frac{dy}{dx}(x) \equiv \frac{1}{h}[y(x+h) - y(x)] \tag{9.6}$$

or the *backward-difference* formula

$$y'(x) = \frac{dy}{dx}(x) \equiv \frac{1}{h}[y(x) - y(x-h)]. \tag{9.7}$$

Approximation by the finite element method. For simplicity we will consider the domain $\Omega \subset I\!R$ to be polygonal, that is the boundary $\partial\Omega$ of Ω is made up of straight segments. More general domain $\Omega \subset I\!R^n$, can be approximated as better as by *polyhedral* (polygonal for $n = 2$) domains with $O(h^2)$-accuracy. Here we consider piecewise linear functions on triangles only. For more sophisticated methods making use of higher degree polynomials, quadrilateral elements, curved elements, see Axelsson & Barker [57], Ciarlet [58], among others.

We start by realizing a *triangulation* of domain Ω, denoted in the sequel by T_h. For this, we divide Ω into a finite number nt of triangles T_i, $i = 1, 2, ..., nt$, whose vertices are the nodes of the triangulation

$$T_h = \{T_i \mid T_i \subset \Omega,\ T_i - triangle,\ i = \overline{1, nt}\}. \tag{9.8}$$

By convention, the *mesh parameter* of the family T_h is

$$h = \max_{T \in T_h} h_T$$

where $h_T =$ *leght* of the largest edge of T. We define also $\rho_T =$ *diameter* of the inscribed circle in T.

Definition 9.1. The family T_h in (9.8) is a *triangulation* of $\bar{\Omega}$ iff

i) for every h, T_h is a finite collection of closed triangles T contained in $\bar{\Omega}$;

ii) $\underset{T \in T_h}{\cup} = \bar{\Omega}$;

iii) if T_i, $T_j \in T_h$, $i \neq j$, $i, j = \overline{1, nt}$, we have

$$\overset{\circ}{T_i} \cap \overset{\circ}{T_j} = \emptyset \quad (\overset{\circ}{T} \text{ is interior of } T)$$

and either $T_i \cap T_j = \emptyset$ or T_i and T_j have in common one whole edge or only one vertex.

Definition 9.2. A family T_h is said to be *regular* if there exists a constant $\delta > 0$ such that

$$\frac{h_T}{\rho_T} \leq \delta \quad \text{for all } T \in T_h \quad \text{and for all } h.$$

We now define the simplest finite element space S_h as follows

$$S_h = \{v_h \mid v_h \in C(\bar{\Omega}),\ v_h|_T \in P_1 \ \ \forall T \in T_h\}$$

where P_1 is the space of polynomials in two variables of degree less than or equal to 1 (e.g., $p(x,y) = ax + by + c$, $a, b, c \in I\!R$). We note that any polynomial $p(x,y)$ of P_1 is uniquely defined by its values at the three vertices of a triangle.

Let Σ_h be the finite set of vertices of T_h and let us set $nn =$ card Σ_h. For any nodal point $N_k \in \Sigma_k$ we associate the function φ_k (unique) defined by

$$\varphi_k \in S_h, \quad \varphi_k(N_k) = 1, \quad \varphi_k(N_j) = 0, \quad k \neq j.$$

Such a function is called a *Courant basis function.* Functions $\{\varphi_1, \cdots, \varphi_{nn}\}$ form a basis of the space S_h, i.e.,

$$v_h = \sum_{k=1}^{nn} v_h(N_k)\varphi_k \quad \forall v_h \in S_h. \tag{9.9}$$

The results on the approximation properties of the particular finite element space introduced above can be found in Ciarlet [58].

Cubic spline functions of one variable. Let $y : [a,b] \longrightarrow I\!R$ be a real function and let $\Delta : a = x_1 < x_2 < \cdots < x_N = b$ be some fixed nodes at which the values of the function y, $y_i = y(x_i)$, are given.

Definition 9.3. The function $S_y : [a,b] \longrightarrow I\!R$ is called a *cubic spline* with the set $\Delta : a = x_1 < x_2 < \cdots < x_N = b$ of nodes if there exists $N-1$ cubic polynomials such that:

(i) $S_y|_{[x_i,x_{i+1}]} \in P_3([x_i, x_{i+1}])$, i=1,2,...,N-1;

(ii) $S_y \in C^2(a,b)$.

It is known that the cubic spline $S_y(x)$ has the following reprezentation

$$S_y(x) = M_i\frac{(x_{i+1}-x)^3}{6h_i} + M_{i+1}\frac{(x-x_i)^3}{6h_i} \tag{9.10}$$

$$+\left(y_i - \frac{M_i h_i^2}{6}\right)\frac{x_{i+1}-x}{h_i} + \left(y_{i+1} - \frac{M_{i+1}h_i^2}{6}\right)\frac{x-x_i}{h_i} \qquad x \in [x_i, x_{i+1}],$$

where $M_i = S_y''(x_i)$, $h_i = x_{i+1} - x_i$, $i = 1, 2, ..., N-1$. For detailes we refer to Prenter [59].

1.10 The phase-field system (Caginalp's model)

In this Section we will describe the phase-field model introduced in mathematical literature by G. Caginalp [7]. For this, we consider a material in a region $\Omega \subset I\!R^N$ which may be in either of two phases, e.g., solid and liquid (see Fig. **1**). Let us denote by $u(t,x)$ - *the reduced temperature distribution*, $u(t,x) = \theta(t,x) - \theta_M$, where: $\theta(t,x)$ represent the temperature of the material at $(t,x) \in [0,T] \times \Omega$, $T \in I\!R_+ = (0,\infty)$, and θ_M is *the melting temperature* (the temperature at which solid and liquid may coexist in equilibrium, separated by a planar interface).

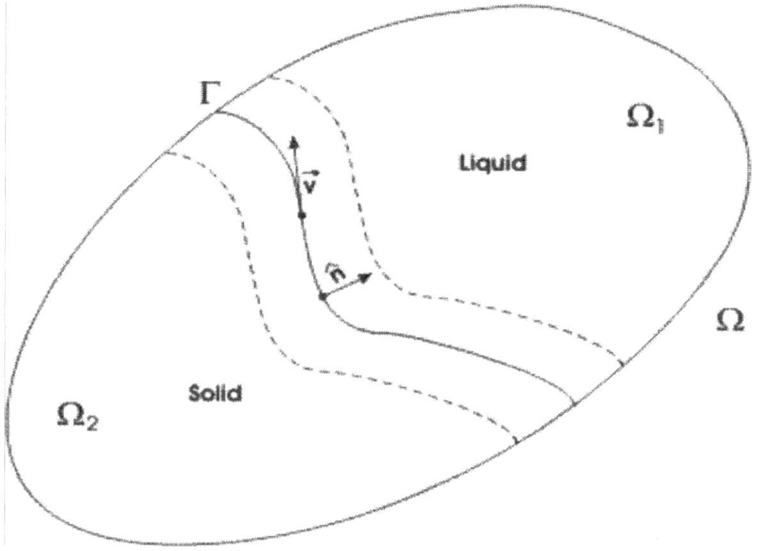

Figure 1 *A material Ω exists in two phases. The dotted lines indicate possible thickness of the interface*

It is known that in the classical Stefan problem in two phases (the first mathematical model of solidification, see L.I. Rubinstein [60]) the *interface* between the solid and liquid (denoted in the sequel by Γ) is considered to be

$$\Gamma(t) = \{x \in \Omega : \quad u(t,x) = 0\}, \tag{10.1}$$

(i.e., $\theta(t,x) = \theta_M, \quad \forall (t,x) \in \Gamma$) while the liquid and solid regions are defined as follows

$$\Omega_1(t) = \{x \in \Omega : \ u(t,x) > 0\}, \quad \Omega_2(t) = \{x \in \Omega : \ u(t,x) < 0\} \tag{10.2}$$

for all $t \in [0, T]$. In Ω_1 and Ω_2, the function $u(t, x)$ must then satisfy the heat diffusion equation

$$u_t = k\Delta u, \tag{10.3}$$

where k is the thermal conductivity divided by heat capacity per unit volume (equal to unity here) assumed to be the same constant in both the phases.

Across the interface Γ, the latent heat of fusion (per unit mass) ℓ must be dissipated or absorbed in accordance with the law of conservation of energy, i.e.

$$\ell\, \vec{v}(t, x) \cdot \hat{n} = k(\nabla^+ u(t, x) - \nabla^- u(t, x)) \cdot \hat{n} \quad x \in \Gamma(t), \tag{10.4}$$

where (see Definition 6.2) $\hat{n}$ is the unit normal at each point of $\Gamma(t)$ (in the direction solid $\to$ lichid), $\vec{v}$ is a local velocity of the interface, and

$$\nabla^+ u(t, x) = \lim_{\bar{x} \to x,\ \bar{x} \in \Omega_1(t)} \nabla u(t, \bar{x}),$$

$$\nabla^- u(t, x) = \lim_{\bar{x} \to x,\ \bar{x} \in \Omega_2(t)} \nabla u(t, \bar{x}).$$

Together with the condition

$$u(t, x) = 0 \quad x \in \Gamma(t), \tag{10.5}$$

(derived from the definition of $\Gamma(t)$, see (10.1)) one must specify initial and boundary conditions for $u(t, x)$, e.g.,

$$u(t, x) = u_\partial(t, x) \quad x \in \partial\Omega, \ \ t > 0, \tag{10.6}$$

$$u(0, x) = u_0(x) \quad x \in \Omega, \tag{10.7}$$

and this complete the mathematical statements of the *classical Stefan model in two phases*. The problem then is to find $u(t, x)$ and $\Gamma(t)$ in suitable function spaces, satisfaing equations (10.3)-(10.7). The interface $\Gamma(t)$ is often called the *free boundary*.

One method for studying the equations (10.3)-(10.7) is the enthalpy or H-method (see O.A Oleinik [61]). The basic idea is to consider the function $H = H(u)$ defined by

$$H(u) = u + \frac{\ell}{2}\varphi(u), \quad \varphi(u) = \begin{cases} +1, & u > 0 \\ -1, & u < 0. \end{cases} \tag{10.8}$$

Then the ecuations (10.3)-(10.4) can be incorporated in a weak sense (see O.A Oleinik [61]) into a single ecuation

$$\frac{\partial}{\partial t}H(u) = k\Delta u \tag{10.9}$$

which represents a balance of heat ecuation. This weak formulation was used by G. Caginalp [7] to introduce a new model which reflects more accurately the physical aspects.

The physical situation is generally more complicated than that described by equations (10.3)-(10.4). One of the physical effects neglected by the classical Stefan model is that of surface tension. If is accepted the Gibbs-Thompson relation

$$u(t,x) = -\frac{\sigma}{\Delta s}\cdot \Xi \quad (t,x) \in \Gamma, \tag{10.10}$$

where σ is the surface tension, Δs is the difference in entropy between solid and liquid, and Ξ is the sum of principal curvatures at a point on the interface, one may consider equations (10.3)-(10.4) together with (10.10) as an alternative to the classical Stefan model. Perhaps the most interesting aspect observed by metallurgists is the presence of liquid at subzero temperatures (the physical phenomena of *supercooling*) and analogously phenomenon for solid (*superheating*). It is clear that the dual role of the temperature $u(t,x)$ is no longer possible. Acceptance of the idea that the temperature need not to be zero at the interface (or negative in the solid, etc) leads to the question of how one distinguishes the two phases.

Let us consider *the interface* as **a continuous region**, more vast (in which the liquid can coexist with the solid), of finite thickness, and in which the change of phase occurs continuously.

Then, it is natural that the function φ in (10.8) to be replaced by a function illustrated in Fig. (**2**), i.e., a smooth function with values from $\varphi = -1$ (solid) to $\varphi = +1$ (liquid). **The key question is how one can determine the function φ?**

In statistical mechanics, a model in which atoms are assumed to interact with a mean field created by the other atoms is known as a mean field theory [7], [62]-[65]. The Landau-Ginzburg theory of phase transition [7] (see and J.W. Cahn & J.E. Hilliard [66]) is such one theory. Thus, the free energy of stationary process in this case may be written as

$$F_u(\varphi) = \int_\Omega \Big(\frac{\xi^2}{2}|\nabla\varphi|^2 - \frac{1}{a}G(\varphi) - 2u\varphi\Big)dx \tag{10.11}$$

where ξ is a length scale, $\frac{1}{2}|\nabla\varphi|^2$ is an extra term of penalty for φ, $G(\varphi)$ is a symmetric double well potential with minima at ± 1, e.g., $G(\varphi) = \frac{1}{8}(\varphi^2 - 1)^2$. This double well potential can be viewed in terms of a probabilistic measure on the individual atoms. The extent to which this measure discriminates against the interfacial region and in favor of the liquid or solid phases depends on how close a is to zero. The last term in (10.11) which introduces the coupling between u and φ, may be understood best as the part of the free energy which corresponds to the component which is generally written as temperature times change in entropy.

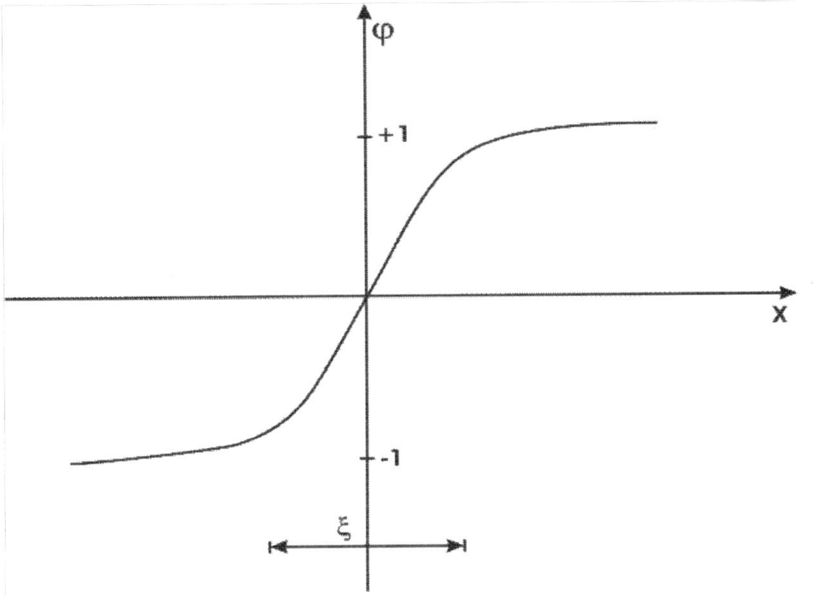

Figure 2 *A possible choice of phase parameter* φ

In equilibrium, one expects φ to be a minimizer of F_u so that $\delta F_u/\delta\varphi = 0$, i.e., φ verify the equation

$$\xi^2\Delta\varphi + \frac{1}{a}g(\varphi) + 2u = 0. \qquad (10.12)$$

When the material is not in equilibrium (the time-dependent situation), φ will not minimize $F_u(\varphi)$ but will differ by a term proportional to φ_t. The Euler-Lagrange equations, i.e., $\tau\varphi_t = -\delta F_u/\delta\varphi$, imply the identity

$$\tau\varphi_t = \xi^2\Delta\varphi + \frac{1}{2a}(\varphi - \varphi^3) + 2u \tag{10.13}$$

where τ is a relaxation time. Unifying now this last equation with the equation (10.9), i.e.,

$$u_t + \frac{\ell}{2}\varphi_t = k\Delta u \tag{10.14}$$

subject to initial conditions

$$u(0, x) = u_0(x) \quad x \in \Omega, \tag{10.15}$$

$$\varphi(0, x) = \varphi_0(x) \quad x \in \Omega \tag{10.16}$$

and appropriate boundary conditions, e.g.,

$$u = u_\partial(x) \quad x \in \partial\Omega, \tag{10.17}$$

$$\varphi = \varphi_\partial(x) \quad x \in \partial\Omega \tag{10.18}$$

then, the system (10.13)-(10.18) represents the mathematical problem to be studied in time-dependence case.

1.11 The continuous-casting process in metallurgy

From the thermo-kinetics point of view, the solidification and cooling, as well as the simultaneous heating in a continuous casting process of steel, represents a very complex problem of non-stationary heat and mass transfer. To solve at present such a problem it is impossible without numerical models of the temperature field and computer technology. In the sequel we will present in short a continuous casting machine, including the key phenomena of interest and a new numerical model that we have used in the settlement of the problem mentioned above.

The casting machine. In a modern steel casting machine (its essential features are illustrated in Fig. (**3**), the molten metal is tapped from a ladle into a copper mold (crystallizer). Here, the water-cooled walls of the mold (the primary cooling zone) extract heat what leads to solidify a shell that contains the liquid pool. Below the mold (the secondary cooling zone),

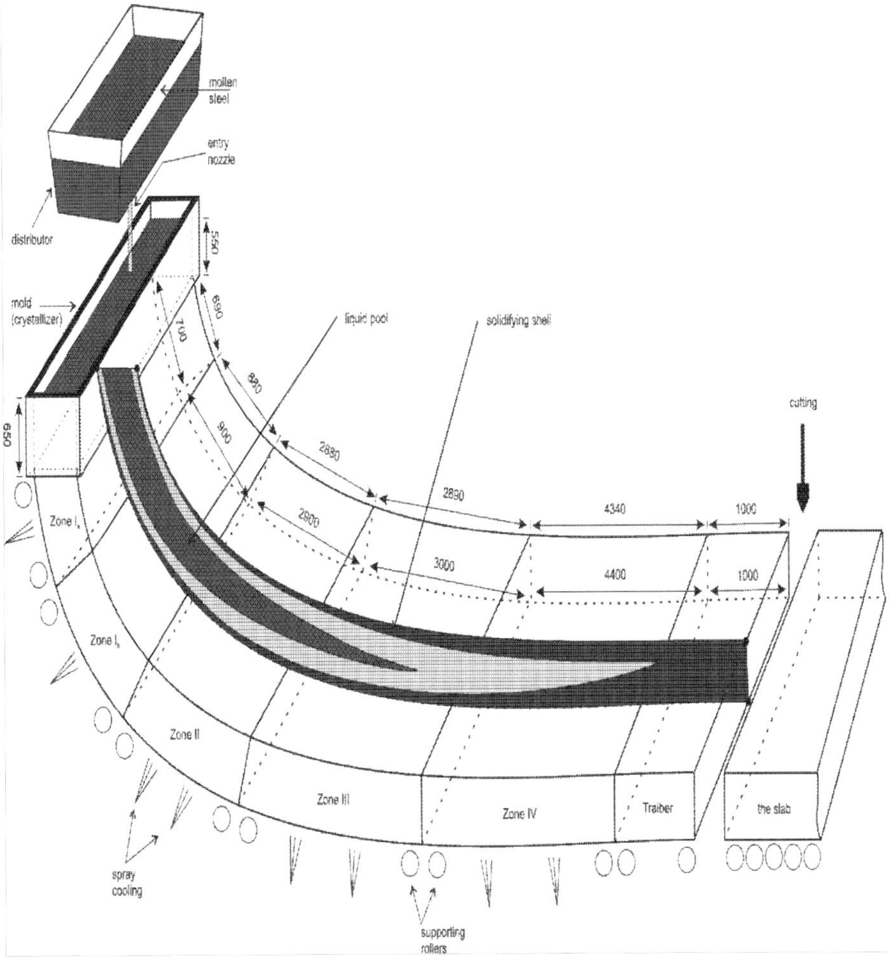

Figure 3 *Schematic representation of a continuous-casting machine*

the product is supported by rollers and is cooled down by water sprays that extract heat from the surface, and, eventually, the core becomes fully solid when the metallurgical length increase at 1213m. After the end of secondary cooling zone the product is cooled only by radiation (traiber). Finally, the continuous-casting product must be cut into the optimum lengths (cutting) to achive a maximum yield of metal.

CHAPTER 2

APPROXIMATION OF THE PHASE FIELD TRANSITION SYSTEM

Costică MOROŞANU

Faculty of Mathematics, "Al.I. Cuza" University of Iasi, Romania

Abstract: In this chapter we will deal, on one hand, with the study of the existence, uniqueness, regularity and estimates of the solution of *phase-field transition system*, and on the other hand, we will analyze the convergence of some *approximating schemes of fractional steps type*, associated to this nonlinear system.

In the first Section the existence, uniqueness and regularity of the solution of the phase-field transition system subject to the non-homogeneous Cauchy-Neumann boundary conditions (**Theorem 2.1**) are studied. Such kind of conditions allow the phase-field system (Caginalp's model) to be considered as *a model of heat transfer* from the surface of product to the environment, when we assume that heat is extracted by convection and conduction in the continuous casting process.

As it is well known, the phase-field system is a refinement of the classical Stefan problem in two phases by adding a new parabolic equation which contains a nonlinear term - *the phase function* $\varphi(t,x)$ used to distinguish between the phases of the material. In order to express more accurately the physical phenomenon (*superheating, supercooling, effects of tension surface, separating zones* etc), different nonlinearities have been proposed by several authors. Thus, the purpose of Section II.2 is to treat the phase-field system with a general nonlinearity in φ. The existence, uniqueness, regularity and estimates of the solution are proved (**Theorem 2.3**) in this relevant case. Basic tools in this approach are the Leray-Schauder degree theory, the L_p-theory of linear parabolic equations, properties of the Nemytskij operator, and a priori estimates in $L^p(Q)$.

The Section 2.4 is devoted to the extension of some results known for m-accretive operators (see, e.g., Barbu [39] and Brézis [40]) to ω-m-accretive operators (**Theorem 2.4**), with a fixed real number ω. As an application of Theorem 2.4, it is indicated that how the nonlinear phase-field transition system (with homogeneous Neumann-Neumann boundary conditions) can be decoupled in two simples systems.

The main idea of this book resides in the construction of *approximation schemes* for nonlinear phase-field transition system. In this respect, we considered different **b**oundary **c**onditions (**bc** shortly), as follows (see and (3')-(3^v) in Preface):

- $u = 0, \quad \varphi = 0$ (homogeneous Dirichlet - Dirichlet **bc**);
- $\frac{\partial}{\partial \nu} u = 0, \quad \frac{\partial}{\partial \nu} \varphi = 0$ (homogeneous Neumann - Neumann **bc**);
- $\frac{\partial}{\partial \nu} u + hu = 0, \ \varphi = 0$ (homogeneous Cauchy - Dirichlet **bc**);
- $\frac{\partial}{\partial \nu} u + hu = w(t, x), \quad \frac{\partial}{\partial \nu} \varphi = 0$ (non-homogeneous Cauchy - Neumann **bc**);
- $\frac{\partial}{\partial \nu} u + hu = w(t) g(x), \quad \varphi = 1$ (non-homogeneous Cauchy - Dirichlet **bc**);

The convergence of each scheme (**Proposition 2.1**, **Theorem 2.5**, **Theorem 2.6**) is proved by different methods. The weak stability of approximating scheme, corresponding to homogeneous Cauchy-Dirichlet boundary conditions, is proved too (**Corollary 2.1**).

On a bounded domain $\Omega \subset I\!\!R^N$, $N \geq 1$, with a C^2 boundary $\partial\Omega$ and for a time $T > 0$, we consider the nonlinear parabolic system ($Q := [0,T] \times \Omega$, $\Sigma = [0,T] \times \partial\Omega$):

$$\begin{cases} \rho c \frac{\partial}{\partial t} u + \frac{\ell}{2} \frac{\partial}{\partial t} \varphi = k\Delta u + f(t,x) & (t,x) \in Q, \\ \tau \frac{\partial}{\partial t} \varphi = \xi^2 \Delta\varphi + \frac{1}{2a}(\varphi - \varphi^3) & \\ \qquad\qquad + 2u + g(t,x) & (t,x) \in Q, \\ \frac{\partial}{\partial \nu} u + hu = w(t,x), \ \frac{\partial}{\partial \nu}\varphi = 0 & \text{on } \Sigma, \\ u(0,x) = u_0(x), \ \varphi(0,x) = \varphi_0(x) & \text{on } \Omega, \end{cases} \tag{2.1}$$

where:

u - represents *the temperature distribution* on Q,

φ - is *the phase function* used to distinguish between the phases of material Ω, at every moment of time $t \in [0,T]$,

- the positive parameters: $\rho, c, \ell, k, \tau, \xi, h, a$ have a physical meaning, namely:
  - ρ - is *the density,*
  - c - is *the heat capacity,*
  - ℓ - denotes *the latent heat,*
  - k - *the heat conductivity,*
  - τ - is *the relaxation time,*
  - ξ - is *the length scale of the interface,*
  - h - *the heat transfer coefficient,*
  - a is a *probabilistic measure* on the individual atoms, (a depends on ξ),
- $f \in L^p(Q)$, $g \in L^q(Q)$ are given functions,
- $w \in W_p^{2-\frac{1}{p}, 1-\frac{1}{2p}}(\Sigma)$ - is a given function (the temperature of the surrounding at $\partial\Omega$), being assimilated as *boundary control* in Chapters 3 and 4,
- $u_0 \in W_p^{2-2/p}(\Omega)$, $\varphi_0 \in W_q^{2-2/q}(\Omega)$, provided $\frac{\partial}{\partial \nu} u_0 + hu_0 = w$, $\frac{\partial}{\partial \nu}\varphi_0 = 0$ on $\partial\Omega$,
- p, q are given numbers which satisfy

$$q \geq p \geq 2. \tag{2.2}$$

As usual, $\frac{\partial}{\partial \nu}$ denotes the outwards unit normal to $\partial\Omega$ (see Definition 6.2 in Chapter 1).

We recall that the embedding theorem of Lions and Peetre (see (5.3) in Chapter 1) ensures the existence of a continuous embedding $W_p^{2,1}(Q) \subset L^\mu(Q)$ (p satisfying (2.2)), where

$$\mu = \begin{cases} \infty, & \text{if } N+2 < 2p, \\ \text{any positive number} > 2p & \text{if } N+2 = 2p, \\ \dfrac{p(N+2)}{N+2-2p} & \text{if } N+2 > 2p. \end{cases} \tag{2.3}$$

Throughout the present Chapter we will denote different positive constants by C for which we will mention possible dependences.

Keywords: nonlinear parabolic systems, abstract approximation theory, nonlinear functional analysis, distributions, operator theory, dissipative operators, degree theory, semigroups of nonlinear operators, numerical analysis, fractional steps method (Lie-Trotter product formula), stability and convergence of numerical methods.

2.1 Existence and regularity result

Within the framework of this Section, we will approach the nonlinear parabolic system (2.1) in the spirit given by

Definition 2.1 (Hadamard's well-posedness conditions). *A problem is well posed if:*

i. *a solution exists,*
ii. *the solution is unique*, and
iii. *the solution depends continuously on the data, in particular the initial and boundary values.*

Consequently, our main result in studying problem (2.1) is the following

Theorem 2.1. *Problem* (2.1) *has a unique solution* (u, φ) *with* $u \in W_p^{2,1}(Q)$ *and* $\varphi \in W_\nu^{2,1}(Q)$, *where* $\nu = \min\{q, \mu\}$. *In addition, the pair function* (u, φ) *satisfies*

$$\|u\|_{W_p^{2,1}(Q)} + \|\varphi\|_{W_\nu^{2,1}(Q)} \tag{2.4}$$

$$\leq C\left(1+\|u_0\|_{W_p^{2-2/p}(\Omega)}+\|\varphi_0\|^{2-1/q}_{W_q^{2-2/q}(\Omega)}\right.$$

$$\left.+\|w\|_{W_p^{2-\frac{1}{p},1-\frac{1}{2p}}(\Sigma)}+\|f\|_{L^p(Q)}+\|g\|_{L^q(Q)}\right)$$

where the constant C depends on $|\Omega|$ (the measure of Ω),T, N, p, q and physical parameters.

Moreover, given any number $M > 0$, if (u_1, φ_1), (u_2, φ_2) are solutions of (2.1) corresponding to the data (f_1, g_1, w_1), $(f_2, g_2, w_2) \in L^p(Q) \times L^q(Q) \times W_p^{2-\frac{1}{p},1-\frac{1}{2p}}(\Sigma)$, respectively, for the same initial conditions, such that $\|\varphi_1\|_{L^\nu(Q)}$, $\|\varphi_2\|_{L^\nu(Q)} \leq M$, then the estimate below holds

$$\|u_1-u_2\|_{W_p^{2,1}(Q)}+\|\varphi_1-\varphi_2\|_{W_\nu^{2,1}(Q)} \tag{2.5}$$

$$\leq C\left(\|f_1-f_2\|_{L^p(Q)}+\|g_1-g_2\|_{L^q(Q)}+\|w_1-w_2\|_{W_p^{2-\frac{1}{p},1-\frac{1}{2p}}(\Sigma)}\right)$$

where $C > 0$ depends on $|\Omega|, T, N, M, p, q$, and physical parameters.

2.1a The nonlinear equation in the phase-field transition system

Before starting the proof of Theorem 2.1, we recall a result that concerns the existence and the uniqueness of the solution to an auxiliary nonlinear parabolic equation derived from (2.1). Then we consider the nonlinear parabolic boundary value problem

$$\begin{cases} \tau\frac{\partial}{\partial t}v-\xi^2\Delta v=\frac{1}{2a}(v-v^3)+\bar{g}(t,x) & (t,x)\in Q,\\ \frac{\partial}{\partial \nu}v=0 & \text{on } \Sigma,\\ v(0,x)=v_0(x) & x\in\Omega, \end{cases} \tag{2.6}$$

where $\bar{g} \in L^p(Q)$ and $v_0 \in W_q^{2-2/q}(\Omega)$ verify $\frac{\partial}{\partial\nu}v_0 = 0$ on $\partial\Omega$.

Theorem 2.2. *There exists a unique solution $v \in W_p^{2,1}(Q)$ for (2.6), and v satisfies*

$$\|v\|_{W_p^{2,1}(Q)}\leq C\left(1+\|v_0\|^{(2p-1)/p}_{W_q^{2-2/q}(\Omega)}+\|\bar{g}\|_{L^p(Q)}\right) \tag{2.7}$$

where $C > 0$ *is a constant which depends on* $|\Omega|, T, N, p,$ *and physical parameters.*

If v_1, v_2 *are solutions of* (2.6) *corresponding to* $v_0^1, v_0^2 \in W_q^{2-2/q}(\Omega)$ *and* $\bar{g}_1$, $\bar{g}_2$, *respectively, with* $\|v_1\|_{W_p^{2,1}(Q)}$, $\|v_2\|_{W_p^{2,1}(Q)} \leq M$ (M as in Theorem 2.1 above), *then*

$$\|v_1 - v_2\|_{W_p^{2,1}(Q)} \leq C\left(\|v_0^1 - v_0^2\|_{W_q^{2-2/q}(\Omega)} + \|\bar{g}_1 - \bar{g}_2\|_{L^p(Q)}\right) \quad (2.8)$$

where $C > 0$ *is a constant which depends on* $|\Omega|, T, N, M, p$ *and physical parameters.*

Proof. Taking $F(t,x,v) = \frac{1}{2a}(v - v^3)$ in Moroşanu & Motreanu [32, problem (P0), pp. 191] (see also Example 2.1 in Section II.2), we may conclude that (2.7)-(2.8) holds true and so the statements made in Theorem 2.2 above are correct. **q.e.d.**

2.1b The properties of the homotopy $H : L^p(Q) \times [0,1] \to L^p(Q)$

We introduce the homotopy $H : L^p(Q) \times [0,1] \to L^p(Q)$ as follows

$$H(v,\lambda) = u, \qquad \forall (v,\lambda) \in L^p(Q) \times [0,1], \quad (2.9)$$

where u is the unique solution of the linear problem

$$\begin{cases} \rho c \frac{\partial}{\partial t} u - k\Delta u = \lambda\left(-\frac{\ell}{2}\frac{\partial}{\partial t}\varphi + f(t,x)\right) & (t,x) \in Q, \\ \frac{\partial}{\partial \nu} u + hu = w(t,x) & \text{on } \Sigma, \\ u(0,x) = u_0(x) & \text{on } \Omega, \end{cases} \quad (2.10)$$

with φ represents the unique solution of the nonlinear parabolic boundary value problem

$$\begin{cases} \tau \frac{\partial}{\partial t}\varphi - \xi^2 \Delta\varphi = \frac{1}{2a}(\varphi - \varphi^3) + 2v(t,x) + g(t,x) & (t,x) \in Q, \\ \frac{\partial}{\partial \nu}\varphi = 0 & \text{on } \Sigma, \\ \varphi(0,x) = \varphi_0(x) & x \in \Omega. \end{cases} \quad (2.11)$$

Since $p \leq q$ (see (2.2)), then $v + g \in L^p(Q)$. Using Theorem 2.2 above we see that there exists a unique solution $\varphi \in W_p^{2,1}(Q)$ of (2.11) and thus $-\frac{\ell}{2}\frac{\partial}{\partial t}\varphi + f(t,x) \in L^p(Q)$. The L_p-theory guarantees that the linear parabolic equation (2.10) has a unique solution $u \in W_p^{2,1}(Q)$. Hence the mapping H introduced in (2.9) is well defined.

We shall prove now the following technical lemmas

Lemma 2.1. *The mapping $H : L^p(Q) \times [0,1] \to L^p(Q)$ in (2.9) has the following properties:*

(i) $H(\cdot,\lambda) : L^p(Q) \to L^p(Q)$ is compact for every $\lambda \in [0,1]$, i.e., it is continuous and maps bounded sets into relatively compact sets;[1]

(ii) for every $\varepsilon > 0$ and every bounded set $A \subset L^p(Q)$ there exists $\delta > 0$ such that

$$\|H(v,\lambda_1) - H(v,\lambda_2)\|_{L^p(Q)} < \varepsilon$$

whenever $v \in A$ and $|\lambda_1 - \lambda_2| < \delta$.

Proof. (i) Let us check the continuity of $H(\cdot,\lambda)$, $\forall\lambda \in [0,1]$, at the point $\bar{v} \in L^p(Q)$. Let $\bar{u} = H(\bar{v},\lambda)$ and for any $v \in L^p(Q)$, $u = H(v,\lambda)$. By (2.9) and (2.10) we derive

$$\begin{cases} \rho c \frac{\partial}{\partial t}(u-\bar{u}) - k\Delta(u-\bar{u}) = -\lambda\frac{\ell}{2}\frac{\partial}{\partial t}(\varphi-\bar{\varphi}) & \text{in } Q, \\ \frac{\partial}{\partial \nu}(u-\bar{u}) + h(u-\bar{u}) = 0 & \text{on } \Sigma, \\ (u-\bar{u})(0,x) = 0 & x \in \Omega. \end{cases} \tag{2.12}$$

Applying the L_p–theory to the linear problem (2.12) it follows that there is a constant $C > 0$ such that

$$\|u-\bar{u}\|_{W_p^{2,1}(Q)} \le C\left\|\frac{\partial}{\partial t}(\varphi-\bar{\varphi})\right\|_{L^p(Q)}. \tag{2.13}$$

Estimate (2.8) in Theorem 2.2 above applied to (2.11) in place of (2.6) choosing $v_1 = \varphi$, $v_2 = \bar{\varphi}$, $v_0^1 = v_0^2 = \varphi_0$ and $\bar{g}_1 = v + g$, $\bar{g}_2 = \bar{v} + g$ insures that

$$\|\varphi-\bar{\varphi}\|_{W_p^{2,1}(Q)} \le C\|v-\bar{v}\|_{L^p(Q)}. \tag{2.14}$$

[1] A relatively compact subspace (or relatively compact subset) Y of a topological space X is a subset whose closure is compact.

We emphasize that we could apply (2.8) because by convergence of v to $\bar{v}$ in $L^p(Q)$, by (2.7) we have that φ runs in a bounded set in $W_p^{2,1}(Q)$ containing $\bar{\varphi}$. Relations (2.13) and (2.14) yield the estimate

$$\|u - \bar{u}\|_{L^p(Q)} \leq C\|v - \bar{v}\|_{L^p(Q)}. \tag{2.15}$$

Hence, by (2.15) the continuity of the map $H(\cdot, \lambda)$ at $\bar{u}$ is true, for each $\lambda \in [0, 1]$.

We have that $H(\cdot, \lambda)$ is expressed as the composition

$$L^p(Q) \longrightarrow W_p^{2,1}(Q) \subset L^\mu(Q) \subset L^p(Q), \tag{2.16}$$

where the second map ($W_p^{2,1}(Q) \subset L^p(Q)$) is a compact inclusion by Lions and Peetre embedding theorem Lions [3] (see relation (5.3) in Section 1.5). Therefore the map $H(\cdot, \lambda)$ is compact. **q.e.d.**

(ii) Fix $\varepsilon > 0$ and a bounded set $A \subset L^p(Q)$. Consider (u_1, λ_1), $(u_2, \lambda_2) \in L^p(Q) \times [0, 1]$ solving (2.10) where we take any $v \in A$ in (2.11) then we have

$$\begin{cases} \rho c \frac{\partial}{\partial t}(u_1 - u_2) - k\Delta(u_1 - u_2) & \\ \quad = -\frac{\ell}{2}\left(\lambda_1 \frac{\partial}{\partial t}\varphi_1 - \lambda_2 \frac{\partial}{\partial t}\varphi_2\right) + (\lambda_1 - \lambda_2) f & \text{in } Q, \\ \frac{\partial}{\partial \nu}(u_1 - u_2) + h(u_1 - u_2) = 0 & \text{on } \Sigma, \\ (u_1 - u_2)(0, x) = 0 & x \in \Omega. \end{cases} \tag{2.17}$$

The same argument as the one to obtain (2.14), relying on (2.7), (2.8) in Theorem 2.2 applied with $v_1 = \varphi_1$, $v_2 = \varphi_2$, $v_0^1 = v_0^2 = \varphi_0$ and $\bar{g}_1 = \lambda_1(v + g)$, $\bar{g}_2 = \lambda_2(v + g)$ insures that

$$\|\varphi_1 - \varphi_2\|_{W_p^{2,1}(Q)} \leq C|\lambda_1 - \lambda_2|\|v + g\|_{L^p(Q)} \leq C(A)|\lambda_1 - \lambda_2|. \tag{2.18}$$

By the L_p-theory applied to the linear problem (2.17) we get

$$\|u_1 - u_2\|_{W_p^{2,1}(Q)}$$

$$\leq C\left[|\lambda_1 - \lambda_2|\left(\left\|\frac{\partial}{\partial t}\varphi_1\right\|_{L^p(Q)} + \|f\|_{L^p(Q)}\right) + \lambda_2\left\|\frac{\partial}{\partial t}(\varphi_1 - \varphi_2)\right\|_{L^p(Q)}\right].$$

Estimate (2.7) assures that $\left\|\frac{\partial}{\partial t}\varphi_1\right\|_{L^p(Q)}$ is uniformly bounded with respect to v, because A is bounded. Then, on the basis of (2.18), the estimate above is of the following form

$$\|u_1 - u_2\|_{W_p^{2,1}(Q)} \leq C(A)|\lambda_1 - \lambda_2|. \tag{2.19}$$

It is clear from (2.19) that assertion (ii) is verified. **q.e.d.**

Lemma 2.2. *There exists a number $r > 0$ with the property*

$$H(u, \lambda) = u \quad \Longrightarrow \|u\|_{L^p(Q)} < r. \tag{2.20}$$

Proof. Using (2.9), the equality $H(u, \lambda) = u$ is equivalent to (2.10), where φ is the unique solution of the nonlinear parabolic boundary value problem (see Theorem 2.2 above)

$$\begin{cases} \tau\frac{\partial}{\partial t}\varphi - \xi^2\Delta\varphi = \frac{1}{2a}(\varphi - \varphi^3) + 2u + g & \text{in } Q, \\ \frac{\partial}{\partial \nu}\varphi = 0 & \text{on } \Sigma, \\ \varphi(0, x) = \varphi_0(x) & x \in \Omega. \end{cases} \tag{2.21}$$

Since $\bar{g} = 2u + g \in L^p(Q)$ (see (2.6) and (2.2)) then by Theorem 2.2 we have a unique solution $\varphi \in W_p^{2,1}(Q)$ that exists and solves (2.21). In addition, estimate (2.7) is valid for φ and $u + g$ replacing v and $\bar{g}$, respectively. Then (2.7) and the continuous inclusion $W_p^{2,1}(Q) \subset L^p(Q)$ imply that

$$\left\|\frac{\partial}{\partial t}\varphi\right\|_{L^p(Q)} \leq \|\varphi\|_{W_p^{2,1}(Q)} \tag{2.22}$$

$$\leq C\left(1 + \|\varphi_0\|_{W_q^{2-2/q}(\Omega)}^{(2p-1)/p} + \|u\|_{L^p(Q)} + \|g\|_{L^p(Q)}\right),$$

where $C > 0$ is a constant which depends on $|\Omega|, T, N, p$ and physical parameters.

Multiplying the first equation in (2.10) by $|u|^{p-2}u$, integrating over Q_t, $(t \in (0, T))$ and using Fubini's theorem, Green's formula and Young's inequality lead to

$$\frac{\rho c}{p}\int_\Omega |u(t,x)|^p dx + k(p-1)\int_{Q_t} |\nabla u|^2|u|^{p-2} ds dx + kh\int_{\Sigma_t} |u|^p ds d\gamma \tag{2.23}$$

$$\leq \frac{\rho c}{p}\|u_0\|^p_{L^p(\Omega)} + \frac{\ell}{2}\frac{1}{p}\int\limits_{Q_t}\left|\frac{\partial}{\partial t}\varphi\right|^p dsdx + \frac{2(p-1)}{p}\int\limits_{Q_t}|u|^p dsdx$$

$$+\frac{1}{p}\int\limits_{Q_t}|f|^p\, dsdx + k\int\limits_{\Sigma_t} w|u|^{p-2}u\, dsd\gamma,$$

for all $t \in (0,T]$. The Holder's and Cauchy's inequality applied to the term $k\int\limits_{\Sigma_t} w|u|^{p-2}u \;\; dsd\gamma$ give us:

$$k\int\limits_{\Sigma_t} w|u|^{p-2}u \;\; dsd\gamma \leq \frac{kh}{p}\int\limits_{\Sigma_t}|u|^p \;\; dsd\gamma + \frac{p-1}{ph}\int\limits_{\Sigma_t}|w|^p \;\; dsd\gamma.$$

Combining (2.22) (Q_t in place of Q) with (2.23) and making use of the last inequality, as well as of the inequality (*) in footnote[2] , it turns out that

$$\frac{\rho c}{p}\int\limits_{\Omega}|u(t,x)|^p dx \tag{2.24}$$

$$+k(p-1)\int\limits_{Q_t}|\nabla u|^2|u|^{p-2}dsdx + kh(1-\frac{1}{p})\int\limits_{\Sigma_t}|u|^p dsd\gamma$$

$$\leq \frac{\rho c}{p}\|u_0\|^p_{L^p(\Omega)} + \frac{\ell}{2}\frac{C}{p}\left(1+\|\varphi_0\|^{2p-1}_{W_q^{2-2/q}(\Omega)} + \|g\|^p_{L^p(Q)}\right)$$

$$+\left[\frac{\ell}{2}\frac{C}{p} + \frac{2(p-1)}{p}\right]\int\limits_{Q_t}|u|^p dsdx$$

$$+\frac{1}{p}\int\limits_{Q_t}|f|^p dsdx + \frac{p-1}{ph}\int\limits_{\Sigma_t}|w|^p dsd\gamma, \;\; \forall t \in (0,T].$$

Now, by Gronwall's lemma, (2.24) and (2.2) we derive the estimate

$$\|u\|_{L_p(Q)} \leq C(|\Omega|, T, N, p, q, \rho, c, \ell, k, h) \tag{2.25}$$

[2] (*) $\quad (a+b+c)^p \leq a^p + b^p + c^p$

$$*\left(1+\|u_0\|_{L^p(\Omega)}+\|\varphi_0\|^{2p-1}_{W_q^{2-2/q}(\Omega)}+\|f\|_{L^p(Q)}+\|g\|_{L^q(Q)}+\|w\|_{L^p(\Sigma)}\right).$$

The a priori estimate (2.25) ensures that a constant $r > 0$ can be found such that the property expressed in (2.20) is true. **q.e.d.**

2.1c Proof of the Theorem 2.1

Denoting $B_r := \{u \in L^p(Q) : \|u\|_{L^p(Q)} < r\}$, Lemma 2.2 assures that there exists $r > 0$ such that

$$H(u, \lambda) \neq u, \quad \text{for all } u \in \partial B_r \text{ and } \lambda \in [0,1].$$

Lemma 2.1 allows to consider (for all $\lambda \in [0,1]$) the Leray-Schauder degree $\deg(Id_{L^p(Q)} - H(\cdot,\lambda), B_r, 0)$,(see Fonseca & Gangbo [67], Definition 7.6). The homotopy invariance of Leray-Schauder degree shows that the equality below holds

$$\deg(\mathrm{Id}_{L^p(Q)} - H(\cdot,0), B_r, 0) = \deg(\mathrm{Id}_{L^p(Q)} - H(\cdot,1), B_r, 0). \quad (2.26)$$

Observing that $H(\cdot, 0) = 0$, it follows that

$$\deg(\mathrm{Id}_{L^p(Q)} - H(\cdot,0), B_r, 0) = \deg(\mathrm{Id}_{L^p(Q)}, B_r, 0) = 1. \quad (2.27)$$

From (2.26), (2.27) we conclude that problem (2.10) has a solution $u \in W_p^{2,1}(Q)$. This solution u is determined by the unique solution φ of (2.21). Since one has $W_p^{2,1}(Q) \subset L^\mu(Q)$ (see (2.3)), we can apply Theorem 2.2 for $\bar{g} = u + g \in L^\nu(Q)$, $\nu = \min\{q, \mu\}$. This ensures the existence of a solution $(u, \varphi) \in W_p^{2,1}(Q) \times W_\nu^{2,1}(Q)$ of problem (2.1). Moreover, estimate (2.7) in Theorem 2.2 yields that

$$\|\varphi\|_{W_\nu^{2,1}(Q)} \le C(|\Omega|, T, N, p, q, \tau, \xi, a) \quad (2.28)$$

$$*\left(1+\|\varphi_0\|^{(2q-1)/q}_{W_q^{2-2/q}(\Omega)} + \|u\|_{L^\nu(Q)} + \|g\|_{L^q(Q)}\right).$$

In writing (2.28) we used the embedding $W_q^{2-2/q}(\Omega) \subset W_\nu^{2-2/\nu}(Q)$.

The L_p-theory applied to (2.1) (with the unknown u), combined with estimate (2.7) expressed by relation (2.22), implies the estimate

$$\|u\|_{W_p^{2,1}(Q)} \quad (2.29)$$

$$\leq C\left(\|u_0\|_{W_p^{2-2/p}(\Omega)} + \frac{\ell}{2}\left\|\frac{\partial}{\partial t}\varphi\right\|_{L^p(Q)} + \|f\|_{L^p(Q)} + \|w\|_{W_p^{2-\frac{1}{p},1-\frac{1}{2p}}(\Sigma)}\right)$$

$$\leq C\Big(1 + \|u_0\|_{W_p^{2-2/p}(\Omega)} + \|\varphi_0\|^{(2p-1)/p}_{W_q^{2-2/q}(\Omega)}$$

$$+\|u\|_{L^p(Q)} + \|g\|_{L^p(Q)} + \|f\|_{L^p(Q)} + \|w\|_{W_p^{2-\frac{1}{p},1-\frac{1}{2p}}(\Sigma)}\Big).$$

On the basis of the interpolation inequality (see Chapter I, relation (5.2)), with $\varepsilon > 0$ sufficiently small for the embedding

$$W_p^{2,1}(Q) \subset L^\nu(Q) \subset L^p(Q), \tag{2.30}$$

relations (2.2), (2.25), (2.28), (2.29) lead to estimate (2.4).

Finally, let us verify estimate (2.5). To this end we consider (u_1, φ_1), (u_2, φ_2) as in the statement of Theorem 2.1. we already established that $u_1, u_2 \in W_p^{2,1}(Q)$ and $\varphi_1, \varphi_2 \in W_\nu^{2,1}(Q)$. Subtracting the equations in (2.1) corresponding to (u_1, φ_1), (u_2, φ_2) and using estimate (2.8) in Theorem 2.2 in conjunction with (2.2) we find

$$\left\|\frac{\partial}{\partial t}(\varphi_1 - \varphi_2)\right\|_{L^p(Q)} \leq C(|\Omega|, T, p, q)\|\varphi_1 - \varphi_2\|_{W_\nu^{2,1}(Q)} \tag{2.31}$$

$$\leq C(|\Omega|, T, M, p, q)\left(\|u_1 - u_2\|_{L^\nu(Q)} + \|g_1 - g_2\|_{L^q(Q)}\right).$$

The L_p-theory combined with (2.31) shows that[3]

$$\|u_1 - u_2\|_{W_p^{2,1}(Q)} \tag{2.32}$$

$$\leq C(|\Omega|, T)\left(\frac{\ell}{2}\left\|\frac{\partial}{\partial t}(\varphi_1 - \varphi_2)\right\|_{L^p(Q)} + \|f_1 - f_2\|_{L^p(Q)} + C_W\right)$$

$$\leq C(|\Omega|, T, M, q)\big(\|u_1 - u_2\|_{L^\nu(Q)} + \|f_1 - f_2\|_{L^p(Q)} + \|g_1 - g_2\|_{L^q(Q)} + C_W\big).$$

On the basis of (2.30) the following interpolation inequality holds

$$\|u_1 - u_2\|_{L^\nu(Q)} \leq \varepsilon\|u_1 - u_2\|_{W_p^{2,1}(Q)} + C(\varepsilon)\|u_1 - u_2\|_{L^p(Q)}. \tag{2.33}$$

[3] $C_W = \|w_1 - w_2\|_{W_p^{2-\frac{1}{p},1-\frac{1}{2p}}(\Sigma)}$, $C_L = \|w_1 - w_2\|_{L^p(\Sigma)}$.

From (2.31)-(2.33) one deduces that

$$(1-2\varepsilon)\|u_1-u_2\|_{W_p^{2,1}(Q)}+\|\varphi_1-\varphi_2\|_{W_\nu^{2,1}(Q)} \tag{2.34}$$
$$\leq C(|\Omega|,T,M,p,q,\ell,a)$$
$$\times\left(\|u_1-u_2\|_{L^p(Q)}+\|f_1-f_2\|_{L^p(Q)}+\|g_1-g_2\|_{L^q(Q)}+C_W\right).$$

In order to estimate $\|u_1-u_2\|_{L^p(Q)}$ we note that u_1-u_2 solves the problem

$$\begin{cases} \rho c\frac{\partial}{\partial t}(u_1-u_2)-k\Delta(u_1-u_2) & \\ \qquad =-\frac{\ell}{2}\frac{\partial}{\partial t}(\varphi_1-\varphi_2)+(f_1-f_2) & \text{in } Q, \\ \frac{\partial}{\partial\nu}(u_1-u_2)+h(u_1-u_2)=w_1-w_2 & \text{on } \Sigma, \\ (u_1-u_2)(0,x)=0 & x\in\Omega. \end{cases} \tag{2.35}$$

From (2.35), the same reasoning for obtaining (2.25) from (2.10) (with the only difference to use estimate (2.8) in place of (2.7)) leads to

$$\|u_1-u_2\|_{W_p^{2,1}(Q)} \tag{2.36}$$
$$\leq C(|\Omega|,T,p,q,\ell,a)\left(\|f_1-f_2\|_{L^p(Q)}+\|g_1-g_2\|_{L^q(Q)}+C_L\right).$$

Taking $\varepsilon>0$ small enough, from (2.34) and (2.36) we get (2.5).

The uniqueness of solution (u,φ) follows from relation (2.5) by taking $f_1=f_2$, $g_1=g_2$, $w_1=w_2$, and so the proof of Theorem 2.1 is finished **q.e.d.**

2.2 The phase-field system with a general nonlinearity

We consider the nonlinear parabolic problem

$$\begin{cases} \frac{\partial}{\partial t}u+\ell\frac{\partial}{\partial t}\varphi=\Delta u+f(t,x) & \text{in } Q, \\ \frac{\partial}{\partial t}\varphi=\Delta\varphi+F(t,x,\varphi)+u+g(t,x) & \text{in } Q, \\ \frac{\partial}{\partial\nu}u=\frac{\partial}{\partial\nu}\varphi=0 & \text{on } \Sigma, \\ u(0,x)=u_0(x), \quad \varphi(0,x)=\varphi_0(x) & x\in\Omega, \end{cases} \tag{P}$$

with $f \in L^p(Q)$, $g \in L^q(Q)$, $u_0 \in W_p^{2-2/p}(\Omega)$, $\varphi_0 \in W_q^{2-2/q}(\Omega)$, provided $\frac{\partial}{\partial \nu}u_0 = \frac{\partial}{\partial \nu}\varphi_0 = 0$ on $\partial\Omega$, a constant $\ell > 0$ and given numbers p, q which satisfy $q \geq p \geq 2$ (see (2.2) in this Chapter).

As we already mentioned at the beginning of this Chapter, the phase-field system (see Section 1.10) is a refinement of the classical Stefan problem in two phases by adding a new parabolic equation which contains a nonlinear term in φ. In this way new aspects of the physical phenomenon are reflected: *superheating*, *supercooling*, *effects of tension surface*, *separating zones* etc. In order to express more accurately the delicate physical features, different nonlinearities have been proposed by several authors (see Bates and Zheng [68], Hoffman and Jiang [44], Moroşanu and Motreanu [31, 34], and Penrose and Fife [69]). It is our purpose to treat problem (P) with the admissible nonlinearities $F(t, x, \varphi)$ of maximal order r in the growth conditions for F (see (H_0), (H_2) below). Under general hypotheses for the nonlinear term $F(t, x, \varphi)$, we prove here the existence, the uniqueness and the estimates of the solution to problem (P) extending the corresponding known results (see, for example, G. Caginalp, X. Chen, and C. Eck [64], P. Colli, F. Luterotti, G. Schimperna, and U. Stefanelli [70], A. Miranvile [71], and U. Stefanelli [72]). Precisely, condition (2.37) below improves the assumption $r < (N+2)/(N-2)$ if $N-2 > 0$ in the growth condition for F in Moroşanu and Motreanu [32] allowing to take a larger growth exponent $r < (N+2)/(N+2-2p)$ if $N+2-2p > 0$. This fact in conjunction with the sharp growth condition for r (cf. (2.37)) leads to better estimates.

Another novelty relative to [32] is the presence of the term $f(t, x)$ in problem (P) which could have weaker regularity properties than $g(t, x)$ (see (2.2)). In addition, we point out that for $F(x, t, \varphi) = a(x, t)\varphi + b(x, t)\varphi^2 - \varphi^3$, with $a, b \in L^\infty(Q)$ and $N = 3$ (so, $r = 3$), one obtains the nonlinear part in the problem studied by Hoffman and Jiang [44] and Liu Xiyuan [73]. The general result also includes other various relevant cases.

Basic tools in treating the problem (P) are the Leray–Schauder degree theory, properties of the Nemytskij operator, and the L_p–theory of linear parabolic equations.

The following hypotheses for the function F entering problem (P) are assumed:

$F : Q \times I\!R \to I\!R$ is a Carathéodory function, i.e., $F(\cdot, \cdot, z) : Q \to I\!R$ is measurable, $\forall z \in I\!R$, and $F(t, x, \cdot) : I\!R \longrightarrow I\!R$ is continuous, $\forall (t, x) \in Q$, with $F(\cdot, \cdot, 0) \in L^\infty(Q)$. Furthermore, the function F is required to satisfy:

(H_0) $F(t,x,z)|z|^{pr-r-1}z \leq \alpha\left(1+|z|^{pr-1}\right) - \beta|z|^{pr}$ for constants $\alpha, \beta > 0$ and $r \geq 1$ provided

$$r < \frac{N+2}{N+2-2p} \quad \text{if } N+2-2p > 0; \tag{2.37}$$

(H_1) There is a constant $a_0 \in I\!R$ such that $(F(t,x,z_1) - F(t,x,z_2))(z_1 - z_2) \leq a_0(z_1-z_2)^2$, $\forall (t,x) \in Q$, $z_1, z_2 \in I\!R$;

(H_2) There is a function $\bar{F} : Q \times I\!R^2 \to I\!R$ and a constant $b_0 > 0$ verifying the relations ($\forall (t,x) \in Q$, $z_1, z_2 \in I\!R$):

$$(F(t,x,z_1) - F(t,x,z_2))^2 \leq \bar{F}(t,x,z_1,z_2)(z_1-z_2)^2,$$
$$\bar{F}(t,x,z_1,z_2) \leq b_0(1+|z_1|^{2r-2}+|z_2|^{2r-2}).$$

A first consequence of the above assumptions gives the growth condition for F.

Lemma 2.3. *Assumption* (H_2) *implies that* F *fulfills the estimate*

$$|F(t,x,z)| \leq b_1(1+|z|^r), \quad \forall (t,x) \in Q, \quad z \in I\!R. \tag{2.38}$$

Proof. Setting $z_1 = z$ and $z_2 = 0$ in (H_2) we get

$$|F(t,x,z)| \leq |F(t,x,0)| + \bar{F}(t,x,z,0)^{1/2}|z|$$
$$\leq |F(t,x,0)| + b_0^{1/2}(1+|z|^{2r-2})^{1/2}|z|, \quad \forall z \in I\!R.$$

Since $F(\cdot,\cdot,0) \in L^\infty(Q)$, relation (2.38) follows. **q.e.d.**

For ease of reading this Section, we recall that the embedding theorem of Lions and Peetre (see and (2.3)) ensures the existence of a continuous embedding $W_p^{2,1}(Q) \subset L^\mu(Q)$ (p satisfying (2.2) and r satisfying (2.37)), where

$$\mu = \begin{cases} \infty, & \text{if } N+2 < 2p, \\ \text{any positive number} > pr & \text{if } N+2 = 2p, \\ \dfrac{p(N+2)}{N+2-2p} & \text{if } N+2 > 2p. \end{cases} \tag{2.3'}$$

Our main result in studying problem (P) is the following

Theorem 2.3. *Under assumptions* (H_0)-(H_2), *problem* (P) *has a unique solution* (u,φ) *with* $u \in W_p^{2,1}(Q)$ *and* $\varphi \in W_\nu^{2,1}(Q)$, *where* $\nu = \min\{q,\mu\}$. *In addition, the solution* (u,φ) *satisfies*

$$\|u\|_{W_p^{2,1}(Q)} + \|\varphi\|_{W_\nu^{2,1}(Q)} \tag{2.39}$$

$$\leq C\left(1 + \|u_0\|_{W_p^{2-2/p}(\Omega)} + \|\varphi_0\|_{W_q^{2-2/q}(\Omega)}^{\frac{qr-r+1}{q}} + \|f\|_{L^p(Q)} + \|g\|_{L^q(Q)}\right)$$

where the constant C *depends on* $|\Omega|$ (*the measure of* Ω),$T, N, p, q, r, \ell, \alpha$, β, b_1.

Moreover, given any number $M > 0$, *if* (u_1,φ_1), (u_2,φ_2) *are solutions of* (P) *corresponding to the data* (f_1,g_1), $(f_2,g_2) \in L^p(Q)\times L^q(Q)$, *respectively, for the same initial conditions, such that* $\|\varphi_1\|_{L^\nu(Q)}$, $\|\varphi_2\|_{L^\nu(Q)} \leq M$, *then the estimate below holds*

$$\|u_1 - u_2\|_{W_p^{2,1}(Q)} + \|\varphi_1 - \varphi_2\|_{W_\nu^{2,1}(Q)} \tag{2.40}$$

$$\leq C\left(\|f_1 - f_2\|_{L^p(Q)} + \|g_1 - g_2\|_{L^q(Q)}\right)$$

where $C > 0$ *depends on* $|\Omega|, T, N, M, p, q, r, \ell, \alpha, \beta, a_0, b_0$.

Now we will indicate a large class of functions $F(t,x,z)$ verifying the assumptions (H_0)–(H_2). For the sake of simplicity we take $F(t,x,z) = F(z)$ because the major difficulty lies in the nonlinearity of $F(t,x,z)$ with respect to z. Examples of F depending explicitly on x and t can be found in Moroşanu and Motreanu [32].

Example 2.1. Fix a number p with

$$p \geq \max\left\{2, \frac{N+2}{4}\right\}. \tag{2.41}$$

Let $F : \mathbb{R} \to \mathbb{R}$ be any function of the following form

$$F(z) = \varphi(z) - a|z|^{r-1}z, \quad \forall z \in \mathbb{R}, \tag{2.42}$$

with constants $a > 0$, $r \geq 2$ satisfying (2.37), and $\varphi \in C^1(\mathbb{R})$ which fulfills the properties

$$|\varphi'(z)| \leq b(1 + |z|^{r-2}), \quad \forall z \in \mathbb{R} \tag{2.43}$$

and

$$(\varphi(z_1) - \varphi(z_2))(z_1 - z_2) \le \gamma(z_1 - z_2)^2, \quad \forall z_1, z_2 \in I\!R, \qquad (2.44)$$

for constants $b > 0$ and $\gamma > 0$. There exists number r as required because due to (2.41), $2 < (N+2)/(N+2-2p)$ whenever $N+2-2p > 0$.

We note that (2.43) implies by integration that $\varphi(z)$ satisfies the growth condition

$$|\varphi(z)| \le \gamma_0(1 + |z|^{r-1}), \quad \forall z \in I\!R, \qquad (2.45)$$

with a constant $\gamma_0 > 0$. By (2.42) and (2.45), one finds some constant $\alpha > 0$ such that (H_0) is satisfied with $\beta = a$. Using (2.42), (2.44) and the fact that the function $f(t) = |t|^{r-1}t$ is increasing, we see that (H_1) is verified for $a_0 = \gamma$. Finally, from (2.42), (2.43) and Mean Value Theorem we have

$$(F(z_1) - F(z_2))^2 \le \frac{1}{2}(\varphi(z_1) - \varphi(z_2))^2 + \frac{a^2}{2}\left(|z_1|^{r-1}z_1 - |z_2|^{r-1}z_2\right)^2$$

$$\le \bar{F}(z_1, z_2)(z_1 - z_2)^2, \quad \forall z_1, z_2 \in I\!R,$$

where

$$\bar{F}(z_1, z_2) = \frac{1}{2}\sup_{0\le t\le 1}(\varphi'((1-t)z_1+tz_2))^2 + \frac{a^2r^2}{2}\sup_{0\le t\le 1}((1-t)z_1+tz_2)^{2(r-1)}$$

$$\le \frac{b^2}{2}(1 + |z_1|^{r-2} + |z_2|^{r-2})^2 + \frac{a^2r^2}{2}(|z_1| + |z_2|)^{2(r-1)}$$

$$\le b_0(1 + |z_1|^{2(r-1)} + |z_2|^{2(r-1)}), \quad \forall z_1, z_2 \in I\!R,$$

for some constant $b_0 > 0$, which ensures that (H_2) is satisfied.

A possible choice in (2.42) is $\varphi(z) = \varphi_1(z) + \varphi_2(z)$, $\forall z \in I\!R$, with $\varphi_1 \in C^1(I\!R)$ is non-increasing, satisfying (2.43), and $\varphi_2 \in C^1(I\!R)$ Lipschitzian on $I\!R$. The classical case of Caginalp's model [7] is obtained for $r = 2$, $\varphi_1(z) = 0$, $\varphi_2(z) = \frac{1}{2}z$, $a = \frac{1}{2}$, i.e., $F(z) = \frac{1}{2a}(z - z^3)$ (see equation $(2.1)_2$).

Proof of Theorem 2.3. Denoting

$$B_\rho := \{u \in L^p(Q) : \quad \|u\|_{L^p(Q)} < \rho\},$$

Lemma 3.2 in [34, pp. 199] assures that there exists $\rho > 0$ such that

$$H(u,\lambda) \neq u, \quad \text{for all } u \in \partial B_\rho \quad \text{and } \lambda \in [0,1].$$

Lemma 3.1 in [34, pp. 198] allows to consider, for all $\lambda \in [0,1]$, the Leray-Schauder degree $\deg(Id_{L^p(Q)} - H(\cdot,\lambda), B_\rho, 0)$, (see Fonseca & Gangbo [67], Definition 7.6). The homotopy invariance of Leray-Schauder degree shows that the equality below holds

$$\deg(\mathrm{Id}_{L^p(Q)} - H(\cdot,0), B_\rho, 0) = \deg(\mathrm{Id}_{L^p(Q)} - H(\cdot,1), B_\rho, 0). \quad (2.46)$$

Observing that $H(\cdot,0) = 0$, it follows that

$$\deg(\mathrm{Id}_{L^p(Q)} - H(\cdot,0), B_\rho, 0) = \deg(\mathrm{Id}_{L^p(Q)}, B_\rho, 0) = 1. \quad (2.47)$$

From (2.46), (2.47) we conclude that problem (P1) below

$$\begin{cases} \frac{\partial}{\partial t}u - \Delta u = \lambda\left(-\ell\frac{\partial}{\partial t}\varphi + f(t,x)\right) & \text{in } Q, \\ \frac{\partial}{\partial \nu}u = 0 & \text{on } \Sigma, \\ u(0,x) = \lambda u_0(x) & x \in \Omega, \end{cases} \quad (P1)$$

has a solution $u \in W_p^{2,1}(Q)$. This solution u is determined by the unique solution φ of (P2):

$$\begin{cases} \frac{\partial}{\partial t}\varphi - \Delta\varphi = F(t,x,\varphi) + u(t,x) + g(t,x) & \text{in Q}, \\ \frac{\partial}{\partial \nu}\varphi = 0 & \text{on } \Sigma, \\ \varphi(0,x) = \varphi_0(x) & x \in \Omega. \end{cases} \quad (P2)$$

Since one has $W_p^{2,1}(Q) \subset L^\mu(Q)$ (see (2.3′)), we can apply Theorem 2.1 in [34, pp. 191] for $\bar{g} = u + g \in L^\nu(Q)$. This ensures the existence of a solution $(u,\varphi) \in W_p^{2,1}(Q) \times W_\nu^{2,1}(Q)$ of problem (P). Moreover, estimate (2.1) in [34, pp. 191] yields that

$$\|\varphi\|_{W_\nu^{2,1}(Q)} \leq C(|\Omega|, T, N, p, q, r, \alpha, \beta, b_1) \quad (2.48)$$

$$\times\left(1+\|\varphi_0\|_{W_q^{2-2/q}(\Omega)}^{\frac{qr-r+1}{q}}+\|u\|_{L^\nu(Q)}+\|g\|_{L^q(Q)}\right).$$

In writing (2.48) we used the embedding $W_q^{2-2/q}(\Omega)\subset W_\nu^{2-2/\nu}(Q)$.

The L_p–theory applied to (P) (with the unknown u), combined with estimate (2.1) in [34, pp. 191] expressed by:

$$\left\|\frac{\partial}{\partial t}\varphi\right\|_{L^p(Q)}\leq\|\varphi\|_{W_p^{2,1}(Q)}$$

$$\leq C\left(1+\|\varphi_0\|_{W_q^{2-2/q}(\Omega)}^{\frac{pr-r+1}{p}}+\|u\|_{L^p(Q)}+\|g\|_{L^p(Q)}\right),$$

implies the estimate

$$\|u\|_{W_p^{2,1}(Q)}\leq C\left(\|u_0\|_{W_p^{2-2/p}(\Omega)}+\ell\left\|\frac{\partial}{\partial t}\varphi\right\|_{L^p(Q)}+\|f\|_{L^p(Q)}\right)\qquad(2.49)$$

$$\leq C\left(1+\|u_0\|_{W_p^{2-2/p}(\Omega)}+\|\varphi_0\|_{W_q^{2-2/q}(\Omega)}^{\frac{pr-r+1}{p}}\right.$$

$$\left.+\|u\|_{L^p(Q)}+\|g\|_{L^p(Q)}+\|f\|_{L^p(Q)}\right).$$

On the basis of the interpolation inequality (see, e.g., (2.33)), relations (2.2), (3.13) in [34, pp. 200]), (2.48), (2.49) lead to estimate (2.39).

Let us verify estimate (2.40). For this we consider (u_1,φ_1), (u_2,φ_2) as in the statement of Theorem 2.3. Subtracting the equations in (P) corresponding to (u_1,φ_1), (u_2,φ_2) and using estimate (2.8) in Theorem 2.2 in conjunction with (2.2), we find

$$\left\|\frac{\partial}{\partial t}(\varphi_1-\varphi_2)\right\|_{L^p(Q)}\leq C(|\Omega|,T,p,q)\|\varphi_1-\varphi_2\|_{W_\nu^{2,1}(Q)}\qquad(2.50)$$

$$\leq C(|\Omega|,T,N,M,p,q,r,a_0,b_0)\left(\|u_1-u_2\|_{L^\nu(Q)}+\|g_1-g_2\|_{L^q(Q)}\right).$$

The L_p-theory combined with (2.50) shows that

$$\|u_1-u_2\|_{W_p^{2,1}(Q)}\qquad(2.51)$$

$$\leq C(|\Omega|, T, N) \left(\ell \left\| \frac{\partial}{\partial t}\varphi_1 - \frac{\partial}{\partial t}\varphi_2 \right\|_{L^p(Q)} + \|f_1 - f_2\|_{L^p(Q)} \right)$$

$$\leq C(|\Omega|, T, N, M, p, q, r, a_0, b_0)$$

$$\times \left(\|u_1 - u_2\|_{L^\nu(Q)} + \|f_1 - f_2\|_{L^p(Q)} + \|g_1 - g_2\|_{L^q(Q)} \right).$$

From (2.50), (2.51) and (2.33) one deduces that

$$(1 - 2\varepsilon)\|u_1 - u_2\|_{W_p^{2,1}(Q)} + \|\varphi_1 - \varphi_2\|_{W_\nu^{2,1}(Q)} \tag{2.52}$$

$$\leq C(|\Omega|, T, N, M, p, q, r, \ell, a_0, b_0)$$

$$\times \left(\|u_1 - u_2\|_{L^p(Q)} + \|f_1 - f_2\|_{L^p(Q)} + \|g_1 - g_2\|_{L^q(Q)} \right).$$

In order to estimate $\|u_1 - u_2\|_{L^p(Q)}$ we note that $u_1 - u_2$ solves the problem

$$\begin{cases} \frac{\partial}{\partial t}(u_1 - u_2) - \Delta(u_1 - u_2) & \\ \quad = -\ell \frac{\partial}{\partial t}(\varphi_1 - \varphi_2) + (f_1 - f_2) & \text{in } Q, \\ \frac{\partial}{\partial \nu}(u_1 - u_2) = 0 & \text{on } \Sigma, \\ (u_1 - u_2)(0, x) = 0 & x \in \Omega. \end{cases} \tag{P3}$$

From (P3), the same reasoning for obtaining (3.13) in [34, pp. 200] leads to

$$\|u_1 - u_2\|_{W_p^{2,1}(Q)} \tag{2.53}$$

$$\leq C(|\Omega|, T, p, q, r, \ell, \alpha, \beta) \left(\|f_1 - f_2\|_{L^p(Q)} + \|g_1 - g_2\|_{L^q(Q)} \right).$$

Taking $\varepsilon > 0$ small enough, from (2.52) and (2.53) we get (2.40).

The uniqueness of solution (u, φ) follows from relation (2.40) by taking $f_1 = f_2$ and $g_1 = g_2$ and so the proof of Theorem 2.3 is finished **q.e.d.**

2.3 Convergence of the approximating scheme of fractional steps type in the case of homogeneous Dirichlet-Dirichlet boundary conditions

We consider the phase-field system in 1D dimension[4]

$$u_t + \frac{\ell}{2}\varphi_t = k\Delta u \qquad \text{in} \quad Q, \tag{2.54}$$

$$\tau\varphi_t = \xi^2\Delta\varphi + \frac{1}{2a}(\varphi - \varphi^3) + 2u \quad \text{in} \quad Q, \tag{2.55}$$

subject to the Dirichlet homogeneous boundary conditions

$$u = 0, \qquad \varphi = 0 \qquad \text{on} \quad \Sigma, \tag{2.56}$$

and initial conditions

$$u(0,x) = u_0(x), \qquad \varphi(0,x) = \varphi_0(x) \qquad x \in \Omega. \tag{2.57}$$

Setting $y = u + \frac{\ell}{2}\varphi$, the system (2.54)-(2.57) takes the form

$$y_t - k\Delta y + \frac{k\ell}{2}\Delta\varphi = 0 \qquad \text{in } Q, \tag{2.58}$$

$$\varphi_t - \frac{\xi^2}{\tau}\Delta\varphi + \frac{1}{\tau}\left(\ell - \frac{1}{2a}\right)\varphi + \frac{1}{2a\tau}\varphi^3 - \frac{2}{\tau}y = 0 \qquad \text{in } Q, \tag{2.59}$$

$$y = 0, \qquad \varphi = 0, \qquad \text{on } \Sigma, \tag{2.60}$$

$$y(0,x) = y_0(x) = u_0(x) + \frac{\ell}{2}\varphi_0(x), \quad \varphi(0,x) = \varphi_0(x), \quad x \in \Omega. \tag{2.61}$$

Let $X = L^2(\Omega) \times L^2(\Omega)$ be the real Banach space with the norm

$$\left\| \begin{pmatrix} y \\ \varphi \end{pmatrix} \right\|_X = \|y\|_{L^2(\Omega)} + \|\varphi\|_{L^2(\Omega)}.$$

We define the operator $A : D(A) \subset X \longrightarrow X$ by

$$A\begin{pmatrix} y \\ \varphi \end{pmatrix} = \begin{pmatrix} -k\Delta y + \frac{k\ell}{2}\Delta\varphi \\ -\frac{\xi^2}{\tau}\Delta\varphi + \frac{1}{\tau}\left(\ell - \frac{1}{2a}\right)\varphi \end{pmatrix},$$

---

[4] $u_t \overset{\text{not}}{=} \frac{\partial}{\partial t}u$

$$D(A) = \begin{pmatrix} H_0^1(\Omega) \cap H^2(\Omega) \\ H_0^1(\Omega) \cap H^2(\Omega) \end{pmatrix},$$

and the operator $B : D(B) \subset X \longrightarrow X$ by

$$B\begin{pmatrix} y \\ \varphi \end{pmatrix} = \begin{pmatrix} 0 \\ (2a\tau)^{-1}\varphi^3 - \dfrac{2}{\tau}y \end{pmatrix},$$

$$D(B) = \left\{ \begin{pmatrix} y \\ \varphi \end{pmatrix} \in X, \quad (2a\tau)^{-1}\varphi^3 - \frac{2}{\tau}y \in L^6(\Omega) \right\}.$$

Then the system (2.58)-(2.59) can be rewritten in the form[5]

$$\frac{d}{dt}\begin{pmatrix} y \\ \varphi \end{pmatrix} + A\begin{pmatrix} y \\ \varphi \end{pmatrix} + B\begin{pmatrix} y \\ \varphi \end{pmatrix} = 0. \tag{S}$$

We associate to system (S) the following approximating scheme[6]

$$\begin{cases} \begin{pmatrix} y^\varepsilon(t) \\ \varphi^\varepsilon(t) \end{pmatrix}' + A\begin{pmatrix} y^\varepsilon(t) \\ \varphi^\varepsilon(t) \end{pmatrix} = 0 & \text{in } (i\varepsilon, (i+1)\varepsilon), \\ \varphi^\varepsilon(i\varepsilon) = z(\varepsilon) & i = 0, 1, ..., M_\varepsilon - 1, \end{cases} \tag{2.62}$$

$$z'(t) + B\begin{pmatrix} y^\varepsilon(t) \\ \varphi^\varepsilon(t) \end{pmatrix} = 0 \quad \text{in } (0, \varepsilon), \tag{2.63}$$

$$z(0) = \varphi_-^\varepsilon(i\varepsilon), \quad \varphi_-^\varepsilon(0) = \varphi_0(x), \tag{2.64}$$

where $0 < \varepsilon < \cdots < M_\varepsilon \leq T$ is a partition of the time-interval $[0, T]$, $\varphi_+^\varepsilon(i\varepsilon)$ is the right limit of $\varphi^\varepsilon(t)$ at $i\varepsilon$ and $\varphi_-^\varepsilon(i\varepsilon)$ is the left limit of $\varphi^\varepsilon(t)$ at $i\varepsilon$.

We will treat the convergence of the numerical scheme (2.62)-(2.64) on the basis of the abstract approximation result due to Barbu & Iannelli [38] (see Section 1.8). This result is not applicable directly to the problem (2.62)-(2.64) because we cannot find a subset C as in hypothesis (H2) such that the operator B be continuous on C. Therefore, we

---

5 for other abstract framework of the phase-field equations (2.54)-(2.57) see, e.g., P.W. Bates & Zheng Songmu [68].

6 we will often denote $\frac{\partial}{\partial t}u$ by $\frac{d}{dt}u$ or u' when discussing u as a function of t

will replace the operator B with another one having all the properties required by Barbu & Iannelli theorem.

Taking into account the estimations (3.27), (3.28) and the Observation 3.1 in 3.1, as well as the Theorem 3.1 in Hoffmann & Jiang [44], we have for $u_0, \varphi_0 \in W^{2,\infty}(\Omega)$, the problem (2.54)-(2.57) (in particular the problem (2.58)-(2.61)) has a solution $u, \varphi \in W_2^{2,1}(Q)$ and, by Lions-Peetre embedding theorem (see Section 1.5, relation (5.3)), we have $u, \varphi \in L^\infty(Q)$.

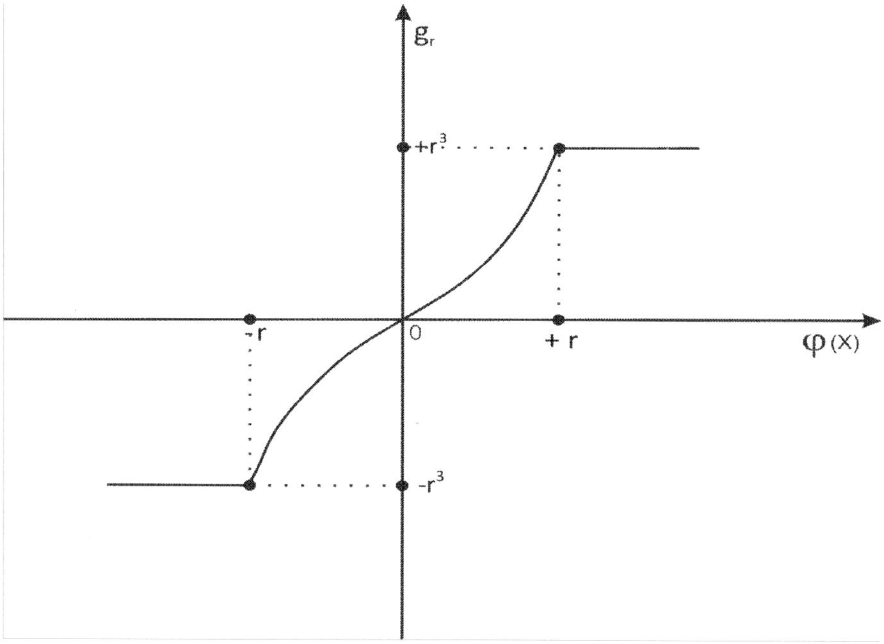

Figure 1 *The truncate operator B_r*

We consider the operator B_r defined by (see Fig. **1**)

$$B_r : X \ \rightarrow \ L^2(\Omega) \times L^\infty(\Omega) \subset X,$$

$$B_r \begin{pmatrix} y \\ \varphi \end{pmatrix} = \begin{pmatrix} 0 \\ (2a\tau)^{-1} g_r(\varphi) - \dfrac{2}{\tau} y \end{pmatrix},$$

where

$$g_r(\varphi) = \begin{cases} \varphi^3, & |\varphi| < r, \\ +r^3, & \varphi > +r, \\ -r^3, & \varphi < -r, \end{cases}$$

Substituting B in (S) and in (2.63) with B_r, we get

$$\frac{d}{dt}\begin{pmatrix} y \\ \varphi \end{pmatrix} + A\begin{pmatrix} y \\ \varphi \end{pmatrix} + B_r\begin{pmatrix} y \\ \varphi \end{pmatrix} = 0, \tag{S'}$$

$$z'(t) + B_r\begin{pmatrix} y^\varepsilon(t) \\ \varphi^\varepsilon(t) \end{pmatrix} = 0, \quad \text{on } (0, \varepsilon). \tag{2.65}$$

We associate now to system (S′) the approximating scheme given by the relations (2.62) and (2.65) coupled with (2.64). The convergence of the foregoing approximation scheme is given by the next result

Proposition 2.1. *Let $u_0, \varphi_0 \in W^{2,\infty}(\Omega)$. If $k\ell^2 < 16\xi^2/\tau$ and $\ell > 1/2a$ then the operators A, B_r together with $Y = C = K = L^2(\Omega) \times L^2(\Omega)$ satisfy all hypotheses of* Barbu & Iannelli's theorem (see Section 1.8).

We shall prove first the following lemma

Lemma 2.4. *If $k\ell^2 < 16\xi^2/\tau$ and $\ell > 1/2a$ then A is ω-accretive operator and satisfies the range condition*

$$R(I + \lambda A) \supset \overline{D(A)} \qquad \forall\, \lambda \in (0, \lambda_0). \tag{2.66}$$

Proof. Using the Definition 7.1 in Section 1.7, we must show that, for every $\binom{y}{\varphi} \in X$, (A is linear and J is univalued, i.e., $J\binom{y}{\varphi} = \binom{y}{\varphi}$)

$$\left(A\begin{pmatrix} y \\ \varphi \end{pmatrix}, \begin{pmatrix} y \\ \varphi \end{pmatrix}\right) \geq 0.$$

Using Green formula and Cauchy-Schwarz's inequality, the above inequality yields

$$\left(\begin{pmatrix} -k\Delta y + \dfrac{k\ell}{2}\Delta\varphi \\ -\dfrac{\xi^2}{\tau}\Delta\varphi + \dfrac{1}{\tau}\Big(\ell - \dfrac{1}{2a}\Big)\varphi \end{pmatrix}, \begin{pmatrix} y \\ \varphi \end{pmatrix}\right)$$

$$= k\|\nabla y\|^2_{L^2(\Omega)} - \frac{k\ell}{2}(\nabla y, \nabla\varphi) + \frac{\xi^2}{\tau}\|\nabla\varphi\|^2_{L^2(\Omega)} + \frac{1}{\tau}\Big(\ell - \frac{1}{2a}\Big)\|\varphi\|^2_{L^2(\Omega)}$$

$$\geq k\|\nabla y\|^2_{L^2(\Omega)} - \frac{k\ell}{2}\|\nabla y\|_{L^2(\Omega)} \cdot \|\nabla\varphi\|^2_{L^2(\Omega)} + \frac{\xi^2}{\tau}\|\nabla\varphi\|^2_{L^2(\Omega)}$$

$$+\frac{1}{\tau}\Big(\ell - \frac{1}{2a}\Big)\|\varphi\|^2_{L^2(\Omega)}.$$

Since $k\ell^2 < 16\xi^2/\tau$ then $k^2\ell^2/4 - 4k\xi^2/\tau \leq 0$ and then

$$k\|\nabla y\|^2_{L^2(\Omega)} - \frac{k\ell}{2}\|\nabla y\|_{L^2(\Omega)} \cdot \|\nabla\varphi\|_{L^2(\Omega)} + \frac{\xi^2}{\tau}\|\nabla\varphi\|^2_{L^2(\Omega)} \geq 0.$$

Thus, for $\ell > 1/2a$, we deduce

$$\left(A\begin{pmatrix} y \\ \varphi \end{pmatrix}, \begin{pmatrix} y \\ \varphi \end{pmatrix}\right) \geq \frac{1}{\tau}\Big(\ell - \frac{1}{2a}\Big)\|\varphi\|^2_{L^2(\Omega)}$$

$$\geq -\frac{1}{\tau}\Big(\ell - \frac{1}{2a}\Big)\Big(\|y\|^2_{L^2(\Omega)} + \|\varphi\|^2_{L^2(\Omega)}\Big),$$

i.e.,

$$\left(A\begin{pmatrix} y \\ \varphi \end{pmatrix} + \frac{1}{\tau}\Big(\ell - \frac{1}{2a}\Big)\begin{pmatrix} y \\ \varphi \end{pmatrix}, \begin{pmatrix} y \\ \varphi \end{pmatrix}\right) \geq 0.$$

Hence the operator A is ω-accretive with $\omega = \dfrac{1}{\tau}\Big(\ell - \dfrac{1}{2a}\Big)$ (see Definition 7.1.c in Section 1.7).

It is easy to verify that for every $\begin{pmatrix} f \\ g \end{pmatrix} \in L^2(\Omega) \times L^2(\Omega) = \overline{D(A)}$ the system

$$\begin{cases} y - \lambda k\Delta y = f - \lambda\frac{k\ell}{2}\Delta\varphi \in L^2(\Omega), \\ \Big[1 + \lambda\frac{1}{\tau}\Big(\ell - \frac{1}{2a}\Big)\Big]\varphi - \frac{\lambda\xi^2}{\tau}\Delta\varphi = g \in L^2(\Omega), \end{cases}$$

has a unique solution $\begin{pmatrix} y \\ \varphi \end{pmatrix} \in D(A)$ for λ small enough (see S. Agmon, A. Douglis, L. Nirenberg [74], V. Barbu [39], H. Brézis [40]). So (2.66) is true. **q.e.d.**

Proof of the Proposition 2.1. By Proposition 3.9 pp.110 in [74], we see that $A + B_r$ is m-accretive and surjective. Taking into account Lemma 2.4 and because A is single valued and the semigruop e^{-At} is differentiable on $D(A)$ (see Barbu [39]), we remark that all hypotheses of Barbu & Iannelli theorem are fulfilled and so

$$\lim_{\varepsilon\to 0}\begin{pmatrix} y^\varepsilon(t) \\ \varphi^\varepsilon(t)\end{pmatrix} = \begin{pmatrix} y(t) \\ \varphi(t)\end{pmatrix} \qquad \forall\, t \geq 0$$

uniformly on the bounded time interval, where $\binom{y^\varepsilon}{\varphi^\varepsilon}$ is the solution of the problem (2.62) (2.65) and (2.64) while $\binom{y}{\varphi}$ is the solution of the system (S′). **q.e.d.**

Observation 2.1. If we can choose r from definition of $g_r(\varphi)$ such that $\|\varphi^\varepsilon(t,x)\|_{L^2(\Omega)} \leq r$, a.e. $x \in \Omega$ then

$$B\begin{pmatrix} y^\varepsilon(t,x) \\ \varphi^\varepsilon(t,x)\end{pmatrix} = B_r\begin{pmatrix} y^\varepsilon(t,x) \\ \varphi^\varepsilon(t,x)\end{pmatrix}$$

and then the solution of the approximate problem (S′)+(2.65) is in fact the solution of the approximate problem (S)+(2.63).

2.4 An extension of Lie-Trotter product formula

This Section is devoted to the extension of some results known for m-accretive operators (see, e.g., Barbu [39] and Brézis [40]) to ω-m-accretive operators, with a fixed real number ω. We point out that if $\omega \leq 0$, an ω-m-accretive operator is m-accretive. Consequently, throughout the Section we fix a number $\omega > 0$.

Definition 2.2. Let C be a subset of H. A one parameter family $\{S(t) : t \geq 0\}$ of operators from C into itself is said to be a *semigroup on* C if it has the following properties:

a. $S(t+s) = S(t)S(s)$, for all $t, s \geq 0$;
b. $\lim\limits_{t\to t_0} S(t)x = S(t_0)x$, $S(0)x = x$, for all $t_0 \geq 0$ and $x \in C$.

Definition 2.3. A (strongly continuous) C_0 - *semigroup* on a Banach space B is a map $S : I\!R_+ \to L(B)$ such that:

1. $S(0) = I := id_B$ (identity operator on B)
2. $\forall t, s \geq 0$, $S(t+s) = S(t)S(s)$
3. $\forall t_0 \in B$, $\|S(t)t_0 - t_0\| \to 0$, as $t \to 0$.

The *infinitesimal generator* A of a C_0 - semigroup S is defined by

$$Ax = \lim_{t \to 0} \frac{1}{t} (S(t) - I)\, x$$

whenever the limit exists. $S(t)$ may also be denoted by the symbol $S(t) = e^{tA}$.

Consider the Cauchy problem

$$\begin{cases} \dfrac{d}{dt} y(t) + Ay(t) + B(y(t)) = 0 \quad t \in (0, T], \\ y(0) = x. \end{cases}$$

Theorem 2.4 *Let $A : D(A) \subset H \longrightarrow H$ be a multi–valued, m-accretive operator and let $B : D(B) \subset H \longrightarrow H$ be a multi–valued, m-ω-accretive operator for some $\omega \geq 0$ such that $A + B : D(A) \cap D(B) \subset H \longrightarrow H$ is m-ω-accretive.*

If there exists a closed convex set $C \subset \overline{D(A)} \cap \overline{D(B)}$ such that

$$(I + \lambda A)^{-1}(C) \subset C, \quad \forall \lambda > 0,$$

$$(I + \lambda B)^{-1}(C) \subset C, \quad \forall \lambda > 0, \ \lambda\omega < 1,$$

then, for any $x \in C \cap \overline{D(A) \cap D(B)}$, we have the convergence

$$\left[S_B\left(\frac{t}{n}\right) S_A\left(\frac{t}{n}\right) \right]^n x \longrightarrow S_{A+B}(t)x \ \ as \ \ in\ H \qquad (*_{\text{II.4}})$$

uniformly on the bounded intervals of $[0, \infty)$.

Here we denote by $S_A(t)$, $S_B(t)$ and $S_{A+B}(t)$ the *semigroups generated on* H by $-A$, $-B$ and $-(A + B)$, respectively.

Proof. Firstly, we show that the operator B belongs to $\mathcal{A}(\omega)$, where $\mathcal{A}(\omega)$ stands for the class considered in Brézis and Pazy [35]. Indeed, for all $0 < \lambda < \omega^{-1}$ and $x_1, x_2 \in D(A) \cap D(B)$, we have

$$(x_1 + \lambda Bx_1 - (x_2 + \lambda Bx_2), x_1 - x_2) - (1 - \lambda\omega)(x_1 - x_2, x_1 - x_2)$$

$$= \lambda((B + \omega I)x_1 - (B + \omega I)x_2, x_1 - x_2) \geq 0.$$

The last inequality follows from the accretiveness of the operator $B+\omega I$. Then, by the Cauchy–Schwarz inequality, we obtain

$$\|x_1 + \lambda Bx_1 - (x_2 + \lambda Bx_2)\| \geq (1 - \lambda\omega)\|x_1 - x_2\|.$$

This ensures that $B \in \mathcal{A}(\omega)$ (see relation (1.1) in Brézis and Pazy [54].

Sugested by the proof of the corresponding result for m-accretive operators as given in Brézis [40, Proposition 4.4], we denote

$$F(t) = S_B(t)S_A(t) \quad \forall t \geq 0. \tag{2.67}$$

Since we have $S_A(t) : \overline{D(A)} \to \overline{D(A)}$ and $S_B(t) : \overline{D(B)} \to \overline{D(B)}$ (see Brézis and Pazy [54, pp.63]), then the composition $F(t) : \overline{D(A)} \cap \overline{D(B)} \to \overline{D(B)}$ in (2.67) is well defined on $\overline{D(A)} \cap \overline{D(B)}$.

Since $C \subset \overline{D(A)} \cap \overline{D(B)}$, Remark 7.1 in Chapter I guarantees that

$$F(t)(C) \subset C \quad \forall t \geq 0. \tag{2.68}$$

We claim that

$$\lim_{t \longrightarrow 0} \frac{1}{t}(x - F(t)x) = Ax + Bx \quad \forall x \in C \cap D(A) \cap D(B). \tag{2.69}$$

To this end, for a fixed $x \in C \cap D(A) \cap D(B)$, we denote

$$y_t = \frac{1}{t}(S_B(t)x - F(t)x) \quad \forall t \geq 0. \tag{2.70}$$

One see that

$$\|y_t\| \leq e^{\omega t}\|Ax\|. \tag{2.71}$$

Indeed, using (2.67), (2.70), relation (1.5) in Brézis and Pazy [54], we find that

$$\|y_t\| = \frac{1}{t}\|S_B(t)x - S_B(t)S_A(t)x\| \tag{2.72}$$

$$\leq e^{\omega t}\frac{1}{t}\|x - S_A(t)x\| \leq e^{\omega t}\frac{1}{t}\int_0^t \|AS_A(\tau)x\|d\tau.$$

Theorem 3.1, part (6) in Brézis [40, pp. 54], ensures that the mapping of $\tau \Rightarrow AS_A(\tau)x$ is non-increasing. Then from (2.72) we derive (2.71).

In addition, we obtain from (2.70) and (2.67)

$$\left(\frac{\xi - S_B(t)\xi}{t} + \frac{x - S_A(t)x}{t} - \frac{x - S_B(t)x}{t} - y_t, \xi - S_A(t)x\right)$$

$$= \frac{1}{t}(\xi - S_B(t)\xi - (I - S_B(t))S_A(t)x, \xi - S_A(t)x)$$

$$= \frac{1}{t}\left[\|\xi - S_A(t)x\|^2 - (S_B(t)\xi - S_B(t)S_A(t)x, \xi - S_A(t)x)\right]$$

$$\geq \frac{1}{t}\left[\|\xi - S_A(t)x\|^2 - \|S_B(t)\xi - S_B(t)S_A(t)x\|\|\xi - S_A(t)x\|\right],$$

$$\forall \xi \in D(B), \quad \forall x \in C \cap D(A) \cap D(B).$$

Again, by (1.5) in Brézis and Pazy [54] applied to the ω-operator B, the inequality above leads to

$$\left(\frac{\xi - S_B(t)\xi}{t} + \frac{x - S_A(t)x}{t} - \frac{x - S_B(t)x}{t} - y_t, \xi - S_A(t)x\right) \tag{2.73}$$

$$\geq \frac{1}{t}\left[\|\xi - S_A(t)x\|^2 - e^{\omega t}\|\xi - S_A(t)x\|^2\right]$$

$$= -\left(\omega + \frac{o(t)}{t}\right)\|\xi - S_A(t)x\|^2, \quad \forall \xi \in D(B), \quad \forall x \in C \cap D(A) \cap D(B),$$

with $o(t)/t \to 0$ in $I\!R$ as $t \to 0$.

Relation (2.71) implies that y_t is bounded as $t \to 0$, thus there exist the weak limit-points of y_t as $t \to 0$. Then letting $t \to 0$ in (2.73) and taking into account the definition of *infinitesimal generator* (see Definition 2.3 in present Section), we find that any weak limit point y of y_t satisfies

$$(B\xi + Ax - Bx - y, \xi - x) \geq -\omega\|\xi - x\|^2,$$

$$\forall \xi \in D(B), \quad \forall x \in C \cap D(A) \cap D(B).$$

Equivalently, this is expressed as follows

$$((B + \omega I)\xi - (Bx - Ax + y + \omega x), \xi - x) \geq 0, \tag{2.74}$$

$$\forall \xi \in D(B), \quad \forall x \in C \cap D(A) \cap D(B).$$

Since B is m-ω-accretive (i.e., $B+\omega I$ is m-accretive) we derive from (2.74) that $(B+\omega I)x = Bx - Ax + y + \omega x$, that is

$$y = Ax. \tag{2.75}$$

By (2.71) and (2.75) we get

$$\limsup_{t\to 0} \|y_t\| \le \|Ax\| = \|y\|.$$

This combined with the *weak convergence* $y_t \rightharpoonup y$ in H as $t \to 0$ yields that

$$y_t \to Ax \quad \text{in } H \text{ as } t \to 0. \tag{2.75'}$$

Then, by (2.75′) in conjunction with (2.67) and (2.70), one sees that

$$\lim_{t\to 0} \frac{1}{t}(x - F(t)x) = \lim_{t\to 0} \left[\frac{1}{t}(x - S_B(t)x) + y_t\right] = Bx + Ax.$$

Therefore, relation (2.69) is proved.

On the other hand let us point out that since $A+B$ is an ω-accretive operator on H, then

$$A + B \in \mathcal{A}(\omega), \tag{2.76}$$

where $\mathcal{A}(\omega)$, $\omega \ge 0$, stands for the class considered in Brézis-Pazy [54].

Furthermore, we remark that

$$\overline{D(A+B)} = H \tag{2.77}$$

and

$$R(I + \lambda(A+B)) = H, \qquad \forall 0 < \lambda < \frac{1}{\omega}. \tag{2.78}$$

Equality (2.77) is true because $A + B : D(A+B) \subset H \to H$ is an m-ω-accretive operator, so it is densely defined (see Brézis [40, pp. 26], applied for $A + B + \omega I$).

To prove (2.78) we take some $0 < \lambda < \frac{1}{\omega}$ and $x \in H$. Then, with $\mu = \frac{\lambda}{1-\lambda\omega} > 0$, by using the m-ω-accretiveness of $A+B$, there exists a unique $v \in D(A) \cap D(B)$ satisfying $v + \mu(A + B + \omega I)v = (1+\mu\omega)x$. This is equivalent to $v + \frac{\mu}{1+\mu\omega}(A+B)v = x$ which establishes (2.78).

Assumption (i) from Theorem 3.2 in Brézis-Pazy [54] holds true since

$$\|F(t)x - F(t)y\| = \|S_B(t)S_A(t)x - S_B(t)S_A(t)y\| \tag{2.79}$$

$$\leq e^{\omega t}\|S_A(t)x - S_A(t)y\| \leq e^{\omega t}\|x-y\|$$

on the basis of relation (1.5) in Brézis-Pazy [54], Theorem 3.1, part (7) in Brézis [40] and (2.67). Thus we can take

$$M(t) = e^{\omega t} = 1 + \omega t + o(t),$$

with $o(t)/t \to 0$ as $t \to 0$.

We note that there exists

$$(I + \lambda(A+B))^{-1} \quad \text{for } 0 < \lambda < \frac{1}{\omega}. \tag{2.80}$$

To show this we use the fact that $A + B + \omega I$ is m-accretive, so the inverse $(I + \lambda(A + B + \omega I))^{-1}$ is defined for all $\lambda > 0$ (see Brézis [40], Proposition 2.2, part (iii), pp. 23). Since

$$(I + \lambda(A + B + \omega I))^{-1} = (1 + \lambda\omega)^{-1}\left(I + \frac{\lambda}{1+\lambda\omega}(A+B)\right)^{-1}$$

and

$$0 < \frac{\lambda}{1+\lambda\omega} < \frac{1}{\omega}, \qquad \forall \lambda > 0,$$

it follows that the claim above is checked.

From (2.79) we have that $F(t)$ in (2.69), with any $t \geq 0$, is Lipschitz continuous with the Lipschitz constant $e^{\omega t} > 1$. Then, applying Lemma 2.2 part (i) in Brézis-Pazy [54], we know that there exists

$$(I + \frac{\lambda}{t}(I - F(t)))^{-1} \quad \text{if } 0 < \frac{\lambda}{t} < (e^{\omega t} - 1)^{-1}$$

which means that

$$0 < \lambda < \frac{t}{e^{\omega t} - 1} = \frac{1}{\omega + \frac{o(t)}{t}}.$$

Choose $\lambda_0 = \frac{1}{2\omega} > 0$. One can find $t_0 > 0$ such that $\frac{o(t)}{t} < \omega$, $0 < t < t_0$. Then we conclude that there exists $(I + \frac{\lambda}{t}(I - F(t)))^{-1}$ if $0 < \lambda < \lambda_0 = \frac{1}{2\omega}$ because $\frac{1}{2\omega} < \frac{1}{\omega + \frac{o(t)}{t}}$, for all $0 < t < t_0$.

For a fixed $x \in H = \overline{D(A) \cap D(B)}$ (see (2.77)) we set

$$z_t = \left(I + \frac{\lambda}{t}(I - F(t))\right)^{-1} x, \tag{2.81}$$

with $0 < \lambda < \frac{1}{2\omega}$ and $0 < t < t_0$.

We claim that assumption (ii) of Theorem 3.2 in Brézis-Pazy [54] is fulfilled for $T(\rho) = F(\rho)$ given by (2.69) and $\lambda_0 = \frac{1}{2\omega}$. To prove this claim it is sufficient to show that

$$z_t \longrightarrow (I + \lambda(A + B))^{-1}x \quad \text{as } t \longrightarrow 0 \tag{2.82}$$

for any $0 < \lambda < \frac{1}{2\omega}$, where the notation (2.81) has been used.

To this end, taking into account Cauchy-Schwarz inequality and again (1.5) in Brézis-Pazy [54], we compute

$$\left(\frac{1}{t}(z_t - F(t)z_t) - \frac{1}{t}(\xi - F(t)\xi), z_t - \xi\right) \tag{2.83}$$

$$= \frac{1}{t}\|z_t - \xi\|^2 - \frac{1}{t}\left(F(t)z_t - F(t)\xi, z_t - \xi\right)$$

$$\geq \frac{1}{t}\|z_t - \xi\|^2 - \frac{1}{t}\|F(t)z_t - F(t)\xi\|\|z_t - \xi\|$$

$$\geq \frac{1}{t}\|z_t - \xi\|^2 - \frac{1}{t}e^{\omega t}\|z_t - \xi\|^2$$

$$= \frac{1}{t}(1 - e^{\omega t})\|z_t - \xi\|^2 = -\left(\omega + \frac{o(t)}{t}\right)\|z_t - \xi\|^2.$$

The equality in (2.81) yields

$$\frac{1}{t}(z_t - F(t)z_t) = \frac{1}{\lambda}(x - z_t).$$

The relation above and (2.83) imply

$$\left(\omega + \frac{o(t)}{t}\right)\|z_t - \xi\|^2 + \left(\frac{1}{\lambda}(x - \xi + \xi - z_t) - \frac{1}{t}(\xi - F(t)\xi), z_t - \xi\right) \geq 0,$$

$$\forall 0 < t < t_0, \quad 0 < \lambda < \frac{1}{2\omega}, \quad \forall \xi \in D(A) \cap D(B), \quad \forall x \in H.$$

Equivalently, this can be expressed as follows

$$\left(\frac{1}{\lambda}(x-\xi), z_t-\xi\right)-\left(\frac{1}{t}(\xi-F(t)\xi), z_t-\xi\right) \tag{2.84}$$

$$\geq\left(\frac{1}{\lambda}-\omega-\frac{o(t)}{t}\right)\|z_t-\xi\|^2.$$

Next, using the Cauchy-Schwarz inequality, we arrive at

$$\frac{1}{\lambda}\|x-\xi\|+\frac{1}{t}\|\xi-F(t)\xi\| \geq\left(\frac{1}{\lambda}-\omega-\frac{o(t)}{t}\right)\|z_t-\xi\|, \tag{2.85}$$

$$\forall 0<t<t_0,\ \ 0<\lambda<\frac{1}{2\omega},\ \ \forall\xi\in D(A)\cap D(B),\ \ \forall x\in H.$$

Let us fix $0<\lambda<\frac{1}{2\omega}$ and $\xi\in D(A)\cap D(B)$ in (2.85). Then we can see that

$$\frac{1}{\lambda}-\omega-\frac{o(t)}{t}>\frac{1}{\lambda}-2\omega>0, \qquad \forall 0<t<t_0.$$

According to (2.69) we know that

$$\frac{1}{t}(\xi-F(t)\xi) \quad \text{is bounded in } H,\ \ 0<t<t_0 \tag{2.86}$$

for a possibly smaller t_0. Then, combaining (2.85) and (2.86) we deduce that

$$z_t \quad \text{is bounded in } H,\ \ 0<t<t_0. \tag{2.87}$$

Let $z_t\to z$ weakly in H, as $t\to 0$, for some $z\in H$. The existence of the weak limit point z is guaranteed by (2.87). Passing to the limit as $t\to 0$ in (2.84) and making use of (2.69), one finds that

$$\left(\frac{1}{\lambda}-\omega\right)\|z-\xi\|^2 \tag{2.88}$$

$$\leq\left(\frac{1}{\lambda}-\omega\right)\liminf_{t\to 0}\|z_t-\xi\|^2\leq\limsup_{t\to 0}\|z_t-\xi\|^2$$

$$\leq\limsup_{t\to 0}\left[\left(\frac{1}{\lambda}(x-\xi), z_t-\xi\right)-\frac{1}{t}\left(\xi-F(t)\xi, z_t-\xi\right)\right]$$

$$\leq \left(\frac{1}{\lambda}(x-\xi), z-\xi\right) - (A\xi + B\xi, z-\xi).$$

It turns out that

$$\left((A+B+\omega I)\xi + \frac{1}{\lambda}(z-x) - \omega z, \xi - z\right) \geq 0,$$

with fixed data $0 < \lambda < \frac{1}{2\omega}$, $\xi \in D(A) \cap D(B)$, $x \in H$. In view of m-ω-accretiveness of operator $A+B$ that is $A+B$ is m-accretive, and since $\xi \in D(A) \cap D(B)$ is arbitary it follows $z \in D(A) \cap D(B)$ and

$$Az + Bz + \omega z = \frac{1}{\lambda}(x-z) + \omega z.$$

On the basis of (2.80) the formula above can be expressed as follows

$$z = (I + \lambda(A+B))^{-1} x. \tag{2.89}$$

Consequently, relation (2.89) allows us to say that

$$z_t \to (I + \lambda(A+B))^{-1} x \quad \text{weakly in } H \text{ as } t \to 0 \tag{2.90}$$

for every $0 < \lambda < \frac{1}{2\omega}$ and $x \in H$ (see (2.81)).

Formula (2.89) ensures that the element $z \in D(A) \cap D(B)$ does not depend on $\xi \in D(A) \cap D(B)$ that we fixed previously. In addition, from (2.81) we observe that z_t is independent of $\xi \in D(A) \cap D(B)$. Thus, we may choose $\xi = z$ in (2.88). Property (2.88) enables us to write that

$$0 \leq \left(\frac{1}{\lambda} - \omega\right) \limsup_{t \to 0} \|z_t - z\|^2 \leq 0. \tag{2.91}$$

From (2.90) and (2.91) it results that the claim in (2.82) is checked.

All the hypotheses of Theorem 3.2 of Brézis-Pazy [54] are verified. We are now in a position to conclude that the convergence result stated in ($*_{II.4}$) is valid. This completes the proof of Theorem 2.4. **q.e.d.**

Remark 2.1. Proposition 4.4 in Brézis [40], with $C = H$ is obtained from Theorem 2.4 for $\omega = 0$.

Theorem 2.4 does not cover the multivalued situation for ω-accretive operators. When $\omega = 0$ the corresponding set-valued result was treated in Kobayashi [56].

2.4a Convergence of the approximating scheme of fractional steps type in the case of homogeneous Neumann-Neumann boundary conditions

As an application of Theorem 2.4 in Section 2.4, we treat the Caginalp's model, i.e.:

$$\begin{cases} \frac{\partial}{\partial t}u + \frac{\ell}{2}\frac{\partial}{\partial t}\varphi = k\Delta u & \text{in } Q, \\ \frac{\partial}{\partial t}\varphi = \frac{\xi^2}{\tau}\Delta\varphi + \frac{1}{2a\tau}(\varphi - \varphi^3) + \frac{2}{\tau}u & \text{in } Q, \\ \frac{\partial}{\partial \nu}u = \frac{\partial}{\partial \nu}\varphi = 0 & \text{on } \Sigma, \\ u(0,x) = u_0(x), \quad \varphi(0,x) = \varphi_0(x) & x \in \Omega, \end{cases} \tag{2.92}$$

where Ω is an open bounded subset of $I\!R^n$, $n = 1, 2, 3$, whose boundary $\partial\Omega$ is sufficiently smooth (for instance of class C^2), u, φ and the positive parameters τ, ξ, ℓ, k, a were described at the beginning of this Chapter.

Setting $y = u + \dfrac{\ell}{2}\varphi$, the system (2.92) takes the form

$$\begin{cases} \frac{\partial}{\partial t}y - k\Delta y + \frac{k\ell}{2}\Delta\varphi = 0 & \text{in } Q, \\ \frac{\partial}{\partial t}\varphi - \frac{\xi^2}{\tau}\Delta\varphi + \frac{1}{\tau}\left(\ell - \frac{1}{2a}\right)\varphi + \frac{1}{2a\tau}\varphi^3 - \frac{2}{\tau}y = 0 & \text{in } Q, \\ \frac{\partial}{\partial \nu}y = \frac{\partial}{\partial \nu}\varphi = 0 & \text{on } \Sigma, \\ y(0,x) = u_0(x) + \frac{\ell}{2}\varphi_0(x), \quad \varphi(0,x) = \varphi_0(x) & x \in \Omega. \end{cases} \tag{2.93}$$

Next, let us put (2.93) in an abstract framework. To this end we consider the space $H = L^2(\Omega) \times L^2(\Omega)$ endowed with the norm $\|\cdot\|$ defined by

$$\left\| \begin{pmatrix} y \\ \varphi \end{pmatrix} \right\|_H = \|y\|_{L^2(\Omega)} + \|\varphi\|_{L^2(\Omega)}.$$

We define the operator

$$A : D(A) = \{(y, \varphi) \in H^2(\Omega)^2 : \ \tfrac{\partial}{\partial \nu}y = \tfrac{\partial}{\partial \nu}\varphi = 0\} \subset H \to H$$

by

$$A\begin{pmatrix} y \\ \varphi \end{pmatrix} = \begin{pmatrix} -k\Delta y + \dfrac{k\ell}{2}\Delta\varphi \\ -\dfrac{\xi^2}{\tau}\Delta\varphi \end{pmatrix},$$

and the operator

$$B : D(B) = L^2(\Omega) \times L^6(\Omega) \subset H \to H$$

by

$$B\begin{pmatrix} y \\ \varphi \end{pmatrix} = \begin{pmatrix} 0 \\ \dfrac{1}{\tau}\left(\ell - \dfrac{1}{2a}\right)\varphi + \dfrac{1}{2a\tau}\varphi^3 - \dfrac{2}{\tau}y \end{pmatrix}.$$

Thus the system (2.93) becomes

$$\frac{d}{dt}\begin{pmatrix} y \\ \varphi \end{pmatrix} + A\begin{pmatrix} y \\ \varphi \end{pmatrix} + B\begin{pmatrix} y \\ \varphi \end{pmatrix} = 0. \tag{2.94}$$

Lemma 2.5. *If $k\ell^2 < 16\xi^2/\tau$ then the operator A is m-accretive.*

Proof. Using Green formula and Cauchy-Schwarz's inequality, we get

$$\left(\begin{pmatrix} -k\Delta y + \frac{k\ell}{2}\Delta\varphi \\ -\frac{\xi^2}{\tau}\Delta\varphi \end{pmatrix}, \begin{pmatrix} y \\ \varphi \end{pmatrix}\right)_H$$

$$= k\|\nabla y\|^2_{L^2(\Omega)} - \frac{k\ell}{2}\langle \nabla y, \nabla\varphi\rangle_{L^2(\Omega;I\!R^n)} + \frac{\xi^2}{\tau}\|\nabla\varphi\|^2_{L^2(\Omega)}$$

$$\geq k\|\nabla y\|^2_{L^2(\Omega)} - \frac{k\ell}{2}\|\nabla y\|_{L^2(\Omega)} \cdot \|\nabla\varphi\|_{L^2(\Omega)} + \frac{\xi^2}{\tau}\|\nabla\varphi\|^2_{L^2(\Omega)}.$$

Since $k\ell^2 < 16\xi^2/\tau$ we have

$$k\|\nabla y\|^2_{L^2(\Omega)} - \frac{k\ell}{2}\|\nabla y\|_{L^2(\Omega)} \cdot \|\nabla\varphi\|_{L^2(\Omega)} + \frac{\xi^2}{\tau}\|\nabla\varphi\|^2_{L^2(\Omega)} \geq 0,$$

thus

$$\left(A\begin{pmatrix} y \\ \varphi \end{pmatrix}, \begin{pmatrix} y \\ \varphi \end{pmatrix}\right)_H \geq 0.$$

Hence A is accretive.

It is well-known that for every $\binom{f}{g} \in H = \overline{D(A)}$ the system

$$\begin{cases} y - \lambda k \Delta y = f - \lambda \dfrac{k\ell}{2} \Delta \varphi \in L^2(\Omega), \\ \varphi - \lambda \dfrac{\xi^2}{\tau} \Delta \varphi = g \in L^2(\Omega), \end{cases}$$

has a unique solution $\binom{y}{\varphi} \in D(A)$ for every $\lambda > 0$ (see Barbu [39] p. 80), so the operator A is m-accretive. **q.e.d.**

Lemma 2.6. *The operator B is m-ω-accretive.*

Proof. Let us choose

$$\omega \geq \max \left\{ \frac{1}{\tau}, -\frac{1}{\tau}\left(\ell - \frac{1}{2a} - 1\right) \right\}. \tag{2.95}$$

We have

$$\left((B + \omega I)\begin{pmatrix} y_1 \\ \varphi_1 \end{pmatrix} - (B + \omega I)\begin{pmatrix} y_2 \\ \varphi_2 \end{pmatrix}, \begin{pmatrix} y_1 \\ \varphi_1 \end{pmatrix} - \begin{pmatrix} y_2 \\ \varphi_2 \end{pmatrix} \right)_H$$

$$= \omega \|y_1 - y_2\|^2_{L^2(\Omega)} + \left(\frac{1}{\tau}\left(\ell - \frac{1}{2a}\right) + \omega \right) \|\varphi_1 - \varphi_2\|^2_{L^2(\Omega)}$$

$$+ \frac{1}{2a\tau} \langle \varphi_1^3 - \varphi_2^3, \varphi_1 - \varphi_2 \rangle_{L^2(\Omega)} - \frac{2}{\tau} \langle y_1 - y_2, \varphi_1 - \varphi_2 \rangle_{L^2(\Omega)}$$

$$\geq \left(\omega - \frac{1}{\tau} \right) \|y_1 - y_2\|^2_{L^2(\Omega)} + \left(\frac{1}{\tau}\left(\ell - \frac{1}{2a} - 1\right) + \omega \right) \|\varphi_1 - \varphi_2\|^2_{L^2(\Omega)} \geq 0,$$

where (2.95) has been used.

It remains to be checked that $R(B + (\lambda + \omega)I) = H, \forall \lambda > 0$. Given $\alpha, \beta \in L^2(\Omega)$, the system

$$\begin{cases} (\lambda + \omega) y = \alpha \\ \dfrac{1}{\tau}\left(\ell - \dfrac{1}{2a}\right)\varphi + \dfrac{1}{2a\tau}\varphi^3 - \dfrac{2}{\tau} y + (\lambda + \omega)\varphi = \beta \end{cases}$$

has a unique solution $(y, \varphi) \in L^2(\Omega) \times L^6(\Omega)$ for every $\lambda > 0$. This expresses that the operator B is m-ω-accretive. **q.e.d.**

The lemmas above show that the conditions of Theorem 2.4 are satisfied. Theorem 2.4 suggests the following approximating scheme for solving system (2.93)

$$\begin{cases} \dfrac{d}{dt}\begin{pmatrix} y_\varepsilon \\ \varphi_\varepsilon \end{pmatrix} + A\begin{pmatrix} y_\varepsilon \\ \varphi_\varepsilon \end{pmatrix} = 0, & \text{in } [i\varepsilon, (i+1)\varepsilon], \\ y_\varepsilon(i\varepsilon) = \psi_\varepsilon((i+1)\varepsilon), & \\ \varphi_\varepsilon(i\varepsilon) = z_\varepsilon((i+1)\varepsilon), & i = 0, 1, \ldots, M-1, \end{cases}$$

$$\begin{cases} \dfrac{d}{dt}\begin{pmatrix} \psi_\varepsilon \\ z_\varepsilon \end{pmatrix} + B\begin{pmatrix} \psi_\varepsilon \\ z_\varepsilon \end{pmatrix} = 0, & \text{in } [i\varepsilon, (i+1)\varepsilon], \\ \psi_\varepsilon(i\varepsilon) = y_\varepsilon^+(i\varepsilon), \quad y_\varepsilon^+(0) = y(0,x), & \\ z_\varepsilon(i\varepsilon) = \varphi_\varepsilon^+(i\varepsilon), \quad \varphi_\varepsilon^+(0) = \varphi_0(x), & i = 0, 1, \ldots, M-1, \end{cases}$$

where $0 < \varepsilon < \cdots < M\varepsilon = T$ is a partition of the time-interval $[0, T]$, and $y_\varepsilon^+(i\varepsilon)$, $\varphi_\varepsilon^+(i\varepsilon)$ are the right limits of y_ε, φ_ε, respectively, at $i\varepsilon$. Theorem 2.4 justifies the convergence of this approximating scheme to the unique solution of system (2.93) (or (2.92)).

2.5 Convergence of the approximating scheme of fractional steps type in the case of homogeneous Cauchy-Dirichlet boundary conditions

The aim of this Section is to investigate the convergence of an iterative scheme of fractional steps type for phase-field transition system

$$\frac{\partial}{\partial t}u + \frac{\ell}{2}\frac{\partial}{\partial t}\varphi = k\Delta u + f \qquad \text{in } \; Q, \tag{2.96}$$

$$\tau\frac{\partial}{\partial t}\varphi = \xi^2\Delta\varphi + \frac{1}{2a}(\varphi - \varphi^3) + 2u + g \qquad \text{in } \; Q, \tag{2.97}$$

subject to the boundary conditions

$$\frac{\partial}{\partial \nu}u + hu = 0 \qquad \text{on } \; \Sigma, \tag{2.98}$$

$$\varphi = 0 \qquad \text{on } \; \Sigma, \tag{2.99}$$

and initial conditions

$$u(0,x) = u_0(x), \qquad \varphi(0,x) = \varphi_0(x), \qquad x \in \Omega. \tag{2.100}$$

An iterative scheme of fractional steps type for the system (2.96)-(2.100) together with the boundary condition of Neumann type, i.e.,

$$\frac{\partial}{\partial \nu} u\Big|_{\Sigma} = \frac{\partial}{\partial \nu}\varphi\Big|_{\Sigma} = 0,$$

has been analyzed by Moroşanu and Motreanu [35] (see also Subsection 2.4a).

We associate to problem (2.96)-(2.100) the following approximating scheme ($Q_i^\varepsilon = [i\varepsilon, (i+1)\varepsilon] \times \Omega$, $\Sigma_i^\varepsilon = [i\varepsilon, (i+1)\varepsilon] \times \partial\Omega$):

$$\begin{cases} \frac{\partial}{\partial t} u^\varepsilon + \frac{\ell}{2}\frac{\partial}{\partial t}\varphi^\varepsilon - k\Delta u^\varepsilon = f & \text{in } Q_i^\varepsilon, \\ \frac{\partial}{\partial \nu} u^\varepsilon + h u^\varepsilon = 0 & \text{on } \Sigma_i^\varepsilon, \end{cases} \tag{2.101}$$

$$\begin{cases} \tau\frac{\partial}{\partial t}\varphi^\varepsilon - \xi^2\Delta\varphi^\varepsilon = \frac{1}{2a}\varphi^\varepsilon + 2u^\varepsilon + g & \text{in } Q_i^\varepsilon, \\ \varphi^\varepsilon = 0 & \text{on } \Sigma_i^\varepsilon, \\ \varphi_+^\varepsilon(i\varepsilon) = z(\varepsilon, \varphi_-^\varepsilon(i\varepsilon)), \end{cases} \tag{2.102}$$

where $z(t, \varphi_-^\varepsilon(i\varepsilon))$ is the solution of the Cauchy problem

$$z'(s) + \frac{1}{2a} z(s)^3 = 0 \qquad s \in [0,T], \tag{2.103}$$

$$z(0) = \varphi_-^\varepsilon(i\varepsilon), \qquad \varphi_-^\varepsilon(0,x) = \varphi_0(x), \tag{2.104}$$

computed at $s = \varepsilon$, for $i = \overline{0, M_\varepsilon - 1}$, where $M_\varepsilon = \left[\frac{T}{\varepsilon}\right]$ and $Q_{M_\varepsilon - 1}^\varepsilon = ((M_\varepsilon - 1)\varepsilon, T) \times \Omega$. Here $\varphi_+^\varepsilon(i\varepsilon) = \lim_{t \downarrow i\varepsilon} \varphi^\varepsilon(t)$ and $\varphi_-^\varepsilon(i\varepsilon) = \lim_{t \uparrow i\varepsilon} \varphi^\varepsilon(t)$.

Definition 2.4. By weak solution to the problem (2.96)-(2.100) we mean a pair of functions $(u, \varphi) \in W \times W_0$, where

$$W = L^2(0,T; H^1(\Omega)) \cap W^{1,2}([0,T]; (H^1(\Omega))'),$$

$$W_0 = L^2(0,T;H_0^1(\Omega)) \cap W^{1,2}([0,T];H^{-1}(\Omega)),$$

which satisfy (2.96)-(2.100) in the following sense

$$\int_Q \frac{\partial}{\partial t}(u + \frac{\ell}{2}\varphi)\psi\ dxdt + k\int_Q \nabla u \nabla \psi\ dxdt + kh\int_\Sigma u\psi\ dxdt \tag{2.105}$$

$$= \int_Q f\psi\ dxdt,$$

$$\int_Q \left(\tau\frac{\partial}{\partial t}\varphi\ \zeta + \xi^2\nabla\varphi\nabla\zeta - \frac{1}{2a}(\varphi - \varphi^3)\zeta - 2u\zeta\right)\ dxdt \tag{2.106}$$

$$= \int_Q g\zeta\ dxdt,$$

for all $(\psi,\zeta) \in L^2(0,T;H^1(\Omega)) \times L^2(0,T;H_0^1(\Omega))$, together with

$$u(0,x) = u_0(x), \quad \varphi(0,x) = \varphi_0(x), \quad x \in \Omega. \tag{2.107}$$

In (2.105), (2.106) we have denoted by the same symbol $\int_Q$ the duality between $L^2(0,T;H^1(\Omega))$, $L^2(0,T;(H^1(\Omega))')$ and $L^2(0,T;H_0^1(\Omega))$, $L^2(0,T;H^{-1}(\Omega))$, respectively.

The main result of this section is the following

Theorem 2.5. *Assume that* $f, g \in L^p(Q)$, $p > 2$ *and* $u_0, \varphi_0 \in W^{1,\infty}(\Omega)$ *satisfy* $\frac{\partial u_0}{\partial \nu} + hu_0 = 0$. *Let* $(u^\varepsilon, \varphi^\varepsilon)$ *be the solution of the approximating scheme* (2.101)-(2.104). *Then, for* $\varepsilon \to 0$, *we have*

$$(u^\varepsilon(t), \varphi^\varepsilon(t)) \quad \to \quad (u^*(t), \varphi^*(t)) \qquad \textit{strongly in } L^2(\Omega) \tag{2.108}$$

for any $t \in [0,T]$, *where* $u^*, \varphi^* \in W^{2,1}([0,T];L^2(\Omega)) \cap L(0,T;H^2(\Omega))$ *is the weak solution to the problem* (2.96)-(2.100).

We shall prove first the following lemmas concerning the Cauchy problem (2.103)-(2.104).

Lemma 2.7. *If* $\varphi_-^\varepsilon(i\varepsilon, x) \in L^\infty(\Omega)$, $i = 0, 1, ..., M_\varepsilon$, *then* $z(\varepsilon) \in L^\infty(\Omega)$.

Proof. We observe that the Cauchy problem (2.103)-(2.104) can be solved directly, by separation of variables. Indeed, we write (2.103) in the form $\left(\frac{1}{2z^2}\right)' = \frac{1}{2a}$ and, integrating on $[0, \varepsilon]$, we obtain

$$z^2(\varepsilon) = \frac{(\varphi_-^\varepsilon(i\varepsilon, x))^2}{1 + \dfrac{\varepsilon}{a}(\varphi_-^\varepsilon(i\varepsilon, x))^2}, \tag{2.109}$$

whence

$$z^2(\varepsilon) \leq (\varphi_-^\varepsilon(i\varepsilon, x))^2 \quad \text{a.e. } x \in \Omega. \tag{2.110}$$

Hence $z(\varepsilon) \in L^\infty$ as claimed. **q.e.d.**

Lemma 2.8. *For $i = 0, 1, ..., M_\varepsilon - 1$, the estimate*

$$\|\nabla\varphi_+^\varepsilon(i\varepsilon, x)\|_{L^2(\Omega)} \leq \|\nabla\varphi_-^\varepsilon(i\varepsilon, x)\|_{L^2(\Omega)}. \tag{2.111}$$

holds.

Proof. Denote $\theta(t) = \nabla z(t)$. From (2.103)-(2.104) we obtain

$$\begin{cases} \theta_t + \dfrac{3}{2a}\theta z^2 = 0 & \text{on } (0, \varepsilon), \\ \theta(0) = \nabla\varphi_-^\varepsilon(i\varepsilon, x). \end{cases} \tag{2.112}$$

The solution of problem (2.112) is given by

$$\theta(\varepsilon) = e^{\int_0^\varepsilon -\frac{3}{2a}\, z^2(t,\cdot)dt}\,\theta(0)$$

from which we easily get (2.111). **q.e.d.**

Lemma 2.9. *The following estimate holds*

$$\|\varphi_+^\varepsilon(i\varepsilon, x) - \varphi_-^\varepsilon(i\varepsilon, x)\|_{L^2(\Omega)} \leq L\varepsilon, \tag{2.113}$$

where $L > 0$ is a constant depending on $|\Omega|$, $\|\varphi_-^\varepsilon(i\varepsilon, x)\|_{L^\infty(\Omega)}$ and on the parameter a.

Proof. By (2.103)-(2.104) we have

$$\frac{1}{2}\frac{d}{dt}|z(t) - z(0)|^2 \leq -\frac{1}{2a}z^3(0)(z(t) - z(0))$$

(we have used the inequality $\frac{1}{2a}z^3(t)(z(t)-z(0))-\frac{1}{2a}z^3(0)(z(t)-z(0)) \geq 0$). Integrating the above inequality on $(0,\varepsilon)$ we obtain

$$|z(\varepsilon)-z(0)| \leq \frac{\varepsilon}{2a}|z^3(0)|$$

and therefore the desired inequality is (2.113). **q.e.d.**

Proof of the Theorem 2.5. We observe first that Lemma 2.7 and Lemma 2.8 ensure that $z(\varepsilon,\varphi_0) \in W^{1,\infty}(\Omega)$. Applying now Theorem A.1 in Annexe A to the problem (2.101)-(2.102) for $i=0$ we obtain the existence of a solution $u^\varepsilon, \varphi^\varepsilon \in W_p^{2,1}(Q_0^\varepsilon) \cap L^\infty(Q_0^\varepsilon)$.

Reasoning iteratively after i we may conclude that $\varphi_-^\varepsilon(i\varepsilon,x) \in L^\infty(\Omega)$, $i=1,2,...,M_\varepsilon-1$, $\varphi_-^\varepsilon(T,x) \in L^\infty(\Omega)$, and so the problem (2.101)-(2.102) has the solution $u^\varepsilon(t,x), \varphi^\varepsilon(t,x) \in W_p^{2,1}(Q_i^\varepsilon) \cap L^\infty(Q_i^\varepsilon)$, $i=0,1,...,M_\varepsilon-1$.

Multiplying $(2.101)_1$ by $\frac{4}{\ell}u^\varepsilon$ and $(2.102)_1$ by φ_t^ε,[7] using integration by parts and Green's formula, yields

$$\frac{2}{\ell}\frac{\partial}{\partial t}\int_\Omega (u^\varepsilon)^2dx + 2\int_\Omega u^\varepsilon\varphi_t^\varepsilon dx + \frac{4k}{\ell}\int_\Omega |\nabla u^\varepsilon|^2dx \tag{2.114}$$

$$+\frac{4kh}{\ell}\int_{\partial\Omega}(u^\varepsilon)^2dx = \frac{4}{\ell}\int_\Omega fu^\varepsilon dx,$$

$$\tau\int_\Omega(\varphi_t^\varepsilon)^2dx + \frac{\xi^2}{2}\frac{\partial}{\partial t}\int_\Omega|\nabla\varphi^\varepsilon|^2dx \tag{2.115}$$

$$= 2\int_\Omega u^\varepsilon\varphi_t^\varepsilon dx + \frac{1}{2a}\int_\Omega\varphi^\varepsilon\varphi_t^\varepsilon dx + \int_\Omega g\varphi_t^\varepsilon dx.$$

Adding (2.114) and (2.115), we obtain

$$\frac{2}{\ell}\frac{\partial}{\partial t}\int_\Omega(u^\varepsilon)^2dx + \frac{4k}{\ell}\int_\Omega|\nabla u^\varepsilon|^2dx \tag{2.116}$$

---

[7] $\varphi_t^\varepsilon \overset{\text{not}}{=} \frac{\partial}{\partial t}\varphi^\varepsilon$

$$+\frac{4kh}{\ell}\int\limits_{\partial\Omega}(u^{\varepsilon})^2dx+\tau\int\limits_{\Omega}(\varphi_t^{\varepsilon})^2dx+\frac{\xi^2}{2}\frac{\partial}{\partial t}\int\limits_{\Omega}|\nabla\varphi^{\varepsilon}|^2dx$$

$$=\frac{4}{\ell}\int\limits_{\Omega}fu^{\varepsilon}dx+\frac{1}{2a}\int\limits_{\Omega}\varphi^{\varepsilon}\varphi_t^{\varepsilon}dx+\int\limits_{\Omega}g\varphi_t^{\varepsilon}dx.$$

If we now multiply $(2.102)_1$ by φ^{ε} and then we integrate over Ω, by Green's formula we deduce

$$\frac{\tau}{2}\frac{\partial}{\partial t}\int\limits_{\Omega}(\varphi^{\varepsilon})^2dx+\xi^2\int\limits_{\Omega}|\nabla\varphi^{\varepsilon}|^2dx \qquad (2.117)$$

$$=2\int\limits_{\Omega}u^{\varepsilon}\varphi^{\varepsilon}dx+\frac{1}{2a}\int\limits_{\Omega}(\varphi^{\varepsilon})^2dx+\int\limits_{\Omega}g\varphi^{\varepsilon}dx.$$

Adding (2.116) with (2.117), performing same computation we get

$$\frac{2}{\ell}\frac{\partial}{\partial t}\int\limits_{\Omega}(u^{\varepsilon})^2dx+\frac{\tau}{2}\frac{\partial}{\partial t}\int\limits_{\Omega}(\varphi^{\varepsilon})^2dx \qquad (2.118)$$

$$+\frac{4k}{\ell}\int\limits_{\Omega}|\nabla u^{\varepsilon}|^2dx+\xi^2\int\limits_{\Omega}|\nabla\varphi^{\varepsilon}|^2dx$$

$$+\frac{\tau}{2}\int\limits_{\Omega}(\varphi_t^{\varepsilon})^2dx+\frac{\xi^2}{2}\frac{\partial}{\partial t}\int\limits_{\Omega}|\nabla\varphi^{\varepsilon}|^2dx+\frac{4kh}{\ell}\int\limits_{\partial\Omega}(u^{\varepsilon})dx$$

$$\leq C\Big(\int\limits_{\Omega}(u^{\varepsilon})^2dx+\int\limits_{\Omega}(\varphi^{\varepsilon})^2dx\Big)+C(\|f\|^2_{L^2(\Omega)}+\|g\|^2_{L^2(\Omega)}).$$

Integration on $(0,\varepsilon)$ and by parts gives now

$$\frac{2}{\ell}\|u^{\varepsilon}(\varepsilon)\|^2_{L^2(\Omega)}+\frac{\tau}{2}\|\varphi_-^{\varepsilon}(\varepsilon)\|^2_{L^2(\Omega)}+\frac{4k}{\ell}\int\limits_0^{\varepsilon}\|\nabla u^{\varepsilon}\|^2_{L^2(\Omega)}ds \qquad (2.119)$$

$$+\xi^2\int\limits_0^{\varepsilon}\|\nabla\varphi^{\varepsilon}\|^2_{L^2(\Omega)}ds+\frac{\tau}{2}\int\limits_0^{\varepsilon}\|\varphi_t^{\varepsilon}\|^2_{L^2(\Omega)}ds$$

$$+\frac{\xi^2}{2}\|\nabla\varphi_-^\varepsilon(\varepsilon)\|^2_{L^2(\Omega)}+\frac{4kh}{\ell}\int_0^\varepsilon\int_{\partial\Omega}(u^\varepsilon)^2dxds$$

$$\leq\frac{2}{\ell}\|u_0\|^2_{L^2(\Omega)}+\frac{\tau}{2}\|\varphi_0\|^2_{L^2(\Omega)}+\frac{\xi^2}{2}\|\nabla\varphi_0\|^2_{L^2(\Omega)}$$

$$+C\int_0^\varepsilon(\|u^\varepsilon(s)\|^2_{L^2(\Omega)}+\|\varphi^\varepsilon(s)\|^2_{L^2(\Omega)})ds+C(\|f\|^2_{L^2(Q_0^\varepsilon)}+\|g\|^2_{L^2(Q_0^\varepsilon)}).$$

Similarly, for any $Q_i^\varepsilon,\ i=1,2,...,M_\varepsilon-2$, we obtain

$$\frac{2}{\ell}\|u^\varepsilon((i+1)\varepsilon)\|^2_{L^2(\Omega)}+\frac{\tau}{2}\|\varphi_-^\varepsilon((i+1)\varepsilon)\|^2_{L^2(\Omega)} \tag{2.120}$$

$$+\frac{4k}{\ell}\int_{i\varepsilon}^{(i+1)\varepsilon}\|\nabla u^\varepsilon\|^2_{L^2(\Omega)}ds+\xi^2\int_{i\varepsilon}^{(i+1)\varepsilon}\|\nabla\varphi^\varepsilon\|^2_{L^2(\Omega)}ds$$

$$+\frac{\tau}{2}\int_{i\varepsilon}^{(i+1)\varepsilon}\|\varphi_t^\varepsilon\|^2_{L^2(\Omega)}ds+\frac{\xi^2}{2}\|\nabla\varphi_-^\varepsilon((i+1)\varepsilon)\|^2_{L^2(\Omega)}$$

$$+\frac{4kh}{\ell}\int_{i\varepsilon}^{(i+1)\varepsilon}\int_{\partial\Omega}(u^\varepsilon)^2dxds\leq\frac{2}{\ell}\|u^\varepsilon(i\varepsilon)\|^2_{L^2(\Omega)}+\frac{\tau}{2}\|\varphi_+^\varepsilon(i\varepsilon)\|^2_{L^2(\Omega)}$$

$$+\frac{\xi^2}{2}\|\nabla\varphi_+^\varepsilon(i\varepsilon)\|^2_{L^2(\Omega)}+C\int_{i\varepsilon}^{(i+1)\varepsilon}(\|u^\varepsilon(s)\|^2_{L^2(\Omega)}+\|\varphi^\varepsilon(s)\|^2_{L^2(\Omega)})ds$$

$$+C(\|f\|^2_{L^2(Q_i^\varepsilon)}+\|g\|^2_{L^2(Q_i^\varepsilon)}).$$

On $Q^\varepsilon_{M_\varepsilon-1}$ we have the estimate

$$\frac{2}{\ell}\|u^\varepsilon(T)\|^2_{L^2(\Omega)}+\frac{\tau}{2}\|\varphi_-^\varepsilon(T)\|^2_{L^2(\Omega)} \tag{2.121}$$

$$+\frac{4k}{\ell}\int_{(M_\varepsilon-1)\varepsilon}^{T}\|\nabla u^\varepsilon\|^2_{L^2(\Omega)}ds+\xi^2\int_{(M_\varepsilon-1)\varepsilon}^{T}\|\nabla\varphi^\varepsilon\|^2_{L^2(\Omega)}ds$$

$$+\frac{\tau}{2}\int\limits_{(M_\varepsilon-1)\varepsilon}^{T}\|\varphi_t^\varepsilon\|^2_{L^2(\Omega)}ds+\frac{\xi^2}{2}\|\nabla\varphi_-^\varepsilon(T)\|^2_{L^2(\Omega)}$$

$$+\frac{4kh}{\ell}\int\limits_{(M_\varepsilon-1)\varepsilon}^{T}\int\limits_{\partial\Omega}(u^\varepsilon)^2dxds\le\frac{2}{\ell}\|u^\varepsilon((M_\varepsilon-1)\varepsilon)\|^2_{L^2(\Omega)}$$

$$\frac{\tau}{2}\|\varphi_+^\varepsilon((M_\varepsilon-1)\varepsilon)\|^2_{L^2(\Omega)}+\frac{\xi^2}{2}\|\nabla\varphi_+^\varepsilon((M_\varepsilon-1)\varepsilon)\|^2_{L^2(\Omega)}$$

$$+C\int\limits_{(M_\varepsilon-1)\varepsilon}^{T}(\|u^\varepsilon(s)\|^2_{L^2(\Omega)}+\|\varphi^\varepsilon(s)\|^2_{L^2(\Omega)})ds$$

$$+C(\|f\|^2_{L^2(Q^\varepsilon_{M_\varepsilon-1})}+\|g\|^2_{L^2(Q^\varepsilon_{M_\varepsilon-1})}).$$

From (2.119)-(2.121), taking into account the relations (2.110) (with $\varphi_+^\varepsilon(t)$ in place of $z(t)$) and (2.111), we obtain

$$\frac{2}{\ell}\|u^\varepsilon(T)\|^2_{L^2(\Omega)}+\frac{\tau}{2}\|\varphi_-^\varepsilon(T)\|^2_{L^2(\Omega)}$$

$$+\frac{4k}{\ell}\int\limits_0^T\|\nabla u^\varepsilon\|^2_{L^2(\Omega)}ds+\xi^2\int\limits_0^T\|\nabla\varphi^\varepsilon\|^2_{L^2(\Omega)}ds$$

$$+\frac{\tau}{2}\sum_{i=0}^{M_\varepsilon-1}\int\limits_{i\varepsilon}^{(i+1)\varepsilon}\|\varphi_t^\varepsilon\|^2_{L^2(\Omega)}ds+\frac{\tau}{2}\int\limits_{(M_\varepsilon-1)\varepsilon}^{T}\|\varphi_t^\varepsilon\|^2_{L^2(\Omega)}ds$$

$$+\frac{\xi^2}{2}\|\nabla\varphi_-^\varepsilon(T)\|^2_{L^2(\Omega)}+\frac{4kh}{\ell}\int\limits_0^T\int\limits_{\partial\Omega}(u^\varepsilon)^2dxds$$

$$\le\frac{2}{\ell}\|u_0\|^2_{L^2(\Omega)}+\frac{\tau}{2}\|\varphi_0\|^2_{L^2(\Omega)}+\frac{\xi^2}{2}\|\nabla\varphi_0\|^2_{L^2(\Omega)}$$

$$+C\int\limits_0^T(\|u^\varepsilon(s)\|^2_{L^2(\Omega)}+\|\varphi^\varepsilon(s)\|^2_{L^2(\Omega)})ds+C(\|f\|^2_{L^2(Q)}+\|g\|^2_{L^2(Q)}).$$

We can apply the Gronwall's inequality to the above inequality and, finally, we derive

$$\frac{\tau}{2}\sum_{i=0}^{M_\varepsilon-1}\int_{i\varepsilon}^{(i+1)\varepsilon}\|\varphi_t^\varepsilon\|^2_{L^2(\Omega)}ds+\frac{\tau}{2}\int_{(M_\varepsilon-1)\varepsilon}^{T}\|\varphi_t^\varepsilon\|^2_{L^2(\Omega)}ds \tag{2.122}$$

$$+\frac{4k}{\ell}\int_0^T\|\nabla u^\varepsilon\|^2_{L^2(\Omega)}ds+\xi^2\int_0^T\|\nabla\varphi^\varepsilon\|^2_{L^2(\Omega)}ds$$

$$+\frac{4kh}{\ell}\int_0^T\int_{\partial\Omega}(u^\varepsilon)^2dxds\le C,\quad\forall\varepsilon>0,$$

where $C>0$ **does not depend on** M_ε **and** ε**.**

Also, by (2.101) we have

$$\int_0^\varepsilon\int_\Omega(u_t^\varepsilon)^2dxds+\int_\Omega|\nabla u^\varepsilon(\varepsilon)|^2dx+\int_{\partial\Omega}(u^\varepsilon(\varepsilon))^2dx \tag{2.123}$$

$$\le C\int_0^\varepsilon\|\varphi_t^\varepsilon\|^2_{L^2(\Omega)}ds+\int_\Omega|\nabla u_0|^2dx+\int_{\partial\Omega}(u^\varepsilon(0))^2dx+C\|f\|^2_{L^2(Q_0^\varepsilon)},$$

$$\int_{i\varepsilon}^{(i+1)\varepsilon}\int_\Omega(u_t^\varepsilon)^2dxds \tag{2.124}$$

$$+\int_\Omega|\nabla u^\varepsilon((i+1)\varepsilon)|^2dx+\int_{\partial\Omega}(u^\varepsilon((i+1)\varepsilon))^2dx$$

$$\le C\int_{i\varepsilon}^{(i+1)\varepsilon}\|\varphi_t^\varepsilon\|^2_{L^2(\Omega)}ds+\int_\Omega|\nabla u^\varepsilon(i\varepsilon)|^2dx$$

$$+\int_{\partial\Omega}(u^\varepsilon(i\varepsilon))^2dx+C\|f\|^2_{L^2(Q_i^\varepsilon)},\quad i=1,...,M_\varepsilon-2,$$

$$\int_{(M_\varepsilon-1)\varepsilon}^{T}\int_{\Omega}(u_t^\varepsilon)^2dxds+\int_{\Omega}|\nabla u^\varepsilon(T)|^2dx+\int_{\partial\Omega}(u^\varepsilon(T))^2dx \tag{2.125}$$

$$\leq C\int_{(M_\varepsilon-1)\varepsilon}^{T}\|\varphi_t^\varepsilon\|^2_{L^2(\Omega)}ds+\int_{\Omega}|\nabla u^\varepsilon((M_\varepsilon-1)\varepsilon)|^2dx$$

$$+\int_{\partial\Omega}(u^\varepsilon((M_\varepsilon-1)\varepsilon))^2dx+C\|f\|^2_{L^2(Q^\varepsilon_{M_\varepsilon-1})}.$$

From (2.123)-(2.125), making use of relation (2.122), we get

$$\int_{0}^{T}\|u_t^\varepsilon\|^2_{L^2(\Omega)}ds+\int_{\Omega}|\nabla u^\varepsilon(T)|^2dx \tag{2.126}$$

$$+\int_{\partial\Omega}(u^\varepsilon(T))^2dx\leq C,\quad \forall\varepsilon>0,$$

where $C>0$ **does not depend on** M_ε **and** ε.

By virtue of estimate (2.113) we obtain

$$\sum_{i=0}^{M_\varepsilon-1}\|\varphi_+^\varepsilon(i\varepsilon)-\varphi_-^\varepsilon(i\varepsilon)\|_{L^2(\Omega)}\leq LT=C, \tag{2.127}$$

where $C>0$ **does not depend on** M_ε **and** ε.

Combining (2.122), (2.126) and (2.127), we derive

$$\overset{T}{\underset{0}{V}}\varphi^\varepsilon+\int_{0}^{T}\|u_t^\varepsilon(t)\|^2_{L^2(\Omega)}dt \tag{2.128}$$

$$+\sum_{i=0}^{M_\varepsilon-1}\int_{i\varepsilon}^{(i+1)\varepsilon}\|\varphi_t^\varepsilon(t)\|^2_{L^2(\Omega)}dt+\int_{(M_\varepsilon-1)\varepsilon}^{T}\|\varphi_t^\varepsilon(t)\|^2_{L^2(\Omega)}dt$$

$$+\int_Q (|\nabla u^\varepsilon(t,x)|^2 + |\nabla \varphi^\varepsilon(t,x)|^2)dxdt \leq C \quad \forall \varepsilon > 0, \ \ \forall t \in [0,T],$$

where $\overset{T}{\underset{0}{V}} \varphi^\varepsilon$ stands for the *total variation* of $\varphi^\varepsilon : [0,T] \rightarrow L_2(\Omega)$ (see Definition 2.8 in Section I.2). Since the injection of $L_2(\Omega)$ into $H^{-1}(\Omega)$ is compact and the set $\{\varphi_t^\varepsilon(t)\}$ is bounded in $L_2(\Omega)$ for every $t \in [0,T]$, by an infinite dimensional version of Helly-Foiaş theorem (see for instance Barbu & Precupanu [50], Remark 3.2, p.60), we conclude that there exists a bounded variation function $\varphi^* \in BV([0,T]; H^{-1}(\Omega))$ such that, on a subsequence also denoted φ^ε, we have

$$\varphi^\varepsilon(t) \rightarrow \varphi^*(t) \text{ strongly in } H^{-1}(\Omega) \quad \forall t \in [0,T]. \tag{2.129}$$

From (2.128) we may assume that

$$\varphi^\varepsilon \rightarrow \varphi^* \quad \text{weakly in } L^2(0,T; H_0^1(\Omega)). \tag{2.130}$$

Now, since the inclusion of $H_0^1(\Omega) \subset L_2(\Omega)$ is compact, by the interpolation inequality (see Chapter 1, relation (5.2)) we have

$$\|\varphi^\varepsilon(t) - \varphi^*(t)\|_{L^2(\Omega)}$$
$$\leq \lambda\|\varphi^\varepsilon(t) - \varphi^*(t)\|_{H_0^1(\Omega)} + C(\lambda)\|\varphi^\varepsilon(t) - \varphi^*(t)\|_{H^{-1}(\Omega)},$$

$\forall \varepsilon > 0$, and $\forall t \in [0,T]$. Together with (2.129) and (2.130), the above inequality yields

$$\varphi^\varepsilon \rightarrow \varphi^* \quad \text{strongly in } L^2(\Omega) \quad \forall t \in [0,T]. \tag{2.131}$$

and if we use the relation (2.128) then, from (2.101)-(2.102) we obtain the following estimates

$$\|u_t^\varepsilon\|_{L^2(0,T;L^2(\Omega))} + \|u^\varepsilon\|_{L^2(0,T;H^2(\Omega))} \leq C, \tag{2.132}$$

$$\|\varphi^\varepsilon\|_{L^2(0,T;H_0^1(\Omega)\cap H^2(\Omega))} \leq C. \tag{2.133}$$

As the inclusion of $H^2(\Omega) \subset H^1(\Omega)$ is compact, it turns out that the sequence $\{u^\varepsilon\}$ is compact in $L^2(0,T;H^1(\Omega))$. Therefore, on a subsequence, again denoted u^ε, we have

$$\begin{aligned} u^\varepsilon &\rightarrow u^* \quad \text{strongly in} \quad L^2(0,T;H^1(\Omega)), \\ u^\varepsilon &\rightarrow u^* \quad \text{weakly in} \quad L^2(0,T;H^2(\Omega)), \\ u_t^\varepsilon &\rightarrow u_t^* \quad \text{weakly in} \quad L^2(0,T;L^2(\Omega)). \end{aligned} \tag{2.134}$$

Relations (2.131)-(2.134) also imply that

$$\Delta u^\varepsilon \;\to\; \Delta u^* \;\text{ weakly in } L^2(0,T;L^2(\Omega)), \tag{2.135}$$

$$\Delta \varphi^\varepsilon \;\to\; \Delta \varphi^* \;\text{ weakly in } L^2(0,T;L^2(\Omega)). \tag{2.136}$$

Finally, from (2.131) and (2.134) we may conclude that (2.108) is valid. **q.e.d**

2.5a Weak stability of approximating scheme (2.101)-(2.104)

Let $s < t$ be two arbitrary points in $[0,T]$ such that $i\varepsilon \le s \le (i+1)\varepsilon < ... < j\varepsilon \le t$. Consider the problem

$$\begin{cases} \tau\varphi_t^\varepsilon - \xi^2\Delta\varphi^\varepsilon - \dfrac{1}{2a}\varphi^\varepsilon = 2u^\varepsilon + g & \text{in } Q_k^\varepsilon, \\ \varphi_+^\varepsilon(k\varepsilon) = z(\varepsilon) & \text{on } \Sigma_k^\varepsilon, \\ \varphi^\varepsilon = 0 & \text{on } \Omega. \end{cases} \tag{2.137}$$

In a usual way, from (2.137) we obtain

$$\int_\Omega |\varphi_-^\varepsilon((k+1)\varepsilon) - \varphi_+^\varepsilon(k\varepsilon)|^2 dx + \frac{\xi^2}{2}\int_{k\varepsilon}^{(k+1)\varepsilon}\int_\Omega |\nabla\varphi^\varepsilon|^2 dxdt$$

$$\le \frac{\xi^2}{2}\int_{k\varepsilon}^{(k+1)\varepsilon}\int_\Omega |\nabla\varphi_+^\varepsilon|^2 dxdt + C\int_{k\varepsilon}^{(k+1)\varepsilon}\int_\Omega (\varphi_+^\varepsilon(k\varepsilon))^2 dxdt$$

$$+C\int_{k\varepsilon}^{(k+1)\varepsilon}\int_\Omega ((u^\varepsilon)^2 + (\varphi^\varepsilon)^2 + g^2) dxdt.$$

Taking into account Lemma 2.7 and Lemma 2.8, the last inequality becomes

$$\sum_{k=i}^{j-1} |\varphi_-^\varepsilon((k+1)\varepsilon) - \varphi_+^\varepsilon(k\varepsilon)|_{L^2(\Omega)}^2 \tag{2.138}$$

$$\leq C\sum_{k=i}^{j-1}\int_{k\varepsilon}^{(k+1)\varepsilon}\int_{\Omega}((u^{\varepsilon})^2+(\varphi^{\varepsilon})^2+g^2)dxdt.$$

On the other hand, using Lemma 2.9 we get the estimate

$$\sum_{k=i}^{j-1}\|\varphi_+^{\varepsilon}(k\varepsilon)-\varphi_-^{\varepsilon}(k\varepsilon)\|_{L^2(\Omega)}\leq L(j-i)\varepsilon. \tag{2.139}$$

But

$$\|\varphi^{\varepsilon}(t)-\varphi^{\varepsilon}(s)\|_{L^2(\Omega)}$$

$$\leq \|\varphi^{\varepsilon}(s)-\varphi_-^{\varepsilon}(i\varepsilon)\|_{L^2(\Omega)}+\|\varphi_-^{\varepsilon}(j\varepsilon)-\varphi^{\varepsilon}(t)\|_{L^2(\Omega)}$$

$$+\sum_{k=i}^{j-1}\|\varphi_+^{\varepsilon}(k\varepsilon)-\varphi_-^{\varepsilon}(k\varepsilon)\|_{L^2(\Omega)}+\sum_{k=i}^{j-1}\|\varphi_-^{\varepsilon}((k+1)\varepsilon)-\varphi_+^{\varepsilon}(k\varepsilon)\|_{L^2(\Omega)}.$$

Along with (2.138) and (2.139) the last inequality implies

$$\|\varphi^{\varepsilon}(t)-\varphi^{\varepsilon}(s)\|_{L^2(\Omega)}\leq \tag{2.140}$$

$$\leq C\Big(|t-s|+|t-s|^{1/2}\Big(\int_s^t\int_{\Omega}((u^{\varepsilon})^2+(\varphi^{\varepsilon})^2+g^2)dxd\tau\Big)^{1/2}\Big)$$

and therefore $\varphi^* : [0,T]\to L^2(\Omega)$ is absolutely continuous and consequently almost everywhere differentiable on $[0,T]$. Hence $\varphi_t^*(t)$ exists a.e. on $(0,T)$.

Let $\nu\in L^6(\Omega)$ be an arbitary but fixed element. From (2.137) we have

$$\tau(\varphi^{\varepsilon}(t)-\varphi^{\varepsilon}(s),\varphi^{\varepsilon}(s)-\nu)+\tau\sum_{k=i+1}^{j}(\varphi_-^{\varepsilon}(k\varepsilon)-\varphi_+^{\varepsilon}(k\varepsilon),\varphi_+^{\varepsilon}(k\varepsilon)-\nu) \tag{2.141}$$

$$\leq \xi^2\int_s^t(\Delta\varphi^{\varepsilon},\varphi^{\varepsilon}-\nu)d\tau+\frac{1}{2a}\int_s^t(\varphi^{\varepsilon},\varphi^{\varepsilon}-\nu)d\tau$$

$$+\int_s^t (f^\varepsilon, \varphi^\varepsilon - \nu)d\tau, \quad \forall \nu \in L^6(\Omega),$$

where $f^\varepsilon := 2u^\varepsilon + g$ and $(\cdot, \cdot)$ stand for the inner product in $L^2(\Omega)$ and also for the duality between $H_0^1(\Omega)$ and $H^{-1}(\Omega)$.

We denote by $F : D(F) = L^6(\Omega) \subset L^2(\Omega) \to L^2(\Omega)$ the operator $z \to -\frac{1}{2a}z^3$. Because $-F = \partial\left[\frac{1}{8a}\int_\Omega z^4 dx\right]$ is m-accretive, then F is m-dissipative. Denoted by e^{-Ft} the semigroup generated by F (see Definition 2.3). Then, implying (2.103)-(2.104), we get

$$z(t) = e^{-Ft}\varphi_-^\varepsilon(i\varepsilon) \qquad \text{for } t \in (0,T), \ \ i = 0, 1, ..., M_\varepsilon - 1.$$

Thus

$$\begin{aligned}&(\varphi_-^\varepsilon(k\varepsilon) - \varphi_+^\varepsilon(k\varepsilon), \varphi_+^\varepsilon(k\varepsilon) - \nu) \\ &= (\varphi_-^\varepsilon(k\varepsilon) - e^{-F\varepsilon}\varphi_-^\varepsilon(k\varepsilon), \varphi_-^\varepsilon(k\varepsilon) - \nu) \\ &+(\varphi_-^\varepsilon(k\varepsilon) - e^{-F\varepsilon}\varphi_-^\varepsilon(k\varepsilon), e^{-F\varepsilon}\varphi_-^\varepsilon(k\varepsilon) - \varphi_-^\varepsilon(k\varepsilon)).\end{aligned} \tag{2.142}$$

We set $S_\varepsilon y := y - e^{-F\varepsilon}y$ and using (2.142) and the monotonicity of S_ε, the relation (2.141) becomes

$$\tau(\varphi^\varepsilon(t) - \varphi^\varepsilon(s), \varphi^\varepsilon(s) - \nu) + \tau\sum_{k=i+1}^{j}(S_\varepsilon, \varphi_-^\varepsilon(k\varepsilon) - \nu) \le \tag{2.143}$$

$$\le \tau\sum_{k=i+1}^{j}\|S_\varepsilon\varphi_-^\varepsilon(k\varepsilon)\|_{L^2(\Omega)}^2 + \xi^2\int_s^t(\Delta\varphi^\varepsilon, \varphi^\varepsilon - \nu)d\tau$$

$$+\frac{1}{2a}\int_s^t(\varphi^\varepsilon, \varphi^\varepsilon - \nu)d\tau + \int_s^t(2u^\varepsilon + g, \varphi^\varepsilon - \nu)d\tau, \quad \forall \nu \in L^6(\Omega).$$

Using Lemma 2.9 we have $\|S_\varepsilon\varphi_-^\varepsilon(k\varepsilon)\|_{L^2(\Omega)}^2 \le \varepsilon^2 L^2$ and therefore

$$\sum_{k=i+1}^{j}\|S_\varepsilon\varphi_-^\varepsilon(k\varepsilon)\|_{L^2(\Omega)}^2 \le L^2\varepsilon^2(j-i) = \delta_\varepsilon,$$

where $\delta_\varepsilon \to 0$ for $\varepsilon \to 0$.

We define now $\bar{\varphi}^\varepsilon(\tau) = \varphi_-^\varepsilon(k\varepsilon)$ for $\tau \in (k\varepsilon, (k+1)\varepsilon)$. Then

$$\varepsilon \sum_{k=i+1}^{j} \Big(\frac{S_\varepsilon \nu}{\varepsilon}, \varphi_-^\varepsilon(k\varepsilon) - \nu\Big) = \int_s^t \Big(\frac{S_\varepsilon \nu}{\varepsilon}, \bar{\varphi}^\varepsilon(\tau) - \nu\Big) d\tau.$$

Since, $\|\varphi_-^\varepsilon(k\varepsilon)\|_{L^2(\Omega)} \leq C$, and $\|\bar{\varphi}^\varepsilon(\tau)\|_{L^2(\Omega)} \leq C$ therefore the above integral is well defined. Using (2.108) in Theorem 2.5 we obtain

$$\lim_{\varepsilon \to 0} \varepsilon \sum_{k=i+1}^{j} \Big(\frac{S_\varepsilon \nu}{\varepsilon}, \varphi_-^\varepsilon(k\varepsilon) - \nu\Big) \tag{2.144}$$

$$= \int_s^t (-F\nu, \varphi^*(\tau) - \nu) d\tau \qquad \forall \nu \in L^6(\Omega).$$

From (2.131) and (2.136) we have

$$\begin{aligned} &\int_0^t (-\Delta\varphi^\varepsilon, \varphi^\varepsilon - \nu) d\tau \to \int_0^t (-\Delta\varphi^*(\tau), \varphi^*(\tau) - \nu) d\tau, \\ &\int_0^t (\varphi^\varepsilon, \varphi^\varepsilon - \nu) d\tau \to \int_0^t (\varphi^*, \varphi^* - \nu) d\tau. \end{aligned} \tag{2.145}$$

Taking into account the last two relations and passing to the limit for $\varepsilon \to 0$ in (2.143), we obtain

$$\tau(\varphi^*(t) - \varphi^*(s), \varphi^*(s) - \nu) + \tau \int_s^t (-F\nu, \varphi^*(\tau) - \nu) d\tau \tag{2.146}$$

$$-\xi^2 \int_s^t (\Delta\varphi^*(\tau), \varphi^*(\tau) - \nu) d\tau - \frac{1}{2a} \int_s^t (\varphi^*(\tau), \varphi^*(\tau) - \nu) d\tau$$

$$\leq \int_s^t (2u^*(\tau) + g, \varphi^*(\tau) - \nu) d\tau \quad \forall \nu \in L^6(\Omega).$$

Dividing (2.146) by $t - s > 0$ and making $s \to t$, we get

$$\left(\tau \frac{\partial}{\partial t} \varphi^*(t) - \xi^2 \Delta \varphi^*(t) - \frac{1}{2a} \varphi^*(t) - F\nu, \varphi^*(t) - \nu\right) \tag{2.147}$$

$$\leq (2u^*(t) + g(t), \varphi^*(t) - \nu) \quad a.e. \ \ t \in [0, T], \ \ \forall \nu \in L^6(\Omega).$$

Using now the maximal monotonicity of F, from (2.147) we infer that

$$\tau \frac{\partial}{\partial t} \varphi^*(t) - \xi^2 \Delta \varphi^*(t) - \frac{1}{2a}\left(\varphi^*(t) - (\varphi^*(t))^3\right) - 2u^*(t) = g \quad a.e. \, t \in [0, T].$$

So $\varphi^*(t, \cdot)$ satisfies (2.97) a.e. $t \in [0, T]$.

By (2.101) we get

$$\frac{\ell}{2} \int_Q (\psi(t), d\varphi^\varepsilon(t)) dx + \int_Q (u_t^\varepsilon - k \Delta u^\varepsilon) \psi dx dt$$

$$= \int_Q f \psi dx dt \quad \forall \psi \in L^2(0, T; H^1(\Omega)).$$

Making use of (2.132) and (2.135), we may pass to the limit for $\varepsilon \to 0$ in the last equality and, by Helly's theorem, we obtain

$$\frac{\ell}{2} \int_Q (\psi(t), d\varphi^*(t)) dx + \int_Q (u_t^* - k \Delta u^*) \psi dx dt \tag{2.148}$$

$$= \int_Q f \psi dx dt \quad \forall \psi \in L^2(0, T; H^1(\Omega)).$$

Since φ^* is absolutely continuous from $[0, T]$ to $L^2(\Omega)$ (see (2.140)) the first Stieltjes integral can by written as $\int_Q (\psi(t), \varphi_t^*(t)) dx dt$ and so (2.148) becomes

$$u_t^*(t) + \frac{\ell}{2} \varphi_t^*(t) - k \Delta u^*(t) = f \quad a.e. \ \ t \in [0, T]. \tag{2.149}$$

By tracing theorem (the map $u^\varepsilon \to u^\varepsilon|_{\partial\Omega}$ is continuous from $H^1(\Omega)$ to $H^{1/2}(\partial\Omega) \subset L^2(\partial\Omega)$) we may conclude from (2.149) that $u^*(t,x)$ satisfies (2.96) a.e. $t \in [0,T]$.

As a matter of fact it turns out from Theorem 2.5 that if $f, g \in L^2(Q)$ and $u_0, \varphi_0 \in L^2(\Omega)$, then the *weak solution* $u^*, \varphi^* \in W^{1,2}([0,T]; L^2(\Omega)) \cap L([0,T]; H^2(\Omega))$ is a *strong solution*, i.e. it is absolutely continuous in t on $[0,T]$ and satisfies a.e the system (2.96)-(2.100). So the Theorem 2.5 can be viewed as a constructive way to prove the existence in the system considered.

Corollary 2.1. *Let* $u_0, \varphi_0 \in W^{1,\infty}(\Omega) = \{y : \ y, \nabla y \in L^\infty(\Omega)\}$ *satisfy* $\frac{\partial u_0}{\partial\nu} + hu_0 = 0$. *Then* $(u^*, \varphi^*) \in W \times W_0$ (see Definition 2.4) *is a weak solution of nonlinear parabolic system* (2.96)-(2.100).

2.6 Convergence of the approximating scheme of fractional steps type in the case of non-homogeneous Cauchy-Neumann boundary conditions

Consider the following nonlinear parabolic system in $Q = (0,T] \times \Omega$ ($T > 0$ and Ω a bounded domain in $I\!R^n$ $(n = 1,2,3)$ whose boundary is smooth enough):

$$\begin{cases} \rho c\; u_t + \dfrac{\ell}{2}\varphi_t = k\Delta u & \text{in } Q, \\ \tau\varphi_t = \xi^2\Delta\varphi + \dfrac{1}{2a}(\varphi - \varphi^3) + 2u & \text{in } Q, \end{cases} \tag{2.150}$$

subject to boundary conditions

$$\begin{cases} \dfrac{\partial u}{\partial \nu} + hu = w(t,x) & \text{on } \Sigma = (0,T] \times \partial\Omega, \\ \dfrac{\partial \varphi}{\partial \nu} = 0 & \text{on } \Sigma, \end{cases} \tag{2.151}$$

and initial conditions

$$u(0,x) = u_0(x) \qquad \varphi(0,x) = \varphi_0(x) \;\text{ on }\; \Omega, \tag{2.152}$$

where: u is the reduced temperature, φ is the phase function used to distinguish between the phases of the material Ω, $u_0, \varphi_0 : \Omega \to I\!R$, $w : Q \to I\!R$ are given functions and the positive parameters: ρ,c, τ, ξ, ℓ, k, h, a have a physical meaning (see (2.1)).

At the moment t the material is considered to be *liquid* if φ is close to $1+\delta_1$ and $u(t,x) \geq \delta_2 - \delta_1$, while it is considered to be *solid* if φ is close to $-1-\delta_1$ and $u(t,x) \leq -\delta_2 + \delta_1$, with δ_1, δ_2 positive numbers. Finaly, we define the *separating region* (the interface at the moment t) the set:

$$\Omega_t = \{x \in \Omega;\ |u(t,x)| \leq -\delta_2,\ |\varphi(t,x)| \leq 1+\delta_1\}.$$

The phase-field transition system (2.150)-(2.151) subject to the homogeneous Neumann boundary conditions for unknown u, e.g.

$$\frac{\partial u}{\partial \nu} + hu = 0 \quad \text{on } \Sigma,$$

has been analyzed in Section above. For other works in this context we refere: V. Arnăutu & C. Moroşanu [8], T. Benincasa & C. Moroşanu [9], C. Moroşanu [15], [18]-[20].

For every $\varepsilon > 0$ we associate to problem (2.150)-(2.152) the following approximating scheme ($Q_i^\varepsilon = (i\varepsilon, (i+1)\varepsilon) \times \Omega$, $\Sigma_i^\varepsilon = (i\varepsilon, (i+1)\varepsilon) \times \partial\Omega$):

$$\begin{cases} \rho c\, u_t^\varepsilon + \dfrac{\ell}{2}\varphi_t^\varepsilon = k\Delta u^\varepsilon & \text{in } Q_i^\varepsilon, \\[2ex] \tau\varphi_t^\varepsilon = \xi^2\Delta\varphi^\varepsilon + \dfrac{1}{2a}\varphi^\varepsilon + 2u^\varepsilon & \text{in } Q_i^\varepsilon, \end{cases} \tag{2.153}$$

$$\begin{cases} \dfrac{\partial u^\varepsilon}{\partial\nu} + hu^\varepsilon = w(t,x) & \text{on } \Sigma_i^\varepsilon, \\[2ex] \dfrac{\partial \varphi^\varepsilon}{\partial\nu} = 0 & \text{on } \Sigma_i^\varepsilon, \end{cases} \tag{2.154}$$

$$\begin{cases} u_+^\varepsilon(i\varepsilon,x) = u_-^\varepsilon(i\varepsilon,x), \quad u_-^\varepsilon(0,x) = u_0(x) & \text{on } \Omega \\[2ex] \varphi_+^\varepsilon(i\varepsilon,x) = z((i+1)\varepsilon, \varphi_-^\varepsilon(i\varepsilon,x)) & \text{on } \Omega \end{cases} \tag{2.155}$$

where $z(\cdot, \varphi_-^\varepsilon(i\varepsilon,x))$ is the solution of Cauchy problem:

$$\begin{cases} z'(s) + \dfrac{1}{2a}z^3(s) = 0 \quad s \in (i\varepsilon, (i+1)\varepsilon), \\[2ex] z(i\varepsilon) = \varphi_-^\varepsilon(i\varepsilon,x) \quad \varphi_-^\varepsilon(0,x) = \varphi_0(x), \end{cases} \tag{2.156}$$

for $i = 0, 1, \cdots, M_\varepsilon - 1$, with $M_\varepsilon = \left[\frac{T}{\varepsilon}\right]$, $Q^\varepsilon_{M_\varepsilon - 1} = [(M_\varepsilon - 1)\varepsilon, T] \times \Omega$, $\varphi^\varepsilon_+(i\varepsilon, x) = \lim_{t\downarrow i\varepsilon} \varphi^\varepsilon(t, x)$ and $\varphi^\varepsilon_-(i\varepsilon, x) = \lim_{t\uparrow i\varepsilon} \varphi^\varepsilon(t, x)$.

Definition 2.5. *By weak solution of the nonlinear system* (2.150)-(2.152) *we mean a pair of functions* $u, \varphi \in W$, $W = L^2([0, T]; H^1(\Omega)) \cap W^{2,1}([0, T]; (H^1(\Omega))')$, (see Barbu [75], pp. 146), *which satisfy* (2.150)-(2.152) *in the following sense:*

$$\int_Q (\rho c\, u + \frac{\ell}{2}\varphi)_t \psi \, dxdt + k \int_Q \nabla u \nabla \psi \, dxdt + kh \int_\Sigma u\psi \, d\gamma dt = k \int_\Sigma w\psi \, d\gamma dt, \tag{2.157}$$

$$\tau \int_Q \varphi_t \zeta \, dxdt + \xi^2 \int_Q \nabla\varphi \nabla\zeta \, dxdt = \frac{1}{2a} \int_Q (\varphi - \varphi^3)\zeta \, dxdt + 2 \int_Q u\zeta \, dxdt, \tag{2.158}$$

$\forall\, \psi, \zeta \in L^2([0, T]; H^1(\Omega))$, and

$$u(0, x) = u_0(x) \quad \varphi(0, x) = \varphi_0(x), \quad \text{in } \Omega. \tag{2.159}$$

Definition 2.6. *By weak solution* $(u^\varepsilon, \varphi^\varepsilon)$ *of* (2.153)-(2.156) *we mean a pair of functions* $u^\varepsilon, \varphi^e \in W$ (see Definition 2.5), *which satisfy* (2.153)-(2.156) *in the following sense:*

$$\int_Q (\rho c\, u^\varepsilon + \frac{\ell}{2}\varphi^\varepsilon)_t \psi \, dxdt + k \int_Q \nabla u^\varepsilon \nabla \psi \, dxdt + kh \int_\Sigma u^\varepsilon \psi \, d\gamma dt = k \int_\Sigma w\psi \, d\gamma dt, \tag{2.160}$$

$$\tau \int_Q \varphi^\varepsilon_t \zeta \, dxdt + \xi^2 \int_Q \nabla\varphi^\varepsilon \nabla\zeta \, dxdt = \frac{1}{2a} \int_Q \varphi^\varepsilon \zeta \, dxdt + 2 \int_Q u^\varepsilon \zeta \, dxdt, \tag{2.161}$$

$\forall\, \psi, \zeta \in L^2([0, T]; H^1(\Omega))$, and

$$u^\varepsilon_-(0, x) = u_0(x) \quad \varphi^\varepsilon_-(0, x) = \varphi_0(x) \quad \text{in } \Omega. \tag{2.162}$$

In (2.157)-(2.162) we have denoted by the same symbol $\int_Q$ the duality between $L^2([0, T]; H^1(\Omega))$ and $L^2([0, T]; (H^1(\Omega))')$.

The main result of this Section is the following

Theorem 2.6. *Assume that* $u_0, \varphi_0 \in W^{1,\infty}(\Omega)$ *satisfies* $\dfrac{\partial u_0}{\partial \nu} + hu_0 = w(t,x)$, $\dfrac{\partial \varphi_0}{\partial \nu} = 0$ *and* $w \in W^1([0,T]; L^2(\partial\Omega))$. *Furthermore,* $\Omega \subset \mathbb{R}^n$ $(n = 1,2,3)$ *is a bounded domain with a smooth boundary. Let* $(u^\varepsilon, \varphi^\varepsilon)$ *be the solution of the approximating scheme* (2.153)-(2.156). *Then, for* $\varepsilon \to 0$, *one has*

$$(u^\varepsilon(t), \varphi^\varepsilon(t)) \to (u^*(t), \varphi^*(t)) \qquad \text{strongly in } L^2(\Omega) \text{ for any } t \in (0,T], \tag{2.163}$$

where $u^*, \varphi^* \in W^{2,1}([0,T]; L^2(\Omega)) \cap L^2([0,T]; H^2(\Omega))$ *is the weak solution of the nonlinear system* (2.150)-(2.152).

Before starting the proof of Theorem 2.6 we recall the existence and regularity result regarding the weak solution in (2.153)-(2.156).

Proposition 2.1. *Let* $u_0, \varphi_0 \in W^{1,\infty}(\Omega)$ *satisfy* $\frac{\partial u_0}{\partial \nu} + hu_0 = w(t,x)$, $\frac{\partial \varphi_0}{\partial \nu} = 0$ *and* $w \in W^1([0,T]; L^2(\partial\Omega))$. *Then the linear system* (2.153)-(2.155) *has a unique solution* $u^\varepsilon, \varphi^\varepsilon \in W^{2,1}(Q_i^\varepsilon) \cap L^\infty(Q_i^\varepsilon)$ *on every time interval* $[i\varepsilon, (i+1)\varepsilon]$, $i = 0, 1, ..., M_\varepsilon - 1$.

The sketch proof of Proposition 2.1 can be found in ANNEXE A.

Concerning the Cauchy problem (2.156), Lemmas 2.7 - 2.9 in Section 2.5 are valid.

Proof of Theorem 2.6. We observe first that Lemma 2.7 and Lemma 2.8 ensure that $z(\varepsilon, x) \in W^{1,\infty}(\Omega)$. Applying now Proposition 2.1 below to the problem (2.153)-(2.156) for $i = 0$ we obtain the existence of a solution $u^\varepsilon, \varphi^\varepsilon \in W^{2,1}(Q_0^\varepsilon) \cap L^\infty(Q_0^\varepsilon)$. Reasoning iteratively after i we may conclude that $\varphi_-^\varepsilon(i\varepsilon, x) \in L^\infty(\Omega)$, $i = 1, 2, ..., M_\varepsilon - 1$, and that the problem (2.153)-(2.156) has the solution u^ε, $\varphi^\varepsilon \in W^{2,1}(Q_i^\varepsilon) \cap L^\infty(Q_i^\varepsilon)$, $i = 0, 1, ..., M_\varepsilon - 1$.

Let us next establish *a priori* estimates for the solution u^ε, φ^ε on to each Q_i^ε, $i = 0, 1, ..., M_\varepsilon - 1$.

Multiplying $(2.153)_1$ by $\frac{4}{\ell}u^\varepsilon$ and $(2.153)_2$ by φ_t^ε, using integration by parts, Green's formula and the relations (2.160)-(2.161), yields

$$\frac{2\rho c}{\ell}\frac{\partial}{\partial t}\Big(\int_\Omega (u^\varepsilon)^2 dx\Big) + 2\int_\Omega u^\varepsilon \varphi_t^\varepsilon dx + \frac{4k}{\ell}\int_\Omega |\nabla u^\varepsilon|^2 dx \tag{2.164}$$

$$+\frac{4kh}{\ell}\int_{\partial\Omega} (u^\varepsilon)^2 d\gamma = \frac{4k}{\ell}\int_{\partial\Omega} u^\varepsilon w(t,x) d\gamma,$$

$$\tau\int\limits_{\Omega}(\varphi_t^{\varepsilon})^2dx+\frac{\xi^2}{2}\frac{\partial}{\partial t}\Big(\int\limits_{\Omega}|\nabla\varphi^{\varepsilon}|^2dx\Big)-2\int\limits_{\Omega}u^{\varepsilon}\varphi_t^{\varepsilon}dx=\frac{1}{2a}\int\limits_{\Omega}\varphi^{\varepsilon}\varphi_t^{\varepsilon}dx. \tag{2.165}$$

The Hölder's and Cauchy's inequalities applied to the term $\frac{4k}{\ell}\int\limits_{\partial\Omega}u^{\varepsilon}wd\gamma$ in (2.164) give us:

$$\frac{4k}{\ell}\int\limits_{\partial\Omega}u^{\varepsilon}w(t,x)d\gamma\le\frac{4kh}{\ell}\int\limits_{\partial\Omega}(u^{\varepsilon})^2d\gamma+\frac{k}{\ell h}\int\limits_{\partial\Omega}w^2(t,x)d\gamma.$$

Thus, adding (2.164)-(2.165) and making use of the last inequality, after some simple calculus we obtain:

$$\frac{2\rho c}{\ell}\frac{\partial}{\partial t}\Big(\int\limits_{\Omega}(u^{\varepsilon})^2dx\Big)+\frac{4k}{\ell}\int\limits_{\Omega}|\nabla u^{\varepsilon}|^2dx+\tau\int\limits_{\Omega}(\varphi_t^{\varepsilon})^2dx \tag{2.166}$$

$$+\frac{\xi^2}{2}\frac{\partial}{\partial t}\Big(\int\limits_{\Omega}|\nabla\varphi^{\varepsilon}|^2dx\Big)\le\frac{k}{\ell h}\int\limits_{\partial\Omega}w^2(t,x)d\gamma+\frac{1}{2a}\int\limits_{\Omega}\varphi^{\varepsilon}\varphi_t^{\varepsilon}dx.$$

If we now multiply $(2.153)_2$ by φ^{ε} and then we integrate over Ω, by Green's formula we get

$$\frac{\tau}{2}\frac{\partial}{\partial t}\Big(\int\limits_{\Omega}(\varphi^{\varepsilon})^2dx\Big)+\xi^2\int\limits_{\Omega}|\nabla\varphi^{\varepsilon}|^2dx=2\int\limits_{\Omega}u^{\varepsilon}\varphi^{\varepsilon}dx+\frac{1}{2a}\int\limits_{\Omega}(\varphi^{\varepsilon})^2dx. \tag{2.167}$$

Adding the relations (2.166)-(2.167) and performing same computations implying Buniakovsky-Schwarz's and Cauchy's inequalities, we deduce that:

$$\frac{2\rho c}{\ell}\frac{\partial}{\partial t}\Big(\int\limits_{\Omega}(u^{\varepsilon})^2dx\Big)+\frac{\tau}{2}\frac{\partial}{\partial t}\Big(\int\limits_{\Omega}(\varphi^{\varepsilon})^2dx\Big) \tag{2.168}$$

$$+\frac{4k}{\ell}\int\limits_{\Omega}|\nabla u^{\varepsilon}|^2dx+\xi^2\int\limits_{\Omega}|\nabla\varphi^{\varepsilon}|^2dx$$

$$+\frac{\tau}{2}\int\limits_{\Omega}(\varphi_t^{\varepsilon})^2dx+\frac{\xi^2}{2}\frac{\partial}{\partial t}\Big(\int\limits_{\Omega}|\nabla\varphi^{\varepsilon}|^2dx\Big)$$

$$\leq \int_{\Omega} (u^\varepsilon)^2 dx + \Big(1 + \frac{1}{2a} + \frac{1}{4a\tau}\Big) \int_{\Omega} (\varphi^\varepsilon)^2 dx + \frac{k}{\ell h} \int_{\partial\Omega} w^2(t,x) d\gamma.$$

Integration over $(0,\varepsilon)$ and by parts of (2.168) gives now:

$$\frac{2\rho c}{\ell}\|u^\varepsilon(\varepsilon)\|^2_{L^2(\Omega)} + \frac{\tau}{2}\|\varphi^\varepsilon_-(\varepsilon)\|^2_{L^2(\Omega)} + \frac{4k}{\ell}\int_0^\varepsilon \|\nabla u^\varepsilon\|^2_{L^2(\Omega)} ds \tag{2.169}$$

$$+\xi^2 \int_0^\varepsilon \|\nabla \varphi^\varepsilon\|^2_{L^2(\Omega)} ds + \frac{\tau}{2}\int_0^\varepsilon \|\varphi^\varepsilon_t\|^2_{L^2(\Omega)} ds + \frac{\xi^2}{2}\|\nabla\varphi^\varepsilon_-(\varepsilon)\|^2_{L^2(\Omega)}$$

$$\leq \frac{2\rho c}{\ell}\|u_0\|^2_{L^2(\Omega)} + \frac{\tau}{2}\|\varphi_0\|^2_{L^2(\Omega)} + \frac{\xi^2}{2}\|\nabla\varphi_0\|^2_{L^2(\Omega)}$$

$$+C\int_0^\varepsilon (\|u^\varepsilon(s)\|^2_{L^2(\Omega)} + \|\varphi^\varepsilon(s)\|^2_{L^2(\Omega)}) ds + C\|w\|^2_{L^2(\Sigma^\varepsilon_0)}.$$

Similarly, for Q^ε_i, $i = 1, 2, ..., M_\varepsilon - 2$, we obtain from (2.168):

$$\frac{2\rho c}{\ell}\|u^\varepsilon((i+1)\varepsilon)\|^2_{L^2(\Omega)} + \frac{\tau}{2}\|\varphi^\varepsilon_-((i+1)\varepsilon)\|^2_{L^2(\Omega)} \tag{2.170}$$

$$+\frac{4k}{\ell}\int_{i\varepsilon}^{(i+1)\varepsilon} \|\nabla u^\varepsilon\|^2_{L^2(\Omega)} ds + \xi^2 \int_{i\varepsilon}^{(i+1)\varepsilon} \|\nabla\varphi^\varepsilon\|^2_{L^2(\Omega)} ds$$

$$+\frac{\tau}{2}\int_{i\varepsilon}^{(i+1)\varepsilon} \|\varphi^\varepsilon_t\|^2_{L^2(\Omega)} ds + \frac{\xi^2}{2}\|\nabla\varphi^\varepsilon_-((i+1)\varepsilon)\|^2_{L^2(\Omega)}$$

$$\leq \frac{2\rho c}{\ell}\|u^\varepsilon(i\varepsilon)\|^2_{L^2(\Omega)} + \frac{\tau}{2}\|\varphi^\varepsilon_+(i\varepsilon)\|^2_{L^2(\Omega)} + \frac{\xi^2}{2}\|\nabla\varphi^\varepsilon_+(i\varepsilon)\|^2_{L^2(\Omega)}$$

$$+C\int_{i\varepsilon}^{(i+1)\varepsilon} (\|u^\varepsilon(s)\|^2_{L^2(\Omega)} + \|\varphi^\varepsilon(s)\|^2_{L^2(\Omega)}) ds + C\|w\|^2_{L^2(\Sigma^\varepsilon_i)}.$$

and for $Q^{\varepsilon}_{M_\varepsilon - 1}$, we get

$$\frac{2\rho c}{\ell}\|u^{\varepsilon}(T)\|^2_{L^2(\Omega)} + \frac{\tau}{2}\|\varphi^{\varepsilon}_{-}(T)\|^2_{L^2(\Omega)} \tag{2.171}$$

$$+\frac{4k}{\ell}\int_{(M_\varepsilon-1)\varepsilon}^{T}\|\nabla u^{\varepsilon}\|^2_{L^2(\Omega)}ds + \xi^2\int_{(M_\varepsilon-1)\varepsilon}^{T}\|\nabla\varphi^{\varepsilon}\|^2_{L^2(\Omega)}ds$$

$$+\frac{\tau}{2}\int_{(M_\varepsilon-1)\varepsilon}^{T}\|\varphi^{\varepsilon}_t\|^2_{L^2(\Omega)}ds + \frac{\xi^2}{2}\|\nabla\varphi^{\varepsilon}_{-}(T)\|^2_{L^2(\Omega)}$$

$$\leq \frac{2\rho c}{\ell}\|u^{\varepsilon}((M_\varepsilon - 1)\varepsilon)\|^2_{L^2(\Omega)}$$

$$+\frac{\tau}{2}\|\varphi^{\varepsilon}_{+}((M_\varepsilon - 1)\varepsilon)\|^2_{L^2(\Omega)} + \frac{\xi^2}{2}\|\nabla\varphi^{\varepsilon}_{+}((M_\varepsilon - 1)\varepsilon)\|^2_{L^2(\Omega)}$$

$$+C\int_{(M_\varepsilon-1)\varepsilon}^{T}\Big(\|u^{\varepsilon}(s)\|^2_{L^2(\Omega)} + \|\varphi^{\varepsilon}(s)\|^2_{L^2(\Omega)}\Big)ds + C\|w\|^2_{L^2(\Sigma^{\varepsilon}_{M_\varepsilon-1})}.$$

Let's remember at this point that, corresponding to Cauchy problem (2.156), we have the following inequality (see (2.110)):

$$\frac{\tau}{2}\|\varphi^{\varepsilon}_{+}(i\varepsilon)\|^2_{L^2(\Omega)} \leq \frac{\tau}{2}\|\varphi^{\varepsilon}_{-}(i\varepsilon)\|^2_{L^2(\Omega)}, \tag{2.172}$$

for $i = 1, 2, \cdots, M_\varepsilon - 1$.

If we now consider the inequalities given by (2.110) and (2.172), from (2.170) we can see that (for $i = 1, 2, \cdots, M_\varepsilon - 2$):

$$B^1_{i\varepsilon} + \frac{\tau}{2}\|\varphi^{\varepsilon}_{+}((i+1)\varepsilon)\|^2_{L^2(\Omega)} + \frac{\xi^2}{2}\|\nabla\varphi^{\varepsilon}_{+}((i+1)\varepsilon)\|^2_{L^2(\Omega)} \tag{2.173}$$

$$\leq B^1_{i\varepsilon} + \frac{\tau}{2}\|\varphi^{\varepsilon}_{-}((i+1)\varepsilon)\|^2_{L^2(\Omega)} + \frac{\xi^2}{2}\|\nabla\varphi^{\varepsilon}_{-}((i+1)\varepsilon)\|^2_{L^2(\Omega)}$$

$$\leq B^2_{i\varepsilon} + \frac{\tau}{2}\|\varphi^{\varepsilon}_{+}(i\varepsilon)\|^2_{L^2(\Omega)} + \frac{\xi^2}{2}\|\nabla\varphi^{\varepsilon}_{+}(i\varepsilon)\|^2_{L^2(\Omega)},$$

where

$$B^1_{i\varepsilon} = \tfrac{2\rho c}{\ell}\|u^\varepsilon((i+1)\varepsilon)\|^2_{L^2(\Omega)} + \tfrac{4k}{\ell}\int\limits_{i\varepsilon}^{(i+1)\varepsilon} \|\nabla u^\varepsilon\|^2_{L^2(\Omega)}ds$$

$$+\xi^2 \int\limits_{i\varepsilon}^{(i+1)\varepsilon} \|\nabla\varphi^\varepsilon\|^2_{L^2(\Omega)}ds + \tfrac{\tau}{2}\int\limits_{i\varepsilon}^{(i+1)\varepsilon} \|\varphi^\varepsilon_t\|^2_{L^2(\Omega)}ds,$$

$$B^2_{i\varepsilon} = \tfrac{2\rho c}{\ell}\|u^\varepsilon(i\varepsilon)\|^2_{L^2(\Omega)}$$

$$+C\int\limits_{i\varepsilon}^{(i+1)\varepsilon} (\|u^\varepsilon(s)\|^2_{L^2(\Omega)} + \|\varphi^\varepsilon(s)\|^2_{L^2(\Omega)})ds + C\|w\|^2_{L^2(\Sigma^\varepsilon_i)}.$$

Adding (2.169), (2.171), (2.173) and performing some simple calculus we derive ($M_\varepsilon \cdot \varepsilon = T$):

$$\frac{2\rho c}{\ell}\|u^\varepsilon(T)\|^2_{L^2(\Omega)} + \frac{\tau}{2}\|\varphi^\varepsilon_-(T)\|^2_{L^2(\Omega)} + \frac{4k}{\ell}\int\limits_0^T \|\nabla u^\varepsilon\|^2_{L^2(\Omega)}ds$$

$$+\xi^2\int\limits_0^T \|\nabla\varphi^\varepsilon\|^2_{L^2(\Omega)}ds + \frac{\tau}{2}\sum_{i=0}^{M_\varepsilon-1}\int\limits_{i\varepsilon}^{(i+1)\varepsilon} \|\varphi^\varepsilon_t\|^2_{L^2(\Omega)}ds + \frac{\xi^2}{2}\|\nabla\varphi^\varepsilon_-(T)\|^2_{L^2(\Omega)}$$

$$\leq \frac{2\rho c}{\ell}\|u_0\|^2_{L^2(\Omega)} + \frac{\tau}{2}\|\varphi_0\|^2_{L^2(\Omega)} + \frac{\xi^2}{2}\|\nabla\varphi_0\|^2_{L^2(\Omega)}$$

$$+C\int\limits_0^T \Big(\|u^\varepsilon(s)\|^2_{L^2(\Omega)} + \|\varphi^\varepsilon(s)\|^2_{L^2(\Omega)}\Big)ds + C\|w\|^2_{L^2(\Sigma)}.$$

Continuing by applying Gronwall-Bellman's inequality, we finally obtain:

$$\frac{\tau}{2}\int\limits_0^T\|\varphi^\varepsilon_t\|^2_{L^2(\Omega)}ds+\frac{4k}{\ell}\int\limits_0^T\|\nabla u^\varepsilon\|^2_{L^2(\Omega)}ds+\xi^2\int\limits_0^T\|\nabla\varphi^\varepsilon\|^2_{L^2(\Omega)}ds \leq C, \quad (2.174)$$

for all $\varepsilon > 0$, where C **does not depend on** M_ε **and** ε.

Multiplying now $(2.153)_1$ by u^ε_t, integrating over $[i\varepsilon, (i+1)\varepsilon]$, $i = 0, 1, \cdots, M_\varepsilon - 1$, and, using Hölder's inequality, Cauchy's inequality as well as Green's formula, we get the estimate:

$$\rho c\int\limits_{Q^\varepsilon_i}(u^\varepsilon_t)^2dxds + \frac{k}{2}\int\limits_\Omega |\nabla u^\varepsilon|^2dx + \frac{kh}{2}\int\limits_{\partial\Omega}(u^\varepsilon)^2d\gamma$$

$$\leq k \int_{\Sigma_i^\varepsilon} u_t^\varepsilon w(t,x) d\gamma ds + \frac{\ell^2}{8\rho c} \int_{Q_i^\varepsilon} (\varphi_t^\varepsilon)^2 \, dx ds + \frac{\rho c}{2} \int_{Q_i^\varepsilon} (u_t^\varepsilon)^2 \, dx ds.$$

Summing after i and making use of (2.174), the last inequality leads to:

$$\frac{\rho c}{2} \int_{Q} (u_t^\varepsilon)^2 \, dx ds + \frac{k}{2} \int_{\Omega} |\nabla u^\varepsilon|^2 dx + \frac{kh}{2} \int_{\partial\Omega} (u^\varepsilon)^2 d\gamma \tag{2.175}$$

$$\leq C + k \int_{\Sigma} u_t^\varepsilon w(t,x) d\gamma ds.$$

But

$$k \int_{\Sigma} u_t^\varepsilon w d\gamma ds = k \int_{\Sigma} \frac{\partial}{\partial t}(u^\varepsilon w) d\gamma ds - k \int_{\Sigma} u^\varepsilon w' d\gamma ds$$

and (using Cauchy-Schwartz's and Hölder's inequalities)

$$k \int_{\partial\Omega} u^\varepsilon w d\gamma \leq \frac{kh}{4} \int_{\partial\Omega} (u^\varepsilon)^2 d\gamma + \frac{k}{h} \int_{\partial\Omega} w^2 d\gamma,$$

$$k \int_{\Sigma} u^\varepsilon w' d\gamma ds \leq \frac{k}{2} \int_{\Sigma} (u^\varepsilon)^2 d\gamma ds + \frac{k}{2} \int_{\Sigma} (w')^2 d\gamma ds.$$

Using now the Gronwall-Bellman's inequality, from (2.175) we find the estimate:

$$\frac{\rho c}{2} \int_0^t \int_{\Omega} (u_t^\varepsilon)^2 dx ds + \frac{k}{2} \int_{\Omega} |\nabla u^\varepsilon|^2 dx + \frac{kh}{4} \int_{\partial\Omega} (u^\varepsilon)^2 d\gamma \leq C, \tag{2.176}$$

for all $\varepsilon > 0$, where the constant $C > 0$ **does not depend on** M_ε **and** ε.

By virtue of estimate (2.113), summarizing for $i = 0, 1, \cdots, M_\varepsilon - 1$, we get

$$\sum_{i=0}^{M_\varepsilon - 1} \|\varphi_+^\varepsilon(i\varepsilon, x) - \varphi_-^\varepsilon(i\varepsilon, x)\|_{L^2(\Omega)} \leq TL = C, \tag{2.177}$$

where C **does not depend on** M_ε **and** ε.

Combining (2.174) and (2.176)-(2.177), we get

$$\overset{T}{\underset{0}{V}}\varphi^\varepsilon + \int_0^T \|u_t^\varepsilon(t)\|^2_{L^2(\Omega)}dt + \int_0^T \|\varphi_t^\varepsilon(t)\|^2_{L^2(\Omega)}dt \tag{2.178}$$

$$+ \int_Q \Big(|\nabla u^\varepsilon(t,x)|^2 + |\nabla\varphi^\varepsilon(t,x)|^2\Big)dxdt \leq C \qquad \forall\,\varepsilon > 0,$$

where $\overset{T}{\underset{0}{V}}\varphi^\varepsilon$ stands for the *total variation* of $\varphi^\varepsilon : [0,T] \to L_2(\Omega)$. Since the injection of $L_2(\Omega)$ into $H^{-1}(\Omega)$ is compact and the set $\{\varphi_t^\varepsilon(t)\}$ is bounded in $L_2(\Omega)$ for every $t \in [0,T]$, by an infinite dimensional version of Helly-Foiaş theorem (see for instance [50], Remark 3.2, pp. 60), we conclude that there exists a bounded variation function $\varphi^*(t) \in BV([0,T]; H^{-1}(\Omega))$ such that, on a subsequence also denoted $\varphi^\varepsilon(t)$, we have

$$\varphi^\varepsilon(t) \to \varphi^*(t) \;\; \text{strongly in } H^{-1}(\Omega) \quad \text{for every } t \in [0,T]. \tag{2.179}$$

From (2.178) we may assume that

$$\varphi^\varepsilon \to \varphi^* \qquad \text{weakly in } L^2(0,T; H^1(\Omega)). \tag{2.180}$$

Now, since the embedding of $H^1(\Omega)$ into $L_2(\Omega)$ is compact, for every $\lambda > 0$ there exists $C(\lambda) > 0$ such that (see (5.2) in Chapter 1)

$$\|\varphi^\varepsilon(t) - \varphi^*(t)\|_{L^2(\Omega)}$$

$$\leq \lambda\|\varphi^\varepsilon(t) - \varphi^*(t)\|_{H^1(\Omega)} + C(\lambda)\|\varphi^\varepsilon(t) - \varphi^*(t)\|_{H^{-1}(\Omega)},$$

$\forall\,\varepsilon > 0$ and $\forall\,t \in [0,T]$, where $C(\lambda) \to 0$ as $\lambda \to 0$.

Together with (2.179) and (2.180), this last inequality leads to

$$\varphi^\varepsilon \to \varphi^* \qquad \text{strongly in } L^2(\Omega) \qquad \text{for any } t \in [0,T]. \tag{2.181}$$

From $(2.153)_1$, $(2.153)_2$ and (2.178) we have, respectively:

$$\int_0^t \int_\Omega (\Delta u^\varepsilon(s,x))^2 dxds \leq C \qquad \forall t \in (0,T],$$

$$\int_0^t \int_\Omega (\Delta\varphi^\varepsilon(s,x))^2 dx ds \le C \qquad \forall t \in (0,T].$$

Taking now into account the elliptic regularity of weak solution (see Barbu [76]), we obtain the estimates:

$$\|u^\varepsilon\|_{L^2(0,T;H^2(\Omega))} \le C, \tag{2.182}$$

$$\|\varphi^\varepsilon\|_{L^2([0,T];H^2(\Omega))} \le C. \tag{2.183}$$

Thus, because the embedding $H^2(\Omega) \subset H^1(\Omega)$ is compact, it turns out that the sequence $\{u^\varepsilon\}$ is compact in $L^2(0,T;H^1(\Omega))$. Therefore, on a subsequence, again denoted u^ε, from (2.178) and (2.182) we have

$$u^\varepsilon \to u^* \ \text{ strongly in } L^2([0,T];H^1(\Omega)), \tag{2.184}$$

$$\text{weakly in } L^2([0,T];H^2(\Omega)),$$

$$u_t^\varepsilon \to u_t^* \ \text{ weakly in } L^2([0,T];L^2(\Omega)),$$

and, by the Ascoli-Arzelà theorem

$$u^\varepsilon \to u^* \ \text{ strongly in } C([0,T];L^2(\Omega)). \tag{2.185}$$

From (2.181) and (2.185) we may conclude that (2.163) holds true and so the Theorem 2.6 is proved. **q.e.d.**

As regarding the pair functions $(u^*(t),\varphi^*(t))$, one can prove (as in Section 2.5.1) that it is absolutely continuous in t on $[0,T]$ and satisfies a.e. the nonlinear system (2.150)-(2.152). Thus, (u^*,φ^*) is a strong solution to problem (2.150)-(2.152) and so, Theorem 2.6 can be viewed as a constructive way to prove the existence in the nonlinear parabolic system (2.150)-(2.152).

Corollary 2.2. *Let* $u_0,\varphi_0 \in W^{1,\infty}(\Omega) = \{y;\quad y, \nabla y \in L^\infty(\Omega)\}$ *satisfy* $\frac{\partial u_0}{\partial\nu} + hu_0 = w(t,x)$, $\frac{\partial\varphi_0}{\partial\nu} = 0$ *and* $w(t,x) \in W^1([0,T];L^2(\partial\Omega))$. *Then* $(u^*,\varphi^*) \in W \times W$ (see Definition 2.5) *is a weak solution of nonlinear system* (2.150)-(2.152).

CHAPTER 3

OPTIMAL CONTROL PROBLEMS GOVERNED BY THE PHASE-FIELD TRANSITION SYSTEM

Costică MOROŞANU

Faculty of Mathematics, "Al.I. Cuza" University of Iasi, Romania

Abstract: Some types of *boundary optimal control problems* governed by the nonlinear phase-field transition system, are introduced and analyzed in this Chapter.

The aim of Section 3.1 is to prove (for later use) a priori estimates in $L^2([0,T]; H^2(\Omega))$ for unknown u, φ in phase-field system, in the presence of following boundary conditions (see $(3')$, $(3'')$ in Preface):

$$\frac{\partial}{\partial \nu}u + hu = w(t,x) \quad \frac{\partial}{\partial \nu}\varphi = 0 \quad \text{on } \Sigma = [0,T] \times \partial\Omega,$$
$$\frac{\partial}{\partial \nu}u + hu = w(t)g(x) \quad \varphi = 1 \quad \text{on } \Sigma,$$

where $w : \Sigma \to I\!R$ $(w : [0,T] \to I\!R)$ represents *the boundary control* (the temperature of the surrounding at $\partial\Omega$). We mentioned that the assumptions on given functions: u_0, φ_0, w, are weaker in Theorem 2.1 in Section 2.1, and considering more natural applications covered by mathematical model. Moreover, in the proof of results (**Proposition 3.1 & 3.2**), a new technique has been used - *elliptic regularity of weak solution* (see Barbu [76]). Both cases (w depending explicitly on t and x or only on t) allow the model of phase-field to be involved in realistic numerical simulations in the metallurgic industry.

In Section 2.2, we shall prove the existence of an optimal control (**Proposition 3.3**) for the problem stated there, proof based on estimates established in the previous Section. The *distributed optimal control problem* governed by phase-field system has been analyzed in many researches, carried out by: Z. Chen & K.-H. Hoffmann [41], M. Heinkenschloss & E.W. Sachs [42], M. Heinkenschloss & F. Troltzsch [43], as well as K.-H. Hoffmann & L. Jiang [44]. *Boundary optimal control* *problem* governed by the classical Stefan problem in two phases was studied by V. Barbu [1], A. Friedman [45], C. Saguez [4], D. Tiba [6].

To the optimal control problem stated in 3.2, we associated an *approximating optimal control problem* for which, first, we proved the existence of an optimal control. Based on the existence of an optimal control, the convergence of the optimal solution of approximating problem to the optimal solution of the original problem (**Theorem 3.1**) was proved in the next Section.

For the approximating problem *necessary optimality conditions* (**Theorem 3.2**) are established in Section 3.4. Such a problem was studied, for an optimal control problem governed by nonlinear and parabolic variational inequalities, by Barbu [46].

Another type of boundary optimal control problem is defined in Section 3.5 (*an inverse problem*). Necessary optimality conditions (*maximum principle*) for such sort of problem are given by **Theorem 3.4**.

An optimal control problem, with *distributed control acting on a subset* $\omega \subset \Omega$ and with *state constraint* in time variable - (P_S), is analyzed in Section 3.6. The necessary optimality conditions for (P_S), **Theorem 3.5**, were obtained by passing to the limit for $\varepsilon \longrightarrow 0$ in the approximating control process considered (the adequate penalty problem P_S^ε, $\varepsilon > 0$).

The last Section of this Chapter is dedicated to a *nonhomogeneous boundary optimal control problem* (P). The main result of this Section (**Theorem 3.6**) amounts to saying that problem (P) can be approximated for $\varepsilon \longrightarrow 0$ by the sequence of problems (P^ε) - consequence of approximation scheme associated to the nonlinear phase-field transition system. The convergence of the approximating process (P^ε) leads to an idea of numerical approximation of the optimal control in problem (P), namely (see step **P2** in algorithm `cpht-2D`, Section 4.5), at every iteration *iter*, the computation of the approximate solution corresponding to the nonlinear phase-field transition system is substituted by computation of the approximate solution for an ordinary equation and a linear system. Hence a large amount of time is saved concerning the computations (see Table **1** in Chapter 5).

The results of this Chapter were developed based on the

researches conducted by: T. Benincasa & C. Moroşanu [10], T. Benincasa, A. Favini & C. Moroşanu [13], C. Moroşanu [14], [19]-[24], [26], [29], and C. Moroşanu & G. Wang [37].

Keywords: Nonlinear parabolic systems, free boundary problems for PDE, inverse problems, necessary conditions for optimality, duality, abstract approximation theory, methods of approximations based on necessary conditions, stability and convergence of numerical methods, computer science, analysis of algorithms, methods of gradient type, phase transitions, applications in engineering and industry.

3.1 A priori estimates in $L^2([0,T];H^2(\Omega))$ for unknown functions u and φ

We consider the following nonlinear parabolic system

$$\begin{cases} \rho V\frac{\partial}{\partial t}\, u + \frac{\ell}{2}\frac{\partial}{\partial t}\, \varphi = k\Delta u \\ \\ \tau\frac{\partial}{\partial t}\, \varphi = \xi^2\Delta\varphi + \frac{1}{2a}(\varphi - \varphi^3) + 2u \end{cases} \quad \text{in } Q, \tag{3.1}$$

with the boundary conditions

$$\frac{\partial}{\partial \nu}\, u + hu = w(t,x), \qquad \frac{\partial}{\partial \nu}\, \varphi = 0 \qquad \text{on } \Sigma, \tag{3.2}$$

and initial conditions

$$u(0,x) = u_0(x), \qquad \varphi(0,x) = \varphi_0(x) \qquad \text{on } \Omega, \tag{3.3}$$

where $w : \Sigma \to I\!R$ is *the boundary control* (the temperature of the surrounding at $\partial\Omega$), $w \in \mathcal{U}$ with

$$\mathcal{U} = \{v \in L^\infty(\Sigma);\ 0 \le v(t,x) \le R,\ \ a.e.\ t \in [0,T],\ x \in \Omega\},$$

u, φ, u_0, φ_0, ρ, τ, ξ, ℓ, k, h, a was described in Chapter 2, and V represent the *casting speed.*

The aim of this Section is to prove, for later use, some estimates for the solution (u,φ) of nonlinear parabolic system (3.1)-(3.3). In this respect, we have

Proposition 3.1. *Let* $u_0, \varphi_0 \in W^{1,\infty}(\Omega) = \{y;\ y, \nabla y \in L^\infty(\Omega)\}$ *satisfying* $\frac{\partial u_0}{\partial \nu} + hu_0 = w(0,x)$, $\frac{\partial \varphi_0}{\partial \nu} = 0$ *and* $w \in W^1([0,T]; L^2(\partial\Omega))$. *Then*

the weak solution of system (3.1)-(3.2) (see Definition 2.5 in Chapter 2) *fulfill the following estimates:*

$$\|\varphi_t\|^2_{L^2([0,T];L^2(\Omega))} + \|u_t\|^2_{L^2([0,T];L^2(\Omega))} \leq C, \tag{3.4}$$

$$\|\varphi(t)\|^2_{H^1(\Omega)} + \|u(t)\|^2_{H^1(\Omega)} \leq C \quad \forall t \in [0,T], \tag{3.5}$$

and

$$\|\varphi\|^2_{L^2([0,T];H^2(\Omega))} + \|u\|^2_{L^2([0,T];H^2(\Omega))} \leq C. \tag{3.6}$$

Proof. Multiplying $(3.1)_1$ with $\frac{4}{\ell}u$ and $(3.1)_2$ by φ_t, using Green's formula, Hölder's inequality, integration by parts and the relations (2.157)-(2.158), we obtain

$$\frac{2\rho V}{\ell}\frac{\partial}{\partial t}\Big(\int_\Omega u^2 dx\Big) + 2\int_\Omega u\varphi_t dx + \frac{4k}{\ell}\int_\Omega |\nabla u|^2 dx + \frac{4kh}{\ell}\int_{\partial\Omega} u^2 d\gamma \tag{3.7}$$

$$= \frac{4k}{\ell}\int_{\partial\Omega} uw(t,x)d\gamma,$$

$$\tau\int_\Omega \varphi_t^2 dx + \frac{\xi^2}{2}\frac{\partial}{\partial t}\Big(\int_\Omega |\nabla\varphi|^2 dx\Big) - 2\int_\Omega u\varphi_t dx \tag{3.8}$$

$$= \frac{1}{2a}\int_\Omega \varphi\varphi_t dx - \frac{1}{8a}\frac{\partial}{\partial t}\Big(\int_\Omega \varphi^4 dx\Big).$$

It follows from (3.7) and (3.8) that (after some simple calculus)

$$\tau\int_\Omega \varphi_t^2 dx + \frac{4k}{\ell}\int_\Omega |\nabla u|^2 dx + \frac{4kh}{\ell}\int_{\partial\Omega} u^2 d\gamma \tag{3.9}$$

$$+\frac{\partial}{\partial t}\Big(\frac{2\rho V}{\ell}\int_\Omega u^2 dx + \frac{\xi^2}{2}\int_\Omega |\nabla\varphi|^2 dx + \frac{1}{8a}\int_\Omega \varphi^4 dx\Big)$$

$$= \frac{1}{8a^2\tau}\int_\Omega \varphi^2 dx + \frac{\tau}{2}\int_\Omega \varphi_t^2 dx + \frac{4k}{\ell}\int_{\partial\Omega} uw(t,x)d\gamma.$$

Using now Hölder's and Cauchy's inequalities, the right hand term $\frac{4k}{\ell}\int\limits_{\partial\Omega} uw(t,x)d\gamma$ in (3.9) leads to:

$$\frac{4k}{\ell}\int\limits_{\partial\Omega} uw(t,x)d\gamma \leq \frac{4kh}{\ell}\int\limits_{\partial\Omega} u^2 d\gamma + \frac{k}{\ell h}\int\limits_{\partial\Omega} w^2(t,x)d\gamma.$$

Substituting the last inequality in (3.9), performing some computation and integrating on $[0,t]$, $t\in(0,T]$, we will deduce that:

$$\frac{2\rho V}{\ell}\int\limits_{\Omega} u^2 dx + \frac{\xi^2}{2}\int\limits_{\Omega}|\nabla\varphi|^2 dx \tag{3.10}$$

$$+\frac{1}{8a}\int\limits_{\Omega}\varphi^4 dx + \frac{\tau}{2}\int\limits_0^t\int\limits_{\Omega}\varphi_t^2 dsdx + \frac{4k}{\ell}\int\limits_0^t\int\limits_{\Omega}|\nabla u|^2 dsdx$$

$$\leq \frac{2\rho V}{\ell}\|u_0\|^2_{L^2(\Omega)} + \frac{\xi^2}{2}\|\nabla\varphi_0\|^2_{L^2(\Omega)} + \frac{1}{8a}\|\varphi_0\|^4_{L^2(\Omega)}$$

$$+\frac{1}{8a^2\tau}\int\limits_0^t\int\limits_{\Omega}\varphi^2 dsdx + \frac{k}{\ell h}\int\limits_0^t\int\limits_{\partial\Omega} w^2(t,x)dsd\gamma.$$

Using Gronwall-Bellman's inequality, from (3.10) we can derive the estimate ($Q_t=[0,t]\times\Omega, \Sigma_t=[0,t]\times\partial\Omega$):

$$\|\varphi(t)\|_{L^2(\Omega)} \leq \Big(\frac{2\rho V}{\ell}\|u_0\|^2_{L^2(\Omega)} + \frac{\xi^2}{2}\|\nabla\varphi_0\|^2_{L^2(\Omega)} + \frac{1}{8a}\|\varphi_0\|^4_{L^2(\Omega)}$$

$$+\frac{k}{\ell h}\|w\|^2_{L^2(\Sigma_t)}\Big)\exp(\int\limits_0^t ds) \quad t\in(0,T],$$

and, by integration on $(0,t)$, we get:

$$\|\varphi\|_{L^2(Q_t)} \leq \Big(\frac{2\rho V}{\ell}\|u_0\|^2_{L^2(\Omega)} + \frac{\xi^2}{2}\|\nabla\varphi_0\|^2_{L^2(\Omega)} \tag{3.11}$$

$$+\frac{1}{8a}\|\varphi_0\|^4_{L^2(\Omega)} + \frac{k}{\ell h}\|w\|^2_{L^2(\Sigma_t)}\Big)e^T T.$$

Substituting (3.11) in (3.10) we find that:

$$\frac{2\rho V}{\ell}\int_{\Omega} u^2 dx + \frac{\xi^2}{2}\int_{\Omega}|\nabla\varphi|^2 dx + \frac{1}{8a}\int_{\Omega}\varphi^4 dx \tag{3.12}$$

$$+\frac{\tau}{2}\int_0^t\int_{\Omega}\varphi_t^2 ds dx + \frac{4k}{\ell}\int_0^t\int_{\Omega}|\nabla u|^2 ds dx$$

$$\leq \frac{1}{8a^2\tau}\Big(\frac{2\rho V}{\ell}\|u_0\|^2_{L^2(\Omega)} + \frac{\xi^2}{2}\|\nabla\varphi_0\|^2_{L^2(\Omega)}$$

$$+\frac{1}{8a}\|\varphi_0\|^4_{L^2(\Omega)} + \frac{k}{\ell h}\|w\|^2_{L^2(\Sigma_t)}\Big)e^T T$$

$$+\frac{2\rho V}{\ell}\|u_0\|^2_{L^2(\Omega)} + \frac{\xi^2}{2}\|\nabla\varphi_0\|^2_{L^2(\Omega)} + \frac{1}{8a}\|\varphi_0\|^4_{L^2(\Omega)} + \frac{k}{\ell h}\|w\|^2_{L^2(Q_t)}$$

$$= C_1\Big(\frac{2\rho V}{\ell}\|u_0\|^2_{L^2(\Omega)} + \frac{\xi^2}{2}\|\nabla\varphi_0\|^2_{L^2(\Omega)}$$

$$+\frac{1}{8a}\|\varphi_0\|^4_{L^4(\Omega)} + \frac{k}{\ell h}\|w\|^2_{L^2(\Sigma_t)}\Big) \overset{not}{=} C.$$

Analyzing the last estimate we may conclude that:

$$\|\varphi_t\|^2_{L^2([0,T];L^2(\Omega))} \leq C. \tag{3.13}$$

Let's denote by

$$A = \frac{\tau}{2}\|\varphi_t\|^2_{L^2(Q_t)} + \frac{4k}{\ell}\|\nabla u\|^2_{L^2(Q_t)} + \frac{2\rho V}{\ell}\|u\|^2_{L^2(\Omega)} + \frac{\xi^2}{2}\|\nabla\varphi\|^2_{L^2(\Omega)}.$$

Then, from (3.12) we have

$$A + \frac{1}{8a}\|\varphi\|^4_{L^4(\Omega)} \leq C. \tag{3.14}$$

The inclusion $L^4(\Omega) \subset L^2(\Omega)$ gives us the inequality

$$\|\varphi\|_{L^2(\Omega)} \leq C\|\varphi\|_{L^4(\Omega)}$$

and thus from (3.14) we can derive

$$A + \|\varphi\|^2_{L^2(\Omega)} \leq A + \frac{C}{8a}\|\varphi\|^4_{L^4(\Omega)} \leq C.$$

The last inequality, combined with (3.12), allow us to conclude that

$$\|\varphi(t)\|^2_{H^1(\Omega)} \leq C \qquad \forall t \in [0,T]. \tag{3.15}$$

If we now multiply $(3.1)_1$ by u_t, integrating over Q_t, $t \in [0,T]$, and than using Hölder's and Cauchy's inequalities as well as Green's formula, we get

$$\rho V \int_{Q_t} u_t^2 dsdx + \frac{k}{2}\int_{\Omega} |\nabla u|^2 dx + \frac{kh}{2}\int_{\partial\Omega} u^2 d\gamma$$

$$\leq k\int_{\Sigma_t} u_t w(t,x) dsd\gamma + \frac{\ell^2}{8\rho V}\int_{Q_t} \varphi_t^2 dsdx + \frac{\rho V}{2}\int_{Q_t} u_t^2 dsdx$$

that imply (making use of (3.13)):

$$\frac{\rho V}{2}\int_{Q_t} u_t^2 dsdx + \frac{k}{2}\int_{\Omega} |\nabla u|^2 dx + \frac{kh}{2}\int_{\partial\Omega} u^2 d\gamma \tag{3.16}$$

$$\leq C + k\int_{\Sigma_t} u_t w(t,x) dsd\gamma.$$

But

$$k\int_{\Sigma_t} u_t w dsd\gamma = k\int_{\Sigma_t} \frac{\partial}{\partial t}(uw) dsd\gamma - k\int_{\Sigma_t} uw' dsd\gamma$$

and (we have used Cauchy-Schwartz's and Hölder's inequalities)

$$k\int_{\partial\Omega} uw d\gamma \leq \frac{kh}{4}\int_{\partial\Omega} u^2 d\gamma + \frac{k}{h}\int_{\partial\Omega} w^2 d\gamma,$$

$$k\int_{\Sigma_t} uw' ds d\gamma \leq \frac{k}{2}\int_{\Sigma_t} u^2 ds d\gamma + \frac{k}{2}\int_{\Sigma_t} (w')^2 ds d\gamma.$$

Due to assumption $w \in W^{1,\infty}(\Sigma)$ and by means of Gronwall-Bellman's inequality, from (3.16) we find now the estimate

$$\frac{\rho V}{2}\int_0^t\int_\Omega u_t^2 ds dx + \frac{k}{2}\int_\Omega |\nabla u|^2 dx + \frac{kh}{4}\int_{\partial\Omega} u^2 d\gamma \leq C \quad \forall t \in (0,T]. \quad (3.17)$$

Together with (3.12), (3.17) implies

$$\|u_t\|^2_{L^2([0,T];L^2(\Omega))} \leq C, \quad (3.18)$$

$$\|u(t)\|^2_{H^1(\Omega)} \leq C \quad \forall t \in [0,T]. \quad (3.19)$$

The relations (3.13) and (3.18) imply estimate (3.4) while relations (3.15) and (3.19) imply estimate (3.5). Also, from $(3.1)_1$ and (3.4) we have:

$$\int_0^t\int_\Omega (\Delta u(s,x))^2 ds dx \leq C \quad \forall t \in (0,T].$$

On the basis of the last inequality and taking into account the elliptic regularity of weak solution (see Barbu [76]), we derive the estimate

$$\|u\|^2_{L^2([0,T];H^2(\Omega))} \leq C. \quad (3.20)$$

The Sobolev's embedding theorem ($H^1(\Omega) \subset L^6(\Omega)$, $\Omega \subset I\!R^n$, $n \leq 3$) together with (3.15) give us

$$\|\varphi^3(t)\|^2_{L^2(\Omega)} \leq C\|\varphi(t)\|^2_{H^1(\Omega)} \leq C \quad \forall t \in [0,T].$$

This, together with $(3.1)_2$, (3.13), (3.15) and (3.19), implies

$$\|\Delta\varphi(t,x)\|^2_{L^2(\Omega)} \leq C \quad \forall t \in [0,T],$$

and, invoking again the elliptic regularity of weak solution (Barbu [76]), we derive the estimate

$$\|\varphi\|^2_{L^2([0,T];H^2(\Omega))} \leq C. \quad (3.21)$$

Estimates (3.20)-(3.21) allow us to conclude that estimate (3.6) is true, what completes the proof in Proposition 3.1. **q.e.d.**

If we change the boundary conditions (3.2) with

$$\frac{\partial}{\partial \nu} u + hu = w(t)g(x), \qquad \varphi = 1 \qquad \text{on } \Sigma, \tag{3.2'}$$

where $w \in W^{1,\infty}(0,T)$, $w(0) = 0$ and $g \in L^6(\partial\Omega)$ are given functions, then we have

Proposition 3.2. *Let* $u_0 \in H^1(\Omega)$, $\varphi_0 \in L^4(\Omega) \cap H^1(\Omega)$ *satisfying* $\frac{\partial}{\partial \nu} u_0 + hu_0 = 0$, $w \in W^{1,\infty}(0,T)$, $w(0) = 0$ *and* $g \in L^6(\partial\Omega)$. *Then the weak solution of system* (3.1)-(3.2′) *fulfill the estimates* (3.4)-(3.6).

Proof. The argument of proof is the same as in Proposition 3.1, except the computer technique used for processing terms $\frac{4k}{\ell} \int\limits_{\partial\Omega} uw(t)g(x)\, d\gamma$ and $k \int\limits_0^t \int\limits_{\partial\Omega} u_t w(t)g(x)\, ds d\gamma$, appearing in inequalities (3.9) and (3.16), respectively. Thus, we have considered the inequalities:

$$\frac{4k}{\ell} \int\limits_{\partial\Omega} uw(t)g(x)\, d\gamma$$

$$\leq \sqrt{\frac{8kh}{\ell}} \Big(\int\limits_{\partial\Omega} u^2\, d\gamma \Big)^{1/2} \sqrt{\frac{2k}{\ell h}} \Big(\int\limits_{\partial\Omega} w(t)^3\, d\gamma \Big)^{1/3} \Big(\int\limits_{\partial\Omega} g(x)^6\, d\gamma \Big)^{1/6}$$

$$\leq \frac{4kh}{\ell} \int\limits_{\partial\Omega} u^2\, d\gamma + \frac{k}{\ell h} \|w\|^2_{L^3(\partial\Omega)} \|g\|^2_{L^6(\partial\Omega)}$$

and

$$k \int\limits_0^t \int\limits_{\partial\Omega} u_t w(t) g(x)\, ds d\gamma$$

$$= k \int\limits_{\partial\Omega} g(x)\, d\gamma \int\limits_0^t \frac{\partial}{\partial t} uw(t)\, ds - k \int\limits_{\partial\Omega} g(x)\, d\gamma \int\limits_0^t uw'(t)\, ds$$

$$= k \int\limits_{\partial\Omega} uw(t)g(x)\, d\gamma - k \int\limits_0^t \int\limits_{\partial\Omega} uw'(t)g(x)\, ds d\gamma,$$

$$k\int_{\partial\Omega} uw(t)g(x)\ dsd\gamma \leq \frac{kh}{4}\int_{\partial\Omega} u^2\ d\gamma + \frac{k}{h}\|w\|^2_{L^3(\partial\Omega)}\|g\|^2_{L^6(\partial\Omega)},$$

$$k\int_0^t\int_{\partial\Omega} uw'(t)g(x)\ dsd\gamma \leq \frac{k}{2}\int_0^t\int_{\partial\Omega} u^2\ dsd\gamma + \frac{k}{2}\|w'\|^2_{L^3(\Sigma)}\|g\|^2_{L^6(\Sigma)},$$

respectively. **q.e.d.**

Observation 3.1. Proposition 3.1 remains true for the system (3.1)-(3.2) if the boundary conditions are substituted with the following one

$$u\Big|_{\Sigma} = \varphi\Big|_{\Sigma} = 0$$

or

$$\frac{\partial u}{\partial \nu}\Big|_{\Sigma} = \frac{\partial \varphi}{\partial \nu}\Big|_{\Sigma} = 0.$$

The argument of the proof is exactly the same.

3.2 A boundary optimal control problem. Existence result

Consider the following boundary optimal control problem

$$(P_0) \qquad \textit{Minimize} \int_Q \chi_0(\varphi(t,x)-1)^2\ dtdx + \int_0^T w^2(t)\ dt,$$

on all $(u, \varphi, w, v) \in (C([0,T]; H^1(\Omega)))^2 \times W^{1,\infty}(0,T) \times \mathcal{U}$

subject to

$$\begin{cases} u_t + \frac{\ell}{2}\varphi_t = k\Delta u & \text{in } Q, \\ \frac{\partial u}{\partial \nu} + hu = w(t)g(x) & \text{on } \Sigma, \\ u(0,x) = u_0(x) & x \in \Omega, \end{cases} \tag{3.22}$$

$$\begin{cases} w'(t) = \beta w(t) + v(t) & t \in [0,T], \\ w(0) = 0, \end{cases} \tag{3.23}$$

$$\begin{cases} \tau\varphi_t = \xi^2\Delta\varphi + \dfrac{1}{2a}(\varphi - \varphi^3) + 2u & \text{in } Q, \\ \varphi = 1 & \text{on } \Sigma, \\ \varphi(0,x) = \varphi_0(x) & x \in \Omega. \end{cases} \tag{3.24}$$

Here w is the temperature of the surrounding at $\partial\Omega$, $v \in \mathcal{U}$ (the *boundary control*)

$$\mathcal{U} = \{v \in L^\infty(0,T), \quad 0 \le v(t) \le R \quad a.e. \ \ t \in [0,T]\},$$

$g \in L^6(\partial\Omega)$, χ_0 is the characteristic function of $\Omega_0 \subset \Omega$ and u, φ, u_0, φ_0, ℓ, k, h, τ, ξ, a was described in Section 2.1.

At the moment t the material is liquid if φ is close to $+1$ and $u(t,x) \ge \delta_1$ and it is solid if φ is close to -1 and $u(t,x) \le \delta_1$, with $\delta_1 \ge 0$. We define *interface at the moment* t (or simply *interface*) the set

$$\{x \in \Omega, \quad |u(t,x)| \le \delta_1, \quad |\varphi(t,x)| \le 1 - \delta_2, \quad \delta_2 \ge 0\}.$$

It is more convenient to reformulate problem (P_0) as

$$(P) \qquad \textit{Minimize} \quad \int_Q \chi_0\varphi^2(t,x)\,dtdx + \int_0^T w^2(t)\,dt,$$

where (u,φ,w,v) is the solution of (3.22)-(3.23) and (3.25)

below:

$$\begin{cases} \tau\varphi_t - \xi^2\Delta\varphi = \dfrac{1}{2a}\varphi - \dfrac{1}{2a}(\varphi+1)^3 + 2u + \dfrac{1}{2a} & \text{in } Q, \\ \varphi = 0 & \text{on } \Sigma, \\ \varphi(0,x) = \varphi_0(x) - 1 & x \in \Omega. \end{cases} \tag{3.25}$$

First of all, we recall the notion of weak solution for (3.22)-(3.23) and (3.25).

Definition 3.1. By weak solution to (3.22)-(3.23) and (3.25) we mean a pair of function $(u,\varphi) \in W \times W_0$, where

$W = L^2([0,T]; H^1(\Omega)) \cap W^{2,1}([0,T]; (H^1(\Omega))')$,
$W_0 = L^2([0,T]; H_0^1(\Omega)) \cap W^{2,1}([0,T]; H^{-1}(\Omega))$,

which satisfy (3.22)-(3.23) and (3.25) in the following sense

$$\int_Q (u + \frac{\ell}{2}\varphi)_t \cdot \psi \; dtdx + k \int_Q \nabla u \cdot \nabla \psi \; dtdx + kh \int_\Sigma u \cdot \psi \; dtd\gamma \tag{3.26}$$

$$= k \int_\Sigma wg\psi \; dtd\gamma,$$

$$\int_Q (\tau\varphi_t \cdot \zeta + \xi^2 \nabla\varphi \cdot \nabla\zeta - \frac{1}{2a}(\varphi - (\varphi+1)^3) \cdot \zeta - 2u \cdot \zeta) dtdx \tag{3.27}$$

$$= \frac{1}{2a} \int_Q \zeta \; dtdx,$$

for all $(\psi, \zeta) \in L^2([0,T]; H^1(\Omega)) \times L^2([0,T]; H_0^1(\Omega))$, and

$$u(0,x) = u_0(x), \qquad \varphi(0,x) = \varphi_0(x) - 1 \qquad x \in \Omega. \tag{3.28}$$

We remember that the symbol $\int_Q$ in relations (3.26)-(3.27) has the same significance as in the Definition 2.4 in Chapter 2.

Relative to the existence in problem (P), we have

Proposition 3.3. *For u_0, φ_0 as in* Proposition 3.2, *problem (P) has at least one solution $(u^*, \varphi^*, w^*, v^*)$.*

Proof. Consider $j : L^2(Q) \to I\!R$ given by

$$j(v) = \int_Q \chi_0 (\varphi^v(t,x))^2 \; dtdx + \int_0^T w^2(t) \; dt, \tag{3.29}$$

where φ^v is the solution to (3.22)-(3.23) and (3.25) corresponding to $v \in \mathcal{U}$.

Let $0 \leq d = \inf\{j(v), \; v \in \mathcal{U}\}$ and let $\{v_n\} \subset \mathcal{U}$, $\{w_n\} \subset W^{1,\infty}(0,T)$ be such that

$$d \leq j(v_n) \leq d + \frac{1}{n}. \tag{3.30}$$

The sequence $\{v_n\}$ being bounded in $L^\infty(0,T)$ we may conclude that there exists $v^* \in \mathcal{U}$ and $(w^*)' \in L^\infty(0,T)$ such that on a subsequence

$$v_n \longrightarrow v^* \text{ weak star in } \quad L^\infty(0,T),$$

$$w_n' \longrightarrow (w^*)' \text{ weak star in } \quad L^\infty(0,T).$$

Denote by (u_n, φ_n) the solution to problem (P) corresponding to v_n, i.e.,

$$\begin{cases} \frac{\partial}{\partial t} u_n + \frac{\ell}{2} \frac{\partial}{\partial t} \varphi_n = k \Delta u_n & \text{in } Q, \\ \tau \frac{\partial}{\partial t} \varphi_n = \xi^2 \Delta \varphi_n + \frac{1}{2a} \varphi_n & \\ \qquad -\frac{1}{2a} (\varphi_n + 1)^3 + 2u_n + \frac{1}{2a} & \text{in } Q, \\ \frac{\partial}{\partial \nu} u_n + h u_n = w_n(t) g(x), \quad \varphi_n = 0 & \text{on } \Sigma, \\ u_n(0,x) = u_0(x), \quad \varphi_n(0,x) = \varphi_0(x) - 1 & x \in \Omega, \\ w_n'(t) = \beta w_n(t) + v_n(t) \quad t \in [0,T], & \\ w_n(0) = 0. & \end{cases} \tag{3.31}$$

By (3.4)-(3.6) and the Sobolev embedding theorem, we have (on a subsequence again denoted $\{n\}$)

$$\begin{aligned} u_n \longrightarrow u^* \quad & \text{weakly in} && L^2([0,T]; H^2(\Omega)) \\ & \text{and strongly in} && L^2([0,T]; H^1(\Omega)), \\ \varphi_n \longrightarrow \varphi^* \quad & \text{weakly in} && L^2([0,T]; H^2(\Omega)) \\ & \text{and strongly in} && L^2([0,T]; H_0^1(\Omega)), \end{aligned}$$

$$\frac{\partial}{\partial t}u_n \longrightarrow \frac{\partial}{\partial t}u^* \quad \text{weakly in} \quad L^2([0,T];L^2(\Omega)),$$

$$\frac{\partial}{\partial t}\varphi_n \longrightarrow \frac{\partial}{\partial t}\varphi^* \quad \text{weakly in} \quad L^2([0,T];L^2(\Omega)),$$

$$\varphi_n^3 \longrightarrow (\varphi^*)^3 \quad \text{strongly in} \quad L^2([0,T];L^2(\Omega)),$$

$$\Delta u_n \longrightarrow \Delta u^* \quad \text{weakly in} \quad L^2([0,T];L^2(\Omega)),$$

$$\Delta \varphi_n \longrightarrow \Delta \varphi^* \quad \text{weakly in} \quad L^2([0,T];L^2(\Omega)),$$

and, by the Ascoli-Arzéla theorem

$$u_n \longrightarrow u^* \text{ strongly in } C([0,T];L^2(\Omega)),$$

$$\varphi_n \longrightarrow \varphi^* \text{ strongly in } C([0,T];L^2(\Omega)).$$

Letting $n \to \infty$ in (3.31), we get

$$\begin{cases} \frac{\partial}{\partial t}u^* + \frac{\ell}{2}\frac{\partial}{\partial t}\varphi^* = k\Delta u^* & \text{in } Q, \\ \tau\frac{\partial \varphi^*}{\partial t} = \xi^2\Delta\varphi^* + \frac{1}{2a}\varphi^* & \\ \qquad -\frac{1}{2a}(\varphi^*+1)^3 + 2u^* + \frac{1}{2a} & \text{in } Q, \\ \frac{\partial u^*}{\partial \nu} + hu^* = w^*(t)g(x), \quad \varphi^* = 0 & \text{on } \Sigma, \\ u^*(0,x) = u_0(x), \quad \varphi^*(0,x) = \varphi_0(x) - 1 & \text{in } \Omega, \\ (w^*)'(t) = \beta w^*(t) + v^*(t), \quad t \in [0,T], & \\ w^*(0) = 0. & \end{cases} \tag{3.32}$$

Thus, the uniqueness of solution for (3.32) (see Theorem 2.1 in Chapter 2) implies that (u^*, φ^*) is the solution of problem (3.22)-(3.23) and (3.25) corresponding to $v^* \in \mathcal{U}$.

Since j is continuous, letting $n \to \infty$ in (3.30), we obtain $d = j(v^*)$ and then the proof of Proposition 3.3 is completed. **q.e.d.**

3.3 A convergence result for the approximating optimal control problem

The aim of this Section is the study of the convergence for a iterative scheme of fractional steps type associated to problem (P) introduced in Section 3.2. In this seance, for $\forall\,\varepsilon > 0$, we associate to the system (3.22)-(3.23) and (3.25) the following approximating scheme:

$$\begin{cases} u_t^\varepsilon + \dfrac{\ell}{2}\varphi_t^\varepsilon - k\Delta u^\varepsilon = 0 & \text{in} \quad Q_i^\varepsilon = (i\varepsilon, (i+1)\varepsilon) \times \Omega, \\ \dfrac{\partial}{\partial\nu}u^\varepsilon + hu^\varepsilon = w(t)g(x) & \text{on} \quad \Sigma_i^\varepsilon = (i\varepsilon, (i+1)\varepsilon) \times \partial\Omega, \\ u^\varepsilon(0,x) = u_0(x) & x \in \Omega, \end{cases} \tag{3.33}$$

$$\begin{cases} w'(t) = \beta w(t) + v(t) \quad t \in [0,T], \\ w(0) = 0, \end{cases} \tag{3.34}$$

$$\begin{cases} \tau\varphi_t^\varepsilon - \xi^2\Delta\varphi^\varepsilon = \dfrac{1}{2a}\varphi^\varepsilon + 2u^\varepsilon + \dfrac{1}{2a} & \text{in} \quad Q_i^\varepsilon, \\ \varphi^\varepsilon = 0 & \text{on} \quad \Sigma_i^\varepsilon, \\ \varphi_+^\varepsilon(i\varepsilon, x) = z(\varepsilon, \varphi_-^\varepsilon(i\varepsilon, x)), \end{cases} \tag{3.35}$$

where $z(\cdot, \varphi_-^\varepsilon(i\varepsilon, x))$ is the solution of

$$\begin{cases} z'(s) + \dfrac{1}{2a}(z(s)+1)^3 = 0 & s \in [0,T], \\ z(0) = \varphi_-^\varepsilon(i\varepsilon, x)) \quad \varphi_-^\varepsilon(0,x)) = \varphi_0(x) - 1, \end{cases} \tag{3.36}$$

computed at $s = \varepsilon$, for $i = \overline{0, M_\varepsilon - 1}$, with $M_\varepsilon = \left[\dfrac{T}{\varepsilon}\right]$ and $Q_{M_\varepsilon - 1}^\varepsilon = ((M_\varepsilon - 1)\varepsilon, T) \times \Omega$. As until now, $\varphi_+^\varepsilon(i\varepsilon) = \lim\limits_{t \downarrow i\varepsilon} \varphi^\varepsilon(t)$ and $\varphi_-^\varepsilon(i\varepsilon) = \lim\limits_{t \uparrow i\varepsilon} \varphi^\varepsilon(t)$.

Consequently, for every $\varepsilon > 0$, we consider the approximating optimal control problem

$$(P^\varepsilon) \qquad \textit{Minimize} \int_Q \chi_0(\varphi^\varepsilon(t,x))^2\, dtdx + \int_0^T w^2(t)\, dt,$$

on all $(u^\varepsilon, \varphi^\varepsilon, w, v)$ subject to (3.33)-(3.36).

For future use, we set

$$j^{\varepsilon}(v) = \int_{Q} \chi_0(\varphi_{\varepsilon}^{v}(t,x))^2 \, dtdx + \int_{0}^{T} w^2(t) \, dt \tag{3.37}$$

where $\varphi_{\varepsilon}^{v}$ is the solution to the system (3.33)-(3.36) corresponding to $v \in \mathcal{U}$.

With respect to the existence in problem (P^{ε}), we have

Proposition 3.4. *For u_0, φ_0 as in* Proposition 3.2, *problem (P^{ε}) has at least one solution $(u_{\varepsilon}^*, \varphi_{\varepsilon}^*, w_{\varepsilon}^*, v_{\varepsilon}^*)$.*

The main result is the following theorem

Theorem 3.1. *If $\{v_{\varepsilon}^*\}$ is a sequence of optimal controllers for* (P^{ε}) *then*

$$\lim_{\varepsilon \to 0} (\inf \; j^{\varepsilon}(v)) = \inf \;\; \{j(v); \;\; v \in \mathcal{U}\} \qquad \textit{and} \tag{3.38}$$

$$\lim_{\varepsilon \to 0} j(v_{\varepsilon}^*) = \inf \;\; \{j(v); \;\; v \in \mathcal{U}\}. \tag{3.39}$$

Moreover, every weak limit point of $\{v_{\varepsilon}^*\}$ is an optimal controller for problem (P).

In other words, Theorem 3.1 amounts to saying that (P^{ε}) approximates problem (P) and an optimal controller $\{v_{\varepsilon}^*\}$ of (P^{ε}) is a suboptimal controller for problem (P).

The main ingredient of the proof of Theorem 3.1 is the following lemma

Lemma 3.1. *If $\{v_{\varepsilon}^*\}$ is a sequence of optimal controller for problem* (P^{ε}) *then there exists $\{\varepsilon_n\} \longrightarrow 0$ such that*

$$\begin{array}{llll} v_{\varepsilon_n}^* \longrightarrow v^* & \text{weak star in} & L^{\infty}(0,T), & \\ w_{\varepsilon_n}^* \longrightarrow w^* & \text{strongly in} & C(0,T), & \\ (w_{\varepsilon_n}^*)' \longrightarrow (w^*)' & \text{weak star in} & L^{\infty}(0,T), & \\ \varphi_{\varepsilon_n}^*(t) \longrightarrow \varphi^*(t) & \text{strongly in} & L^2(\Omega) \quad \forall t \in [0,T], & \\ u_{\varepsilon_n}^* \longrightarrow u^* & \text{strongly in} & L^2([0,T]; H^1(\Omega)), & \end{array} \tag{3.40}$$

where $(u_{\varepsilon_n}^*, \varphi_{\varepsilon_n}^*, w_{\varepsilon_n}^*) = (u_{\varepsilon_n}^{v_{\varepsilon_n}^*}, \varphi_{\varepsilon_n}^{v_{\varepsilon_n}^*}, w_{\varepsilon_n}^{v_{\varepsilon_n}^*})$ is the solution to (3.33)-(3.36) *corresponding to $v = v_{\varepsilon_n}^*$ and $(u^*, \varphi^*, w^*) = (u^{v^*}, \varphi^{v^*}, w^{v^*})$ is the solution to* (3.22)-(3.23) *and* (3.25) *corresponding to $v = v^*$.*

Proof. For $\{v_\varepsilon^*\}$ independent of ε this lemma was proved in Chapter 2, Theorem 2.5 in the case of homogeneous Cauchy-Dirichlet boundary conditions. We shall adapt the arguments from there to this new case (see also Barbu [46]). Last but not least, to obtain the necessary estimates in this new case, a new approach is used. Hence, let $\{v_\varepsilon^*\}$ be an optimal controller for problem (P^ε) and let $(u_\varepsilon^*, \varphi_\varepsilon^*, w_\varepsilon^*)$ be the corresponding solution of (3.33)-(3.36) with $v = v_\varepsilon^*$. Since $\{v_\varepsilon^*\}$ is bounded in $L^\infty(0,T)$ there exist $v^* \in L^\infty(0,T)$ and $\{\varepsilon_n\}$ such that (3.40) holds.

We set $v_{\varepsilon_n}^* = v_n^*$, $w_{\varepsilon_n} = w_n$, $u^{\varepsilon_n} = u^n$ and $\varphi^{\varepsilon_n} = \varphi^n$. Then (3.33)-(3.36) becomes

$$\begin{cases} u_t^n + \dfrac{\ell}{2}\varphi_t^n - k\Delta u^n = 0 & \text{in } Q_i^\varepsilon, \\[2mm] \dfrac{\partial}{\partial \nu}u^n + hu^n = w_n(t)g(x) & \text{on } \Sigma_i^\varepsilon, \\[2mm] u^n(0,x) = u_0(x), & x \in \Omega, \end{cases} \tag{3.41}$$

$$\begin{cases} w_n'(t) = \beta w_n(t) + v_n^*(t) & t \in [0,T], \\ w_n(0) = 0, \end{cases} \tag{3.42}$$

$$\begin{cases} \tau\varphi_t^n - \xi^2\Delta\varphi^n = \dfrac{1}{2a}\varphi^n + 2u^n + \dfrac{1}{2a} & \text{in } Q_i^\varepsilon, \\[2mm] \varphi^n = 0 & \text{on } \Sigma_i^\varepsilon, \\[2mm] \varphi_+^n(i\varepsilon, x) = z(\varepsilon, \varphi_-^n(i\varepsilon, x)), \end{cases} \tag{3.43}$$

where $z(\cdot, \varphi_-^n(i\varepsilon, x))$ is the solution of

$$\begin{cases} z'(s) + \dfrac{1}{2a}(z(s)+1)^3 = 0 & s \in [0,T], \\ z(0) = \varphi_-^n(i\varepsilon, x) \quad \varphi_-^n(0,x) = \varphi_0(x) - 1, \end{cases} \tag{3.44}$$

computed at $s = \varepsilon$, for $i = \overline{0, M_\varepsilon - 1}$.

As regards the problem (3.44) we have that if $\varphi_-^n(i\varepsilon, x) \in L^\infty(\Omega)$, then $z(\cdot, \varphi_-^n(i\varepsilon, x)) \in L^\infty(\Omega)$ and the following estimates hold

$$\|\nabla\varphi_+^n(i\varepsilon, x)\|_{L^2(\Omega)} \le \|\nabla\varphi_-^n(i\varepsilon, x)\|_{L^2(\Omega)}, \tag{3.45}$$

$$\|\varphi_+^n(i\varepsilon, x) - \varphi_-^n(i\varepsilon, x)\|_{L^2(\Omega)} \le L\varepsilon, \tag{3.46}$$

for $i = 0, 1, ..., M_\varepsilon - 1$, where $L > 0$ is a constant depending on $|\Omega|$, $\|\varphi_-^n(i\varepsilon, x)\|_{L^2(\Omega)}$ and on the parameter model a (see Lemmas 2.7 - 2.9 in Chapter 2).

Again C will designate a positive constant, independent of ε.

Multiplying $(3.41)_1$ by $\frac{4a}{\ell}u^n$ and $(3.43)_1$ by $a\varphi_t^n$, using integration by parts and Green's formula, yields

$$\frac{2a}{\ell}\frac{\partial}{\partial t}\int_\Omega (u^n)^2 dx + 2a\int_\Omega u^n\varphi_t^n dx + \frac{4ak}{\ell}\int_\Omega |\nabla u^n|^2 dx \tag{3.47}$$

$$+\frac{4akh}{\ell}\int_{\partial\Omega}(u^n)^2 dx = \frac{4ak}{\ell}\int_{\partial\Omega} u^n w_n(t)g(x)dx,$$

$$a\tau\int_\Omega (\varphi_t^n)^2 dx + \frac{a\xi^2}{2}\frac{\partial}{\partial t}\int_\Omega |\nabla\varphi^n|^2 dx = 2a\int_\Omega u^n\varphi_t^n dx \tag{3.48}$$

$$+\frac{1}{2}\int_\Omega \varphi^n\varphi_t^n dx + \frac{1}{2}\int_\Omega \varphi_t^n dx.$$

If we now multiply $(3.43)_1$ by $\frac{1}{\tau}\varphi^n$, and then we integrate over Ω, by Green's formula we get

$$\int_\Omega \varphi_t^n\varphi^n dx + \frac{\xi^2}{\tau}\int_\Omega |\nabla\varphi^n|^2 dx \tag{3.49}$$

$$= \frac{2}{\tau}\int_\Omega u^n\varphi^n dx + \frac{1}{2a\tau}\int_\Omega (\varphi^n)^2 dx + \frac{1}{2a\tau}\int_\Omega \varphi^n dx.$$

Adding (3.47)-(3.49), performing some computation involving Cauchy inequality, we derive

$$\frac{2a}{\ell}\frac{\partial}{\partial t}\int_\Omega (u^n)^2 dx + \frac{1}{2}\frac{\partial}{\partial t}\int_\Omega (\varphi^n)^2 dx \tag{3.50}$$

$$+\frac{4ak}{\ell}\int_\Omega |\nabla u^n|^2 dx + \frac{\xi^2}{\tau}\int_\Omega |\nabla\varphi^n|^2 dx$$

$$+\frac{a\tau}{2}\int\limits_{\Omega}(\varphi_t^n)^2dx+\frac{a\xi^2}{2}\frac{\partial}{\partial t}\int\limits_{\Omega}|\nabla\varphi^n|^2dx+\frac{2akh}{\ell}\int\limits_{\partial\Omega}(u^n)^2dx$$

$$\leq C\Big(\int_{\Omega}(u^n)^2dx+\int\limits_{\Omega}(\varphi^n)^2dx\Big)$$

$$+\frac{1}{2h}\|w_n(t)\|^2_{L^3(\partial\Omega)}\|g(x)\|^2_{L^6(\partial\Omega)}+|\Omega|\frac{1}{8a\tau}.$$

Integration over $(0,\varepsilon)$ and by parts of (3.50) give now

$$\frac{2a}{\ell}\|u^n(\varepsilon)\|^2_{L^2(\Omega)}+\frac{1}{4}\|\varphi^n_-(\varepsilon)\|^2_{L^2(\Omega)} \tag{3.51}$$

$$+\frac{4ak}{\ell}\int\limits_0^{\varepsilon}\|\nabla u^n\|^2_{L^2(\Omega)}ds+\frac{\xi^2}{\tau}\int\limits_0^{\varepsilon}\|\nabla\varphi^n|^2_{L^2(\Omega)}ds$$

$$+\frac{a\tau}{2}\int\limits_0^{\varepsilon}\|\varphi_t^n\|^2_{L^2(\Omega)}ds+\frac{a\xi^2}{2}\|\nabla\varphi^n_-(\varepsilon)|^2_{L^2(\Omega)}+\frac{2akh}{\ell}\int\limits_0^{\varepsilon}\int\limits_{\partial\Omega}(u^n)^2dsdx$$

$$\leq\frac{2a}{\ell}\|u_0\|^2_{L^2(\Omega)}+\frac{1}{4}\|\varphi_0\|^2_{L^2(\Omega)}+\frac{a\xi^2}{2}\|\nabla\varphi_0|^2_{L^2(\Omega)}$$

$$+C\int\limits_0^{\varepsilon}(\|u^n(s)\|^2_{L^2(\Omega)}+\|\varphi^n(s)\|^2_{L^2(\Omega)})ds$$

$$+\frac{1}{2h}\int\limits_0^{\varepsilon}\|w_n(t)\|^2_{L^3(\partial\Omega)}\|g(x)\|^2_{L^6(\partial\Omega)}ds+|\Omega|\frac{\varepsilon}{8a\tau}.$$

Similarly, for any Q_i^{ε}, $i=1,2,...,M_{\varepsilon}-2$, we obtain

$$\frac{2a}{\ell}\|u^n((i+1)\varepsilon)\|^2_{L^2(\Omega)}+\frac{1}{4}\|\varphi^n_-((i+1)\varepsilon)\|^2_{L^2(\Omega)} \tag{3.52}$$

$$+\frac{4ak}{\ell}\int\limits_{i\varepsilon}^{(i+1)\varepsilon}\|\nabla u^n\|^2_{L^2(\Omega)}ds+\frac{\xi^2}{\tau}\int\limits_{i\varepsilon}^{(i+1)\varepsilon}\|\nabla\varphi^n|^2_{L^2(\Omega)}ds$$

$$+\frac{a\tau}{2}\int_{i\varepsilon}^{(i+1)\varepsilon}\|\varphi_t^n\|^2_{L^2(\Omega)}ds+\frac{a\xi^2}{2}\|\nabla\varphi_-^n((i+1)\varepsilon)|^2_{L^2(\Omega)}$$

$$+\frac{2akh}{\ell}\int_{i\varepsilon}^{(i+1)\varepsilon}\int_{\partial\Omega}(u^n)^2dsdx\le\frac{2a}{\ell}\|u^n(i\varepsilon)\|^2_{L^2(\Omega)}+\frac{1}{4}\|\varphi_+^n(i\varepsilon)\|^2_{L^2(\Omega)}$$

$$+\frac{a\xi^2}{2}\|\nabla\varphi_+^n(i\varepsilon)|^2_{L^2(\Omega)}+C\int_{i\varepsilon}^{(i+1)\varepsilon}(\|u^n(s)\|^2_{L^2(\Omega)}+\|\varphi^n(s)\|^2_{L^2(\Omega)})ds$$

$$+\frac{1}{2h}\int_{i\varepsilon}^{(i+1)\varepsilon}\|w_n(t)\|^2_{L^3(\partial\Omega)}\|g(x)\|^2_{L^6(\partial\Omega)}ds+|\Omega|\frac{\varepsilon}{8a\tau}.$$

On $Q^\varepsilon_{M_\varepsilon-1}$ we have the estimate

$$\frac{2a}{\ell}\|u^n(T)\|^2_{L^2(\Omega)}+\frac{1}{4}\|\varphi_-^n(T)\|^2_{L^2(\Omega)} \tag{3.53}$$

$$+\frac{4ak}{\ell}\int_{(M_\varepsilon-1)\varepsilon}^{T}\|\nabla u^n\|^2_{L^2(\Omega)}ds+\frac{a\xi^2}{\tau}\int_{(M_\varepsilon-1)\varepsilon}^{T}\|\nabla\varphi^n|^2_{L^2(\Omega)}ds$$

$$+\frac{a\tau}{2}\int_{(M_\varepsilon-1)\varepsilon}^{T}\|\varphi_t^n\|^2_{L^2(\Omega)}ds+\frac{a\xi^2}{2}\|\nabla\varphi_-^n(T)|^2_{L^2(\Omega)}$$

$$+\frac{2akh}{\ell}\int_{(M_\varepsilon-1)\varepsilon}^{T}\int_{\partial\Omega}(u^n)^2dsdx\le\frac{2a}{\ell}\|u^n((M_\varepsilon-1)\varepsilon)\|^2_{L^2(\Omega)}$$

$$+\frac{1}{4}\|\varphi_+^n((M_\varepsilon-1)\varepsilon)\|^2_{L^2(\Omega)}+\frac{a\xi^2}{2}\|\nabla\varphi_+^n((M_\varepsilon-1)\varepsilon)|^2_{L^2(\Omega)}$$

$$+C\int_{(M_\varepsilon-1)\varepsilon}^{T}(\|u^n(s)\|^2_{L^2(\Omega)}+\|\varphi^n(s)\|^2_{L^2(\Omega)})ds$$

$$+\frac{1}{2h}\int_{(M_\varepsilon-1)\varepsilon}^{T}\|w_n(t)\|^2_{L^3(\partial\Omega)}\|g(x)\|^2_{L^6(\partial\Omega)}ds+|\Omega|\frac{T-(M_\varepsilon-1)\varepsilon}{8a\tau}.$$

Adding (3.51)-(3.53) and taking into account (3.45), we obtain

$$\frac{2a}{\ell}\|u^n(T)\|^2_{L^2(\Omega)}+\frac{1}{4}\|\varphi^n_-(T)\|^2_{L^2(\Omega)}$$

$$+\frac{4ak}{\ell}\int_0^T\|\nabla u^n\|^2_{L^2(\Omega)}ds+\frac{\xi^2}{\tau}\int_0^T\|\nabla\varphi^n|^2_{L^2(\Omega)}ds$$

$$+\frac{a\tau}{2}\sum_{i=0}^{M_\varepsilon-1}\int_{i\varepsilon}^{(i+1)\varepsilon}\|\varphi^n_t\|^2_{L^2(\Omega)}ds+\frac{a\tau}{2}\int_{(M_\varepsilon-1)\varepsilon}^{T}\|\varphi^n_t\|^2_{L^2(\Omega)}ds$$

$$+\frac{a\xi^2}{2}\|\nabla\varphi^n_-(T)|^2_{L^2(\Omega)}+\frac{2akh}{\ell}\int_0^T\int_{\partial\Omega}(u^n)^2dsdx$$

$$\leq\frac{2a}{\ell}\|u_0\|^2_{L^2(\Omega)}+\frac{1}{4}\|\varphi_0\|^2_{L^2(\Omega)}+\frac{a\xi^2}{2}\|\nabla\varphi_0|^2_{L^2(\Omega)}$$

$$+C\int_0^T(\|u^n(s)\|^2_{L^2(\Omega)}+\|\varphi^n(s)\|^2_{L^2(\Omega)})ds$$

$$+\frac{1}{2h}\int_0^T\|w_n(t)\|^2_{L^3(\partial\Omega)}\|g(x)\|^2_{L^6(\partial\Omega)}ds+|\Omega|\frac{T}{8a\tau}.$$

By Gronwall's inequality we may derive finally the estimate

$$\sum_{i=0}^{M_\varepsilon-1}\int_{i\varepsilon}^{(i+1)\varepsilon}\|\varphi^n_t\|^2_{L^2(\Omega)}ds+\int_{(M_\varepsilon-1)\varepsilon}^{T}\|\varphi^n_t\|^2_{L^2(\Omega)}ds \tag{3.54}$$

$$\int_0^T(\|\nabla u^n\|^2_{L^2(\Omega)}+\|\nabla\varphi^n\|^2_{L^2(\Omega)})ds+\int_\Sigma(u^n)^2dxds\leq C_1\quad\forall\,\varepsilon>0,$$

where $C_1 > 0$ **does not depend on** M_ε **and** ε (C_1 depends on τ, k, ℓ, ξ, a, h, T, $|\Omega|$, $\|u_0\|_{L^2(\Omega)}$, $\|\varphi_0\|_{L^2(\Omega)}$, $\|\nabla\varphi_0\|_{L^2(\Omega)}$, $\|w_n\|_{L^3(\partial\Omega)}$ and on $\|g\|_{L^6(\partial\Omega)}$).

Also, from $(3.41)_1$ we have

$$\frac{1}{2}\int_0^\varepsilon\int_\Omega (u_t^n)^2 dsdx + \frac{k}{2}\int_\Omega |\nabla u^n(\varepsilon)|^2 dx + \frac{kh}{2}\int_{\partial\Omega}(u^n(\varepsilon))^2 dx \tag{3.55}$$

$$\leq \frac{\ell^2}{8}\int_0^\varepsilon \|\varphi_t^n\|^2_{L^2(\Omega)} ds + \frac{k}{2}\int_\Omega |\nabla u_0|^2 dx + \frac{kh}{2}\int_{\partial\Omega} u_0^2 dx$$

$$+k\int_0^\varepsilon\int_{\partial\Omega} u_t^n w_n(t) g(x) dxds,$$

$$\frac{1}{2}\int_{i\varepsilon}^{(i+1)\varepsilon}\int_\Omega (u_t^n)^2 dxds + \frac{k}{2}\int_\Omega |\nabla u^n((i+1)\varepsilon)|^2 dx \tag{3.56}$$

$$+\int_{\partial\Omega}(u^n((i+1)\varepsilon))^2 dx \leq \frac{\ell^2}{8}\int_{i\varepsilon}^{(i+1)\varepsilon} \|\varphi_t^n\|^2_{L^2(\Omega)} ds$$

$$+\frac{k}{2}\int_\Omega |\nabla u^n(i\varepsilon)|^2 dx + \frac{kh}{2}\int_{\partial\Omega}(u^n(i\varepsilon))^2 dx$$

$$+k\int_{i\varepsilon}^{(i+1)\varepsilon}\int_{\partial\Omega} u_t^n w_n(t) g(x) dxds \qquad i = 1, ..., M_\varepsilon - 2,$$

$$\frac{1}{2}\int_{(M_\varepsilon-1)\varepsilon}^{T}\int_\Omega (u_t^n)^2 dsdx + \frac{k}{2}\int_\Omega |\nabla u^n(T)|^2 dx + \int_{\partial\Omega}(u^n(T))^2 dx \tag{3.57}$$

$$\leq \frac{\ell^2}{8}\int_{(M_\varepsilon-1)\varepsilon}^{T} \|\varphi_t^n\|^2_{L^2(\Omega)} ds + \frac{k}{2}\int_\Omega |\nabla u^n((M_\varepsilon-1)\varepsilon)|^2 dx$$

$$+\frac{kh}{2}\int\limits_{\partial\Omega}(u^n((M_\varepsilon-1)\varepsilon))^2dx+k\int\limits_{(M_\varepsilon-1)\varepsilon}^{T}\int_{\partial\Omega}u_t^n w_n(t)g(x)dsdx.$$

Adding (3.55)-(3.57) and making use of relation (3.54), we obtain

$$\frac{1}{2}\int\limits_0^T\int\limits_\Omega(u_t^n)^2dxds+\frac{k}{2}\int\limits_\Omega|\nabla u^n(T)|^2dx+\frac{kh}{2}\int\limits_{\partial\Omega}(u^n(T))^2dx \tag{3.58}$$

$$\leq\frac{\ell^2}{8}\sum_{i=0}^{M_\varepsilon-1}\int\limits_{i\varepsilon}^{(i+1)\varepsilon}\|\varphi_t^n\|^2_{L^2(\Omega)}ds+\frac{\ell^2}{8}\int\limits_{(M_\varepsilon-1)\varepsilon}^{T}\|\varphi_t^n\|^2_{L^2(\Omega)}ds$$

$$+\frac{k}{2}\|\nabla u_0\|^2_{L^2(\Omega)}+\frac{kh}{2}\|u_0\|^2_{L^2(\Omega)}+k\int\limits_0^T\int\limits_{\partial\Omega}u_t^n w_n(t)g(x)dxds$$

$$\leq\frac{\ell^2}{8}\bar{C}_1+k\int\limits_0^T\int\limits_{\partial\Omega}u_t^n w_n(t)g(x)dxds,$$

where $\bar{C}_1>0$ depends on C_1 and $\|\nabla u_0\|_{L^2(\Omega)}$. Using now Fubini's theorem, the trace theorem ($\|u\|_{L^2(\partial\Omega)}\leq C\|u\|_{H^1(\Omega)}$), Hölder's inequality and Cauchy's inequality, the relation (3.58) becomes

$$\frac{1}{2}\int\limits_0^T\int\limits_\Omega(u_t^n)^2dsdx+\frac{k}{2}\int\limits_\Omega|\nabla u^n(T)|^2dx+\frac{kh}{2}\int\limits_{\partial\Omega}(u^n(T))^2dx \tag{3.59}$$

$$\leq\frac{\ell^2}{8}\bar{C}_1+\bar{C}_2\int\limits_0^T\|u^n(s)\|^2_{H^1(\Omega)}ds,$$

where $\bar{C}_2>0$ depends on $\|g\|_{L^6(\partial\Omega)}$, $\|w_n\|_{L^3(\partial\Omega)}$ and $\|w_n'\|_{L^3(\partial\Omega)}$.

Using now Gronwall's inequality, from (3.59) we obtain

$$\int\limits_0^T\|u_t^n\|^2_{L^2(\Omega)}ds\leq C_2\quad\forall\,\varepsilon>0, \tag{3.60}$$

where $C_2 > 0$ **does not depend on** M_ε **and** ε (C_2 depends on $\bar{C}_1$, $\bar{C}_2$ and on T).

By virtue of estimate (3.46) we get

$$\sum_{i=0}^{M_\varepsilon - 1} \|\varphi_+^n(i\varepsilon, x) - \varphi_-^n(i\varepsilon, x)\|_{L^2(\Omega)} \leq LT = C \tag{3.61}$$

where $C > 0$ **does not depend on** M_ε **and** ε.

Combining (3.54), (3.60) and (3.61), we derive

$$\overset{T}{\underset{0}{V}} \varphi^n + \int_0^T \|u_t^n(t)\|_{L^2(\Omega)}^2 dt \tag{3.62}$$

$$+ \sum_{i=0}^{M_\varepsilon - 1} \int_{i\varepsilon}^{(i+1)\varepsilon} \|\varphi_t^n(t)\|_{L^2(\Omega)}^2 dt + \int_{(M_\varepsilon - 1)\varepsilon}^{T} \|\varphi_t^n(t)\|_{L^2(\Omega)}^2 dt$$

$$+ \int_0^T \|\nabla u^n\|_{L^2(\Omega)}^2 + \int_0^T \|\nabla \varphi^n\|_{L^2(\Omega)}^2 dt \leq C \quad \forall \varepsilon > 0, \ \ \forall t \in [0, T],$$

where $\overset{T}{\underset{0}{V}} \varphi^n$ stands for the *total variation* of the function $\varphi^n : [0, T] \to L_2(\Omega)$.

By continuing to make the same reasoning as in the proof of Theorem 2.5 in Chapter 2, we can conclude that

$$\varphi^n(t) \to \varphi^*(t) \ \text{ strongly in } H^{-1}(\Omega)) \quad \forall t \in [0, T]. \tag{3.63}$$

On the other hand, from (3.62) we can get that

$$\varphi^n \to \varphi^* \quad \text{weakly in} \quad L^2([0, T]; H_0^1(\Omega)). \tag{3.64}$$

From the sequence of inclusions $H_0^1(\Omega) \subset L_2(\Omega) \subset H^{-1}(\Omega)$, we have the following interpolation inequality

$$\|\varphi^n(t) - \varphi^*(t)\|_{L^2(\Omega)}$$

$$\leq \lambda\|\varphi^n(t) - \varphi^*(t)\|_{H_0^1(\Omega)} + C(\lambda)\|\varphi^n(t) - \varphi^*(t)\|_{H^{-1}(\Omega)},$$

$\forall \varepsilon > 0$, $\forall t \in [0,T]$ where $C(\lambda) \to 0$ as $\lambda \to 0$.

Together with (3.63) and (3.64), the last inequality give us

$$\varphi^n \to \varphi^* \quad \text{strongly in} \quad L^2(\Omega) \quad \forall t \in [0,T]. \tag{3.65}$$

Taking into account (3.60) and (3.62), we can obtain from $(3.41)_1$ the estimate

$$\int_0^t \int_\Omega (\Delta u^n(s,x))^2 dxds \leq C \quad \forall t \in (0,T] \tag{3.66}$$

whence, using elliptic boundary regularity, (3.66) implies

$$\|u^n\|_{L^2(0,T;H^2(\Omega))} \leq C. \tag{3.67}$$

Similarly, from $(3.43)_1$ (implying and (3.62)), we get

$$\int_0^t \int_\Omega (\Delta \varphi^n(s,x))^2 dxds \leq C \quad \forall t \in (0,T] \tag{3.68}$$

and by the same reasoning as above, we derive

$$\|\varphi^n\|_{L^2([0,T];H_0^1(\Omega)\cap H^2(\Omega))} \leq C. \tag{3.69}$$

Relations (3.62) and (3.67), together with the fact that the inclusion $H^2(\Omega) \subset H^1(\Omega)$ is compact, guaranty the compactify of the sequence $\{u^n\}$ in $L^2([0,T];H^1(\Omega))$. So, on a subsequence (again denoted by $\{u^n\}$), we have

$$\begin{aligned} u^n \to u^* \quad & \text{weakly in} \quad L^2([0,T];H^2(\Omega)) \\ & \text{strongly in} \quad L^2([0,T];H^1(\Omega)), \\ u_t^n \to u_t^* \quad & \text{weakly in} \quad L^2([0,T];L^2(\Omega)), \end{aligned} \tag{3.70}$$

and, by Ascoli-Arzelà theorem

$$u^n \to u^* \;\text{ strongly in } C([0,T];L^2(\Omega)).$$

Relations (3.66) and (3.68) imply that

$$\begin{aligned} &\Delta u^n \to \Delta u^* \quad \text{weakly in} \quad L^2([0,T];L^2(\Omega)),\\ &\Delta\varphi^n \to \Delta\varphi^* \quad \text{weakly in} \quad L^2([0,T];L^2(\Omega)). \end{aligned} \tag{3.71}$$

From (3.65) and (3.70) we conclude that the last two relations in (3.40) are satisfied and therefore the proof of Lemma 3.1 is ended. **q.e.d.**

Corollary 3.1. *Let* $u_0 \in H^1(\Omega)$, $\varphi_0 \in L^4(\Omega)\cap H^1(\Omega)$ *satisfying* $\frac{\partial u_0}{\partial\nu} + hu_0 = w(t)g(x)$, $w \in W^{1,\infty}(0,T)$, $w(0) = 0$ *and* $g \in L^6(\partial\Omega)$. *Then* $(u^*,\varphi^*) \in W \times W_0$ (see Definition 3.1) *is a weak solution of nonlinear parabolic system* (3.22)-(3.23) *and* (3.25).

Proof of Theorem 3.1. The idea is the same as in Barbu [46]. Let $\{v_\varepsilon^*\}$ be an optimal controller for problem (P^ε) and let $(u_\varepsilon^*,\varphi_\varepsilon^*,w_\varepsilon^*)$ be the solution of (3.33)-(3.36) corresponding to $v = v_\varepsilon^*$. By Lemma 3.1 it results that there exist $v^* \in L^\infty(0,T)$ and $\{\varepsilon_n\}$ such that

$$\begin{aligned} &v_{\varepsilon_n}^* \to v^* && \text{weak star in} \quad L^\infty(0,T),\\ &w_{\varepsilon_n}^{v_{\varepsilon_n}^*} \to w^{v^*} && \text{strongly in} \quad C(0,T),\\ &\varphi_{\varepsilon_n}^{v_{\varepsilon_n}^*}(t) \to \varphi^{v^*}(t) && \text{strongly in} \quad L^2(\Omega) \quad \forall t\in[0,T], \end{aligned}$$

where $(u_{\varepsilon_n}^{v_{\varepsilon_n}^*},\varphi_{\varepsilon_n}^{v_{\varepsilon_n}^*},w_{\varepsilon_n}^{v_{\varepsilon_n}^*})$ is the solution to (3.33)-(3.36) corresponding to $v = v_{\varepsilon_n}^*$ and $(u^{v^*},\varphi^{v^*},w^{v^*})$ is the solution to (3.22)-(3.23) and (3.25) corresponding to $v = v^*$.

Since $\varphi \to \int_Q \chi_0(\varphi(t,x))^2\,dtdx$ and $w \to \int_0^T w^2(t)\,dt$ are convex continuous functions it follows that these are weakly lower semicontinuous functions (from $L^2(Q) \to I\!R$ and $L^2(0,T) \to I\!R$, respectively). Hence

$$j(v^*) \le \liminf_{n\to\infty}\left[\int_Q \chi_0(\varphi_{\varepsilon_n}^{v_{\varepsilon_n}^*}(t,x))^2\,dtdx + \int_0^T (w_{\varepsilon_n}^{v_{\varepsilon_n}^*}(t))^2\,dt\right]. \tag{3.72}$$

Let $\tilde{v}^*$ be an optimal controller for problem (P) introduced in 3.2. Because $v^*_{\varepsilon_n}$ is an optimal controller for problem ($\mathrm{P}^{\varepsilon_n}$), it follows that

$$\int_Q \chi_0 (\varphi_{\varepsilon_n}^{v^*_{\varepsilon_n}}(t,x))^2 \, dtdx + \int_0^T (w_{\varepsilon_n}^{v^*_{\varepsilon_n}}(t))^2 \, dt$$

$$\leq \int_Q \chi_0 (\varphi_{\varepsilon_n}^{\tilde{v}^*}(t,x))^2 \, dtdx + \int_0^T (w_{\varepsilon_n}^{\tilde{v}^*}(t))^2 \, dt.$$

But, by Lemma 3.1, $\varphi_{\varepsilon_n}^{\tilde{v}^*}(t) \longrightarrow \varphi^{\tilde{v}^*}(t)$ strongly in $L^2(\Omega)$ $\forall t \in [0,T]$, and so, the above inequality implies

$$\lim_{n\to\infty} \left[\int_Q \chi_0 (\varphi_{\varepsilon_n}^{\tilde{v}^*}(t,x))^2 \, dtdx + \int_0^T (w_{\varepsilon_n}^{\tilde{v}^*}(t))^2 \, dt \right] \tag{3.73}$$

$$\leq \int_Q \chi_0 (\varphi^{\tilde{v}^*}(t,x))^2 \, dtdx + \int_0^T (w^{\tilde{v}^*}(t))^2 \, dt.$$

From (3.72) and (3.73) we obtain

$$j(v^*) = \int_Q \chi_0 (\varphi^{v^*}(t,x))^2 \, dtdx + \int_0^T (w^{v^*}(t))^2 \, dt$$

$$\leq \liminf_{n\to\infty} \left[\int_Q \chi_0 (\varphi_{\varepsilon_n}^{v^*_{\varepsilon_n}}(t,x))^2 \, dtdx + \int_0^T (w_{\varepsilon_n}^{v^*_{\varepsilon_n}}(t))^2 \, dt \right]$$

$$\leq \limsup_{n\to\infty} \left[\int_Q \chi_0 (\varphi_{\varepsilon_n}^{v^*_{\varepsilon_n}}(t,x))^2 \, dtdx + \int_0^T (w_{\varepsilon_n}^{v^*_{\varepsilon_n}}(t))^2 \, dt \right]$$

$$\leq \int_Q \chi_0 (\varphi^{\tilde{v}^*}(t,x))^2 \, dtdx + \int_0^T (w^{\tilde{v}^*}(t))^2 \, dt.$$

Therefore

$$\lim_{\varepsilon_n \to 0} \inf j^{\varepsilon_n}(v^*_{\varepsilon_n}) = j(\tilde{v}^*) = \inf \ \{j(v), \quad v \in \mathcal{U}\}$$

and then (3.38) in Theorem 3.1 holds.

To prove (3.39), we set $\tilde{\varphi}_\varepsilon = \varphi^{v^*_\varepsilon}$, $\tilde{u}_\varepsilon = u^{v^*_\varepsilon}$, $\tilde{w}_\varepsilon = w^{v^*_\varepsilon}$ (we recall that v^*_ε is optimal in (P^ε)). We have on a subsequence $\{\varepsilon_n\}$

$$\begin{aligned}
v^*_{\varepsilon_n} &\longrightarrow v^0 && \text{weak star in} && L^\infty(0,T),\\
\tilde{w}^{v^*_{\varepsilon_n}}_{\varepsilon_n} &\longrightarrow w^0 && \text{strongly in} && C[0,T],\\
\tilde{u}_{\varepsilon_n} &\longrightarrow u && \text{strongly in} && L^2([0,T];H^1(\Omega)),\\
\tilde{\varphi}_{\varepsilon_n} &\longrightarrow \varphi && \text{strongly in} && L^2([0,T];H^1(\Omega)),
\end{aligned}$$

where (u,φ,w^0,v^0) satisfy (3.22)-(3.23) and (3.25), i.e., $(u,\varphi,w^0) = (u^{v^0},\varphi^{v^0},w^{v^0})$.

Therefore, we have

$$\int_Q \chi_0(\varphi^{v^0}(t,x))^2 \ dtdx + \int_0^T (w^{v^0}(t))^2 \ dt \le \inf P$$

and since $\{\varepsilon_n\}$ was chosen arbitrary, (3.39) follows.

Let us establish now the second part in Theorem 3.1. Since v^*_ε is an optimal controller for problem (P^ε), it follows that

$$\int_Q \chi_0(\varphi^{v^*_\varepsilon}_\varepsilon(t,x))^2 \ dtdx + \int_0^T (w^{v^*_\varepsilon}_\varepsilon(t))^2 \ dt$$

$$\le \int_Q \chi_0(\varphi^{v}_\varepsilon(t,x))^2 \ dtdx + \int_0^T (w^{v}_\varepsilon(t))^2 \ dt \qquad \forall\, v \in \mathcal{U}.$$

But, as we seen above,

$$j(v^*) \le \liminf_{\varepsilon \to 0} \left[\int_Q \chi_0(\varphi^{v^*_\varepsilon}_\varepsilon(t,x))^2 \ dtdx + \int_0^T (w^{v^*_\varepsilon}_\varepsilon(t))^2 \ dt \right]$$

and thus, along with the above inequality, we have

$$j(v^*) \le \lim_{\varepsilon\to 0}\left[\int_Q \chi_0(\varphi_\varepsilon^v(t,x))^2\, dtdx + \int_0^T (w_\varepsilon^v(t))^2\, dt\right] \qquad \forall\, v \in \mathcal{U}.$$

Therefore

$$j(v^*) \le j(v) \quad \forall\, v \in \mathcal{U},$$

i.e., the weak limit point v^* for $v^*_{\varepsilon_n}$ is a suboptimal controller for the problem (P). This completes the proof of Theorem 3.1. **q.e.d.**

3.4 Necessary optimality conditions for the approximating optimal control problem

In order to establish the optimality conditions for problem (P^ε) introduced in 3.3, we consider $(u^\varepsilon, \varphi^\varepsilon, w, v)$ the solution of (3.33)-(3.36) and the corresponding variations:

$$\begin{aligned} &u^\lambda = u^\varepsilon + \lambda\tilde{u}^\varepsilon, \quad \varphi^\lambda = \varphi^\varepsilon + \lambda\tilde{\varphi}^\varepsilon, \quad w^\lambda = w + \lambda\tilde{w},\\ &z^\lambda = z + \lambda\tilde{z}, \qquad v^\lambda = v + \lambda\tilde{v}, \end{aligned} \tag{3.74}$$

where $\tilde{v} \in T_{\mathcal{U}}(v)$, $\lambda > 0$.[1] For $(u^\lambda, \varphi^\lambda, w^\lambda, v^\lambda)$ we have:

$$\begin{cases} u_t^\lambda + \dfrac{\ell}{2}\varphi_t^\lambda = k\Delta u^\lambda & \text{in } Q_i^\varepsilon,\\ \dfrac{\partial}{\partial\nu}u^\lambda + hu^\lambda = w^\lambda(t)g(x) & \text{on } \Sigma_i^\varepsilon,\\ u^\lambda(0,x) = u_0(x), & x \in \Omega, \end{cases} \tag{3.75}$$

$$\begin{cases} (w^\lambda(t))' = \beta w^\lambda(t) + v^\lambda(t) & t \in [0,T],\\ w^\lambda(0) = 0, \end{cases} \tag{3.76}$$

$$\begin{cases} \tau\varphi_t^\lambda = \xi^2\Delta\varphi^\lambda + \dfrac{1}{2a}\varphi^\lambda + 2u^\lambda + \dfrac{1}{2a} & \text{in } Q_i^\varepsilon,\\ \varphi^\lambda = 0 & \text{on } \Sigma_i^\varepsilon,\\ \varphi_+^\lambda(i\varepsilon, x) = z^\lambda(\varepsilon, \varphi_-^\lambda(i\varepsilon, x)), \end{cases} \tag{3.77}$$

[1] $T_{\mathcal{U}}(v)$ is the tangent cone at $\mathcal{U}$ in v

where z^λ is the solution of

$$\begin{cases} (z^\lambda(s))' + \dfrac{1}{2a}(z^\lambda(s)+1)^3 = 0 & s \in [0,T], \\ z^\lambda(0) = \varphi_-^\lambda(i\varepsilon, x), \end{cases} \tag{3.78}$$

computed at $s = \varepsilon$, for $i = \overline{0, M_\varepsilon - 1}$.

From (3.74) we get

$$\begin{aligned} \tilde{u}^\varepsilon &= \lim_{\lambda\to 0} \frac{u^\lambda - u^\varepsilon}{\lambda}, & \tilde{\varphi}^\varepsilon &= \lim_{\lambda\to 0} \frac{\varphi^\lambda - \varphi^\varepsilon}{\lambda}, \\ \tilde{w} &= \lim_{\lambda\to 0} \frac{w^\lambda - w}{\lambda}, & \tilde{z} &= \lim_{\lambda\to 0} \frac{z^\lambda - z}{\lambda}, \\ \tilde{v} &= \lim_{\lambda\to 0} \frac{v^\lambda - v}{\lambda}, \end{aligned} \tag{3.79}$$

$$\begin{aligned} \eta &= \lim_{\lambda\to 0} \frac{z^\lambda(\varepsilon, \varphi_-^\lambda(i\varepsilon, x)) - z(\varepsilon, \varphi_-^\varepsilon(i\varepsilon, x))}{\lambda} = \\ &= z(\varepsilon, \varphi_-^\varepsilon(i\varepsilon, x))\tilde{\varphi}_-^\varepsilon(i\varepsilon, x) + \tilde{z}(\varepsilon, \varphi_-^\varepsilon(i\varepsilon, x)). \end{aligned} \tag{3.80}$$

Subtracting (3.33)-(3.36) from (3.75)-(3.78), letting $\lambda \to 0$ and taking into account the relations (3.79) and (3.80), we get the *system in variations*:

$$\begin{cases} \tilde{u}_t^\varepsilon + \dfrac{\ell}{2}\tilde{\varphi}_t^\varepsilon = k\Delta\tilde{u}^\varepsilon & \text{in } Q_i^\varepsilon, \\ \dfrac{\partial}{\partial\nu}\tilde{u}^\varepsilon + h\tilde{u}^\varepsilon = \tilde{w}(t)g(x) & \text{on } \Sigma_i^\varepsilon, \\ \tilde{u}^\varepsilon(0,x) = 0, & x \in \Omega, \end{cases} \tag{3.81}$$

$$\begin{cases} \tilde{w}'(t) = \beta\tilde{w}(t) + \tilde{v}(t) & t \in [0,T], \\ \tilde{w}(0) = 0, \end{cases} \tag{3.82}$$

$$\begin{cases} \tau\tilde{\varphi}_t^\varepsilon = \xi^2\Delta\tilde{\varphi}^\varepsilon + \dfrac{1}{2a}\tilde{\varphi}^\varepsilon + 2\tilde{u}^\varepsilon & \text{in } Q_i^\varepsilon, \\ \tilde{\varphi}^\varepsilon = 0 & \text{on } \Sigma_i^\varepsilon, \\ \tilde{\varphi}_+^\varepsilon(i\varepsilon, x) = \eta(\varepsilon) & x \in \Omega, \end{cases} \tag{3.83}$$

for $i = \overline{0, M_\varepsilon - 1}$, where $\eta(\varepsilon)$ is the solution of

$$\begin{cases} \eta'(s) + \dfrac{3}{2a}(z+1)^2\eta(s) = 0 \qquad s \in [0,T], \\ \eta(0) = \tilde{\varphi}_-^\varepsilon(i\varepsilon, x) \qquad \tilde{\varphi}_-^\varepsilon(0,x) = 0. \end{cases} \tag{3.84}$$

The solution of problem (3.84) is given by

$$\eta(\varepsilon) = \exp\Big(-\int_0^\varepsilon \frac{3}{2a}(z(t,\cdot)+1)^2 dt\Big)\tilde{\varphi}_-^\varepsilon(i\varepsilon, x) \tag{3.85}$$

and then the relation $(3.83)_3$ is equivalent with

$$\tilde{\varphi}_-^\varepsilon(i\varepsilon, x) = \exp\Big(\int_0^\varepsilon \frac{3}{2a}(z(t,\cdot)+1)^2 dt\Big)\tilde{\varphi}_+^\varepsilon(i\varepsilon, x). \tag{3.86}$$

Let us introduce now the *adjoint state system.* The equations $(3.81)_1$, $(3.83)_1$ can be written as

$$\frac{\partial}{\partial t}\begin{pmatrix} \tilde{u}^\varepsilon \\ \tilde{\varphi}^\varepsilon \end{pmatrix} = A\begin{pmatrix} \tilde{u}^\varepsilon \\ \tilde{\varphi}^\varepsilon \end{pmatrix} \quad \text{in } Q_i^\varepsilon$$

where

$$A = \begin{pmatrix} k\Delta - \dfrac{\ell}{\tau} & -\dfrac{\ell\xi^2}{2\tau}\Delta - \dfrac{\ell}{4a\tau} \\ \dfrac{2}{\tau} & \dfrac{\xi^2}{\tau}\Delta + \dfrac{1}{2a\tau} \end{pmatrix}.$$

Then

$$A^* = \begin{pmatrix} k\Delta - \dfrac{\ell}{\tau} & \dfrac{2}{\tau} \\ -\dfrac{\ell\xi^2}{2\tau}\Delta - \dfrac{l}{4a\tau} & \dfrac{\xi^2}{\tau}\Delta + \dfrac{1}{2a\tau} \end{pmatrix}$$

and the adjoint state system is

$$\frac{\partial}{\partial t}\begin{pmatrix} p \\ q \end{pmatrix} = -A^*\begin{pmatrix} p \\ q \end{pmatrix} + \chi_0\begin{pmatrix} 0 \\ \varphi^\varepsilon \end{pmatrix}, \tag{3.87}$$

i.e., the optimality conditions for problem (P^ε) can be written as

$$\begin{cases} p_t + k\Delta p - \dfrac{\ell}{\tau}p + \dfrac{2}{\tau}q = 0 & \text{in } Q_i^\varepsilon, \\ \frac{\partial}{\partial\nu}p + hp = 0 & \text{on } \Sigma_i^\varepsilon, \\ p^-((i+1)\varepsilon, x) = 0, \quad p^-(T,x) = 0 & x \in \Omega, \end{cases} \tag{3.88}$$

$$\begin{cases} \tau q_t - \dfrac{\ell\xi^2}{2\tau}\Delta p - \dfrac{\ell}{4a\tau}p + \dfrac{\xi^2}{\tau}\Delta q + \dfrac{1}{2a\tau}q = \varphi^\varepsilon \cdot \chi_0 & \text{in } Q_i^\varepsilon, \\ q = \dfrac{\ell}{2}p & \text{on } \Sigma_i^\varepsilon, \\ q^-((i+1)\varepsilon, \cdot) & \\ \quad = \exp\Big(\int\limits_0^\varepsilon \frac{3}{2a}(z(t,\cdot)+1)^2 dt\Big) q^+((i+1)\varepsilon, \cdot) & x \in \Omega, \\ q^-(T, \cdot) = 0, & \end{cases} \tag{3.89}$$

for $i = M_\varepsilon - 2, M_\varepsilon - 3, ..., 1, 0$, where $z(t,\cdot)$ is the solution of (3.36).

Let us introduce the cost functional

$$J^\varepsilon(v) = \frac{1}{2}\int\limits_Q \chi_0(\varphi_\varepsilon^v(t))^2 \, dtdx + \frac{1}{2}\int\limits_0^T w^2(t)\, dt + I_{\mathcal{U}}(v) \tag{3.90}$$

where $I_{\mathcal{U}}(v)$ is the *indicator function* for the set $\mathcal{U}$. If v_ε^* is an optimal controller for problem (P^ε) then

$$\frac{J^\varepsilon(v_\varepsilon^* + \lambda\tilde{v}) - J^\varepsilon(v_\varepsilon^*)}{\lambda} \geq 0 \quad \forall\, \lambda > 0.$$

Letting $\lambda \to 0$ in above inequality, we get

$$\int\limits_{Q_0} \varphi^\varepsilon \tilde{\varphi}^\varepsilon \, dtdx + \int\limits_0^T w\tilde{w}\, dt + I'_{\mathcal{U}}(v_\varepsilon^*, \tilde{v}) \geq 0 \quad \forall\, \tilde{v} \in T_{\mathcal{U}}(v_\varepsilon^*), \tag{3.91}$$

where $Q_0 = [0,T] \times \Omega_0$.

Multiplying $(3.88)_1$ by $\tilde{u}^\varepsilon$ and $(3.89)_1$ by $\tilde{\varphi}^\varepsilon$, using integration by parts and Green's formula, yields

$$\int\limits_{Q_i^\varepsilon} p_t\tilde{u}^\varepsilon \, dtdx + k\int\limits_{Q_i^\varepsilon} p\Delta\tilde{u}^\varepsilon \, dtdx - \frac{\ell}{\tau}\int\limits_{Q_i^\varepsilon} p\tilde{u}^\varepsilon \, dtdx \tag{3.92}$$

$$+\frac{2}{\tau}\int\limits_{Q_i^\varepsilon} q\tilde{u}^\varepsilon\ dtdx + \int\limits_{\Sigma_i^\varepsilon}\Big(\frac{\partial}{\partial\nu}p\tilde{u}^\varepsilon - p\frac{\partial}{\partial\nu}\tilde{u}^\varepsilon\Big)\ dtd\gamma = 0.$$

$$\int\limits_{Q_i^\varepsilon} q_t\tilde{\varphi}^\varepsilon\ dtdx + \frac{\xi^2}{\tau}\int\limits_{Q_i^\varepsilon} q\Delta\tilde{\varphi}^\varepsilon\ dtdx + \frac{\xi^2}{\tau}\int\limits_{\Sigma_i^\varepsilon}\Big(\frac{\partial}{\partial\nu}q\tilde{\varphi}^\varepsilon - q\frac{\partial}{\partial\nu}\tilde{\varphi}^\varepsilon\Big)\ dtd\gamma \quad (3.93)$$

$$+\frac{1}{2a}\int\limits_{Q_i^\varepsilon} q\tilde{\varphi}^\varepsilon\ dtdx - \frac{\ell\xi^2}{2\tau}\int\limits_{Q_i^\varepsilon} p\Delta\tilde{\varphi}^\varepsilon\ dtdx - \frac{\ell\xi^2}{2\tau}\int\limits_{\Sigma_i^\varepsilon}\Big(\frac{\partial}{\partial\nu}p\tilde{\varphi}^\varepsilon - p\frac{\partial}{\partial\nu}\tilde{\varphi}^\varepsilon\Big)\ dtd\gamma$$

$$-\frac{\ell}{4a\tau}\int\limits_{Q_i^\varepsilon} p\tilde{\varphi}^\varepsilon\ dtdx = \int\limits_{Q_i^\varepsilon}\chi_0\varphi^\varepsilon\tilde{\varphi}^\varepsilon\ dtdx.$$

Now we multiply $(3.81)_2$ by p, $(3.88)_2$ by $\tilde{u}^\varepsilon$ and, by subtraction, we get

$$\frac{\partial}{\partial\nu}p\tilde{u}^\varepsilon - p\frac{\partial}{\partial\nu}\tilde{u}^\varepsilon = -p\tilde{w}(t)g(x). \quad (3.94)$$

Adding (3.92)-(3.93) and making use of relations $(3.83)_2$, $(3.89)_2$ as well as of (3.94), after some simple calculations, we obtain

$$\int\limits_{Q_i^\varepsilon} p_t\tilde{u}^\varepsilon\ dtdx + \int\limits_{Q_i^\varepsilon} q_t\tilde{\varphi}^\varepsilon\ dtdx$$

$$+\int\limits_{Q_i^\varepsilon} p\Big[k\Delta\tilde{u}^\varepsilon - \frac{\ell}{\tau}\tilde{u}^\varepsilon - \frac{\ell\xi^2}{2\tau}\Delta\tilde{\varphi}^\varepsilon - \frac{\ell}{4a\tau}\tilde{\varphi}^\varepsilon\Big]dtdx$$

$$+\int\limits_{Q_i^\varepsilon} q\Big[\frac{\xi^2}{\tau}\Delta\tilde{\varphi}^\varepsilon + \frac{1}{2a\tau}\tilde{\varphi}^\varepsilon + \frac{2}{\tau}\tilde{u}^\varepsilon\Big]dtdx - \int\limits_{\Sigma_i^\varepsilon} p\tilde{w}(t)g(x)\ dtd\gamma$$

$$= \int\limits_{Q_i^\varepsilon}\chi_0\varphi^\varepsilon\tilde{\varphi}^\varepsilon\ dtdx$$

i.e., taking into account $(3.81)_1$ and $(3.83)_1$,

$$\int\limits_{Q_i^\varepsilon} p_t\tilde{u}^\varepsilon\ dtdx + \int\limits_{Q_i^\varepsilon} q_t\tilde{\varphi}^\varepsilon\ dtdx + \int\limits_{Q_i^\varepsilon} p\tilde{u}_t^\varepsilon\ dtdx + \int\limits_{Q_i^\varepsilon} q\tilde{\varphi}_t^\varepsilon\ dtdx$$

$$-\int_{\Sigma_i^\varepsilon} p\tilde{w}(t)g(x)\ dtd\gamma = \int_{Q_i^\varepsilon} \chi_0 \varphi^\varepsilon \tilde{\varphi}^\varepsilon\ dtdx.$$

By Fubini's theorem and definition of *distributional derivative*, the last equality implies

$$-\int_{\Sigma_i^\varepsilon} p\tilde{w}(t)g(x)\ dtd\gamma = \int_{Q_i^\varepsilon} \chi_0 \varphi^\varepsilon \tilde{\varphi}^\varepsilon\ dtdx$$

and then (3.91) becomes

$$\int_{\Sigma} (-kpg(x) + w(t))\tilde{w}(t)\ dtd\gamma + I'_{\mathcal{U}}(v_\varepsilon^*, \tilde{v}) \geq 0 \quad \forall\, \tilde{v} \in T_{\mathcal{U}}(v_\varepsilon^*), \quad (3.95)$$

where $\tilde{w}(t) = \int_0^t e^{\beta(t-s)}\tilde{v}(s)ds$. We set

$$r(t) = \int_t^T \Big(\int_{\partial\Omega} (w(s) - kp(s,x)g(x))d\gamma \Big) e^{\beta(s-t)} ds$$

and then (3.95) one rewritten

$$\int_0^T r(t)\tilde{v}(t)dt + I'_{\mathcal{U}}(v_\varepsilon^*, \tilde{v}) \geq 0 \quad \forall\, \tilde{v} \in T_{\mathcal{U}}(v_\varepsilon^*),$$

whence we obtain

$$-r(t) \in \partial I_{\mathcal{U}}(v_\varepsilon^*) \qquad a.e.\ \ t \in [0,T],$$

where by the symbol ∂ we have denoted the *subdifferential*; thus

$$v_\varepsilon^*(t) = \begin{cases} R, & \text{if } r(t) < 0 \\ 0, & \text{if } r(t) > 0. \end{cases} \qquad (3.96)$$

Summing, we have proved the following *maximum principle* for problem (P^ε)

Theorem 3.2. *Let* $(u_\varepsilon^*, \varphi_\varepsilon^*, w_\varepsilon^*, v_\varepsilon^*)$ *be optimal in problem* (P^ε). *Then the optimal control is given by* (3.96) *where* (p, q) *satisfy along with* u_ε^* *and* φ_ε^* *the adjoint state system* (3.88)-(3.89).

3.5 Control of solidification zone in 1D

Denote by $\Omega = (0, b_1] \subset I\!R$, $0 < b_1 < +\infty$. Let $T > 0$ and we set: $Q = [0, T] \times (0, b_1]$, $\Sigma_0 = \{(t, x) \in Q,\ t = f(x)\}$, and $\Sigma = [0, T] \times \{b_1\}$, where $t = f(x)$ is considered to be the equation of the moving boundary separating the liquid and solid phases, $0 = f(b_0)$, $0 < b_0 < b_1$ (see Fig. **1**).

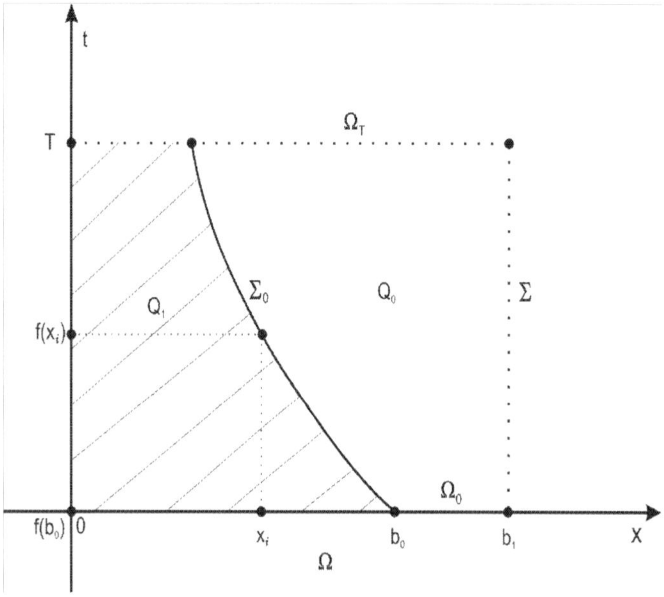

Figure 1 *Geometrical image of the elements in inverse problem* (P_{inv})

Consider the following nonlinear parabolic system in one space dimension:

$$\begin{cases} \rho c\, u_t + \frac{\ell}{2}\varphi_t = k u_{xx} \\ \tau\varphi_t = \xi^2\varphi_{xx} + \frac{1}{2a}(\varphi - \varphi^3) + 2u \end{cases} \quad \text{in } Q, \tag{3.97}$$

subject to non-homogeneous Cauchy-Neumann boundary conditions:

$$u_x + hu = w(t), \qquad \varphi_x = 0 \ \text{ on } \Sigma, \tag{3.98}$$

$$-u_x = 0, \qquad\qquad -\varphi_x = 0 \ \text{ on } \ \Sigma_0, \tag{3.99}$$

and initial conditions:

$$u(0,x) = u_0(x), \quad \varphi(0,x) = \varphi_0(x) \qquad \text{on } \ \Omega_0 = [b_0, b_1], \tag{3.100}$$

where u is the *reduced temperature* distribution, φ is the *phase function* used to distinguish between the phase of Ω, u_0, $\varphi_0 : \Omega \to I\!R$ are given functions, $w : [0,T] \to I\!R$ is the *boundary control* (the temperature surrounding at $x = b_1$), $w \in \mathcal{U}$,

$$\mathcal{U} = \{w \in L^\infty([0,T]), \ \ -R \le w(t) \le 0 \ \ a.e. \ \ t \in [0,T]\}, \tag{3.101}$$

and the positive parameters ρ, c, τ, ξ, ℓ, k, h, a, have the following physical meaning: ρ - is *the density*, c - is *the heat capacity*, τ - is *the relaxation time*, ξ - is *the length scale of the interface*, ℓ - denotes *the latent heat*, k - *the heat conductivity*, h - *the heat transfer coefficient* and a is an *probabilistic measure on the individual atoms* (a depends on ξ).

As regards the existence, it is known that under appropriate conditions on u_0, φ_0 and w, the state system (3.97)-(3.100) has a unique solution u, $\varphi \in W = W_p^{2,1}(Q) \cap L^\infty(Q)$, $p > 3/2$ (see Theorem 2.1 in Chapter 2 with $f = g = 0$).

Given the positive numbers δ_1, δ_2, we define:

- the *pure liquid region*: $\{(t,x) \in Q, \ u(t,x) > \delta_2$ and $\varphi(t,x) \ge 1+\delta_1\}$,
- the *pure solid region*: $\{(t,x) \in Q, \ u(t,x) < -\delta_2$ and $\varphi(t,x) \le -1-\delta_1\}$,
- the *separating region*: $\{(t,x) \in Q, \ |u(t,x)| < \delta_2, \ |\varphi(t,x)| \le 1 + \delta_1\}$,

and we set

$$Q_0 = \big\{(t,x) \in Q, \ f(x) \le t \le T\big\}. \tag{3.102}$$

Consider the following inverse problem:

(P$_{inv}$) *Given Σ_0 find the boundary control $w \in L^\infty([0,T])$ such that Q_0 in (3.102) is in the liquid region, $Q_1 = Q \backslash Q_0$ is in the solid region and a neighborhood of Σ_0 is the separating region between the liquid and the solid region.*

(P$_{inv}$) is in general *"ill posed"* and a common way to treat this inverse problem is to reformulate it as an optimal control problem with an appropriate cost functional. Consequently, we will focus on the present

Section with an optimal control problem associated to the inverse problem (P_{inv}), namely:

$$(\mathrm{P}^0_{inv})\ \textit{Minimize } L_0(w) = \frac{\beta}{2}\int_Q \left[(u(t,x)-\delta_2)^+\right]^2 \cdot \chi_{Q_0}\, dtdx$$

$$+\frac{1}{2}\int_Q (\varphi(t,x)-1-\delta_1)^2\cdot\chi_{Q_0}\, dtdx + \frac{1}{2}\int_0^T w^2(t)\, dt,$$

on all (u,φ) solution of the system (3.97)-(3.100) and for all $w\in\mathcal{U}$ in (3.101). $\beta>0$ is a given constant.

In the statement above we denoted by u^+ the positive part of u:

$$u^+ = \begin{cases} u, & \text{if } u>0, \\ 0, & \text{if } u\le 0. \end{cases}$$

We point out that problem (P^0_{inv}) is an optimal problem with boundary control $w(t)$ depending on time variable $t\in[0,T]$, being dictated by the industrial solidification process like *casting wire.*

Approximating process ($\mathbf{P}^{\varepsilon}_{inv}$). We associate to the nonlinear system (3.97)-(3.100) the following approximating scheme ($\varepsilon>0$):

$$\begin{cases} \rho c\ u^\varepsilon_t + \frac{\ell}{2}\varphi^\varepsilon_t = k\ u^\varepsilon_{xx} \\ \tau\ \varphi^\varepsilon_t = \xi^2\varphi^\varepsilon_{xx} + \frac{1}{2a}\varphi^\varepsilon + 2u^\varepsilon \end{cases} \quad \text{in } Q^\varepsilon_0 = \{(t,x)\in Q,\ \varepsilon\le t\le T\}, \tag{3.103}$$

$$u^\varepsilon_x + hu^\varepsilon = w(t),\quad \varphi^\varepsilon_x = 0 \quad \text{on } \Sigma^\varepsilon = [\varepsilon,T]\times\{b_1\}, \tag{3.104}$$

$$-u^\varepsilon_x = 0,\quad -\varphi^\varepsilon_x = 0 \quad \text{on } \Sigma^\varepsilon_0 = \{(t,x)\in Q,\ \varepsilon\le t\le T\}, \tag{3.105}$$

$$u^\varepsilon(\varepsilon,x) = u_0(x),\qquad \varphi^\varepsilon_+(\varepsilon,x) = z(\varepsilon,\varphi^\varepsilon_-(\varepsilon,x)) \quad \text{on } \Omega_\varepsilon, \tag{3.106}$$

where $z(\varepsilon,\varphi^\varepsilon_-(\varepsilon,x))$ is the solution of the Cauchy problem:

$$\begin{cases} z'(s) + \dfrac{1}{2a}z^3(s) = 0 & s\in(0,\varepsilon), \\ z(0) = \varphi^\varepsilon_-(\varepsilon,x), & \varphi^\varepsilon_-(0,x) = \varphi_0(x), \end{cases} \tag{3.107}$$

and $\varphi_+^\varepsilon(\varepsilon, x) = \lim_{t\downarrow\varepsilon} \varphi^\varepsilon(t, x)$, $\varphi_-^\varepsilon(\varepsilon, x) = \lim_{t\uparrow\varepsilon} \varphi^\varepsilon(t, x)$.

The convergence and weak stability of the approximating scheme (3.103)-(3.106), in a more general case ($w(t, x)$ in place of $w(t)$), was studied in Section 2.6.

Corresponding to the approximating scheme (3.103)-(3.107), we now consider the approximating optimal control problem:

$$(\mathrm{P}_{inv}^\varepsilon) \; Minimize \; L_0^\varepsilon(w) = \frac{\beta}{2} \int\limits_Q \left[(u^\varepsilon(t, x) - \delta_2)^+ \right]^2 \cdot \chi_{Q_0} \, dtdx$$

$$+\frac{1}{2} \int\limits_Q (\varphi^\varepsilon(t, x) - 1 - \delta_1)^2 \cdot \chi_{Q_0} \, dtdx + \frac{1}{2} \int\limits_0^T w^2(t) \, dt,$$

on all $(u^\varepsilon, \varphi^\varepsilon)$ *solution of* (3.103)-(3.107) *corresponding to* $w \in \mathcal{U}$ *in* (3.101).

The main result of this Section (Theorem 3.3) says that problem (P_{inv}^0) can be approximated for $\varepsilon \to 0$ by the sequence of optimal control problems $(\mathrm{P}_{inv}^\varepsilon)$ and so the computation of the approximate boundary control $w(t)$ in (P_{inv}^0) can be substituted by computation of an approximate control of $(\mathrm{P}_{inv}^\varepsilon)$.

3.5a The convergence of problem $(\mathrm{P}_{inv}^\varepsilon)$

The main result of this Subsection is

Theorem 3.3. *Let* $\{w_\varepsilon^*\}$ *be a sequence of optimal controllers for problem* (P_{inv}^ε)*. Then*

$$\lim_{\varepsilon\to 0} \inf \; L_0^\varepsilon(w) = \inf \; \{L_0(w); \;\; w \in \mathcal{U}\} \tag{3.108}$$

and

$$\lim_{\varepsilon\to 0} L_0(w_\varepsilon^*) = \inf \; \{L_0(w); \;\; w \in \mathcal{U}\}. \tag{3.109}$$

Moreover, every weak limit point of $\{w_\varepsilon^*\}$ *is an optimal controller for problem* (P_{inv}^0).

Remark 3.1. Theorem 3.3 amounts to saying that $(\mathrm{P}_{inv}^\varepsilon)$ approximates problem (P_{inv}^0), and an optimal control $\{w_\varepsilon^*\}$ of $(\mathrm{P}_{inv}^\varepsilon)$ is a suboptimal control of problem (P_{inv}^0).

The main ingredient in the proof of the Theorem 3.3 is the following Lemma.

Lemma 3.2. *If* $\{w_\varepsilon^*\}$ *is a sequence of optimal controllers for problems* (P^ε) *then there exists* $\{\varepsilon_n\} \to 0$ *such that*

$$w_{\varepsilon_n}^* \to w^* \quad \text{weak star in } L^\infty(\Sigma), \tag{3.110}$$

$$u_{\varepsilon_n}^* \to u^* \quad \text{strongly in } L^2((0,T); H^1(\Omega)), \tag{3.111}$$

$$\varphi_{\varepsilon_n}^* \to \varphi^* \quad \text{strongly in } L^2((0,T); H^1(\Omega)), \tag{3.112}$$

where $(u_{\varepsilon_n}^*, \varphi_{\varepsilon_n}^*, w_{\varepsilon_n}^*) = (u_{\varepsilon_n}^{w_{\varepsilon_n}^*}, \varphi_{\varepsilon_n}^{w_{\varepsilon_n}^*}, w_{\varepsilon_n}^*)$ *is the solution to* (3.103)-(3.106) *corresponding to* $w = w_{\varepsilon_n}^*$ *and* $(u^*, \varphi^*, w^*) = (u^{w^*}, \varphi^{w^*}, w^*)$ *is the solution to* (3.97)-(3.100) *corresponding to* $w = w^*$.

Proof. Details on the demonstration of this Lemma can be found in Section 3.3, Lemma 3.1. We omit them.

Proof of Theorem 3.3. Let $\{w_\varepsilon^*\}$ be an optimal controller for problem (P_{inv}^ε) and let $(u_\varepsilon^*, \varphi_\varepsilon^*, w_\varepsilon^*)$ be the corresponding solution of (3.103)-(3.106) with $w = w_\varepsilon^*$. Lemma 3.2 above allows us to conclude that there exist $w^* \in L^\infty([0,T])$ and $\{\varepsilon_n\}$ such that relations (3.110)-(3.112) are valid.

Since:

$$u \to \frac{\beta}{2} \int_Q \left((u(t,x) - \delta_2)^+\right)^2 \cdot \chi_{Q_0} \, dtdx,$$

$$\varphi \to \frac{1}{2} \int_Q (\varphi(t,x) - 1 - \delta_1)^2 \cdot \chi_{Q_0} \, dtdx,$$

$$w \to \frac{1}{2} \int_0^T w^2(t) \, dt$$

are convex continuous functions, it follows that these are weakly lower semicontinuous functions. Hence we can write

$$L_0(w^*) \le \liminf_{n\to\infty} L_0^{\varepsilon_n}(w_{\varepsilon_n}^*). \tag{3.113}$$

Let $\bar{w}^*$ be an optimal controller for problem (P_{inv}^0). Since $w_{\varepsilon_n}^*$ is an optimal controller for problem $(P_{inv}^{\varepsilon_n})$ it follows that

$$L_0^{\varepsilon_n}(w_{\varepsilon_n}^*) \le L_0(\bar{w}^\varepsilon).$$

But $u_{\varepsilon_n}^{\bar{w}^*} \to u^{\bar{w}^*}$ $\varphi_{\varepsilon_n}^{\bar{w}^*} \to \varphi^{\bar{w}^*}$ strongly in $L^2((0,T); H^1(\Omega))$ (see (3.111) and (3.112)), and so the latter inequalities implies

$$\lim_{n\to\infty} L_0^{\varepsilon_n}(\bar{w}^\varepsilon) \le L_0(\bar{w}^\varepsilon). \tag{3.114}$$

From (3.113)-(3.114) we get

$$L_0(w^*) \leq \liminf_{n\to\infty} L_0^{\varepsilon_n}(w_{\varepsilon_n}^*) \leq \limsup_{n\to\infty} L_0^{\varepsilon_n}(w_{\varepsilon_n}^*) \leq L_0(\bar{w}^*).$$

Hence

$$\liminf_{\varepsilon_n\to 0} L_0^{\varepsilon_n}(w_{\varepsilon_n}^*) = L_0(\bar{w}^*) = \inf\{L_0(w),\ w \in \mathcal{U}\}$$

and then (3.108) holds.

To prove (3.109) we set: $\bar{u}_\varepsilon = u^{w_\varepsilon^*}$, $\bar{\varphi}_\varepsilon = \varphi^{w_\varepsilon^*}$ (we recall that w_ε^* is chosen to be optimal in (P_{inv}^ε)). On a subsequence $\{\varepsilon_n\}$ we have

$$\begin{aligned} w_\varepsilon^* &\to w^0 \quad \text{weak star in} \quad L^\infty([0,T]),\\ \bar{u}_{\varepsilon_n} &\to u \quad \text{strongly in} \quad L^2((0,T), H^1(\Omega)),\\ \bar{\varphi}_{\varepsilon_n} &\to \varphi \quad \text{strongly in} \quad L^2((0,T), H^1(\Omega)),\end{aligned}$$

where (u, φ, w^0) satisfy (3.97)-(3.100), i.e., $(u, \varphi) = (u^{w^0}, \varphi^{w^0})$. Therefore, we derive

$$L_0(w^0) \leq \inf P_{inv}^0$$

and, because $\{\varepsilon_n\}$ was choose arbitrarily, (3.109) follows.

Now, taking into account that w_ε^* is an optimal controller for problem (P_{inv}^ε), it follows that

$$L_0^\varepsilon(w_\varepsilon^*) \leq L_0^\varepsilon(w) \quad \forall\, w \in \mathcal{U}.$$

On the other hand, on the basis of relation (3.113), we can put

$$L_0(w^*) \leq \liminf_{\varepsilon\to 0} L_0^\varepsilon(w_\varepsilon^*)$$

and thus, along with previous inequality, we may conclude that

$$L_0(w^*) \leq \lim_{\varepsilon\to 0} L_0^\varepsilon(w) \quad \forall\, w \in \mathcal{U}.$$

Consequently

$$L_0(w^*) \leq L_0(w) \quad \forall\, w \in \mathcal{U}$$

i.e., the weak limit point w^* is a suboptimal control for problem (P_{inv}^0). This completes the proof of Theorem 3.3.

3.5b Necessary optimality conditions in $(\mathrm{P}^{\varepsilon}_{inv})$

Let $(u^{\varepsilon}, \varphi^{\varepsilon}, w)$ be the solution of (3.103)-(3.106) and let $\tilde{w} \in L^{\infty}[0,T])$ be arbitrary but fixed and $\lambda > 0$. Set $w^{\lambda} = w + \lambda \tilde{w}$ and let $(u^{\lambda,\varepsilon}, \varphi^{\lambda,\varepsilon})$ be the solution of (3.103)-(3.106) corresponding to w^{λ}, that is:

$$\begin{cases} \rho c u_t^{\lambda,\varepsilon} + \frac{\ell}{2}\varphi_t^{\lambda,\varepsilon} = k u_{xx}^{\lambda,\varepsilon} \\ \tau \varphi_t^{\lambda,\varepsilon} = \xi^2 \varphi_{xx}^{\lambda,\varepsilon} + \frac{1}{2a}\varphi^{\lambda,\varepsilon} + 2u^{\lambda,\varepsilon} \end{cases} \quad \text{in } Q_0^{\varepsilon}, \tag{3.115}$$

subject to non-homogeneous Cauchy-Neumann boundary conditions:

$$u_x^{\lambda,\varepsilon} + h u^{\lambda,\varepsilon} = w^{\lambda}, \qquad \varphi_x^{\lambda,\varepsilon} = 0 \quad \text{on } \Sigma^{\varepsilon}, \tag{3.116}$$

$$u_x^{\lambda,\varepsilon} = 0, \qquad \varphi_x^{\lambda,\varepsilon} = 0 \quad \text{on } \Sigma_0^{\varepsilon}, \tag{3.117}$$

and initial conditions:

$$u^{\lambda,\varepsilon}(\varepsilon, x) = u_0(x), \quad \varphi_+^{\lambda,\varepsilon}(\varepsilon, x) = z^{\lambda}(\varepsilon, \varphi_0(x)) \quad \text{on } \Omega_{\varepsilon}, \tag{3.118}$$

where $z^{\lambda}(\varepsilon, \varphi_0(x))$ is the solution of the Cauchy problem:

$$\begin{cases} \left(z^{\lambda}(s)\right)' + \dfrac{1}{2a}\left(z^{\lambda}(s)\right)^3 = 0 \quad s \in (0, \varepsilon), \\ z^{\lambda}(0) = \tilde{\varphi}_-^{\lambda,\varepsilon}(\varepsilon, x) \quad \tilde{\varphi}_-^{\lambda,\varepsilon}(0, x) = \varphi_0(x). \end{cases} \tag{3.119}$$

Subtracting (3.103)-(3.106) from (3.115)-(3.118) and dividing by $\lambda > 0$, we get

$$\begin{cases} \rho c\left(\frac{u^{\lambda,\varepsilon} - u^{\varepsilon}}{\lambda}\right)_t + \frac{\ell}{2}\left(\frac{\varphi^{\lambda,\varepsilon} - \varphi^{e}}{\lambda}\right)_t = k\left(\frac{u^{\lambda,\varepsilon} - u^{\varepsilon}}{\lambda}\right)_{xx} \\ \tau\left(\frac{\varphi^{\lambda,\varepsilon} - \varphi^{e}}{\lambda}\right)_t = \xi^2\left(\frac{\varphi^{\lambda,\varepsilon} - \varphi^{e}}{\lambda}\right)_{xx} + \frac{1}{2a}\left(\frac{\varphi^{\lambda,\varepsilon} - \varphi^{e}}{\lambda}\right) + 2\left(\frac{u^{\lambda,\varepsilon} - u^{e}}{\lambda}\right) \end{cases} \text{in } Q_0^{\varepsilon}, \tag{3.120}$$

$$\left(\frac{u^{\lambda,\varepsilon} - u^{e}}{\lambda}\right)_x + h\left(\frac{u^{\lambda,\varepsilon} - u^{e}}{\lambda}\right) = \frac{w^{\lambda} - w}{\lambda}, \quad \left(\frac{\varphi^{\lambda,\varepsilon} - \varphi^{e}}{\lambda}\right)_x = 0 \quad \text{on } \Sigma^{\varepsilon}, \tag{3.121}$$

$$\left(\frac{u^{\lambda,\varepsilon} - u^{e}}{\lambda}\right)_x = 0, \quad \left(\frac{\varphi^{\lambda,\varepsilon} - \varphi^{e}}{\lambda}\right)_x = 0 \quad \text{on } \Sigma_0^{\varepsilon}, \tag{3.122}$$

$$\frac{u^{\lambda,\varepsilon}(\varepsilon,x)-u^{\varepsilon}(\varepsilon,x)}{\lambda}=0,\quad \frac{\varphi_+^{\lambda,\varepsilon}(\varepsilon,x)-\varphi_+^{e}(\varepsilon,x)}{\lambda}=\frac{z^{\lambda}(\varepsilon,\varphi_0(x))-z(\varepsilon,\varphi_0(x))}{\lambda}\quad \text{on } \Omega_\varepsilon. \tag{3.123}$$

Letting λ tend to zero in (3.120)-(3.123) we get the system in variation (3.124)-(3.127) below

$$\begin{cases} \rho c\tilde{u}_t^\varepsilon+\frac{\ell}{2}\tilde{\varphi}_t^\varepsilon=k\tilde{u}_{xx}^\varepsilon \\ \\ \tau\tilde{\varphi}_t^\varepsilon=\xi^2\tilde{\varphi}_{xx}^\varepsilon+\frac{1}{2a}\tilde{\varphi}^\varepsilon+2\tilde{u}^\varepsilon \end{cases}\quad \text{in } Q_0^\varepsilon, \tag{3.124}$$

$$\tilde{u}_x^\varepsilon+h\tilde{u}^\varepsilon=\tilde{w},\qquad \tilde{\varphi}_x^\varepsilon=0\qquad \text{on } \Sigma^\varepsilon, \tag{3.125}$$

$$\tilde{u}_x^\varepsilon=0,\qquad \tilde{\varphi}_x^\varepsilon=0\qquad \text{on } \Sigma_0^\varepsilon, \tag{3.126}$$

$$\tilde{u}^\varepsilon(\varepsilon,x)=0,\qquad \tilde{\varphi}_+^\varepsilon(\varepsilon,x)=\eta(\varepsilon,x)\quad \text{on } \Omega_\varepsilon. \tag{3.127}$$

where $\tilde{u}^\varepsilon=\lim\limits_{\lambda\to 0}\dfrac{u^{\lambda,\varepsilon}-u^\varepsilon}{\lambda}$, etc., and

$$\eta(\varepsilon,x)=\lim_{\lambda\to 0}\frac{z^\lambda(\varepsilon,\varphi_0(x))-z(\varepsilon,\varphi_0(x))}{\lambda}=$$

$$=z^\lambda(\varepsilon,\varphi_0(x))\cdot\tilde{\varphi}_-^\varepsilon(\varepsilon,x)+\tilde{z}(\varepsilon,\varphi_0(x))=\tilde{z}(\varepsilon,\varphi_0(x))$$

with $\eta(\cdot,x)$ the solution of Cauchy problem

$$\begin{cases} \eta'(s,x)+\dfrac{3}{2a}z^2(s,x)\eta(s,x)=0 \qquad s\in(0,\varepsilon), \\ \\ \eta(0,x)=\tilde{\varphi}_-^\varepsilon(\varepsilon,x), \end{cases} \tag{3.128}$$

that is

$$\eta(s,x)=\exp\Big(-\int_0^\varepsilon\frac{3}{2a}z(t,\cdot)^2dt\Big)\tilde{\varphi}_-^\varepsilon(\varepsilon,x). \tag{3.129}$$

We now introduce the adjoint state system. For this, the system (3.124) can be written in the abstract form:

$$\begin{pmatrix}\tilde{u}^\varepsilon\\ \tilde{\varphi}^\varepsilon\end{pmatrix}_t=A\begin{pmatrix}\tilde{u}^\varepsilon\\ \tilde{\varphi}^\varepsilon\end{pmatrix}\qquad \text{in } Q_0^\varepsilon$$

where (here $\Delta\varphi = \varphi_{xx}$)

$$A = \begin{pmatrix} \frac{1}{\rho c}\Big(k\Delta - \frac{\ell}{\tau}\Big) & -\frac{\ell}{2\tau\rho c}\Big(\xi^2\Delta + \frac{1}{2a}\Big) \\ \frac{2}{\tau} & \frac{1}{\tau}\Big(\xi^2\Delta + \frac{1}{2a}\Big) \end{pmatrix}.$$

$D(A) = \Big\{(\psi,\gamma) \in H^2(\Omega)\times H^2(\Omega);\ \frac{\partial\psi}{\partial\nu} + h\psi \in L^2(\partial\Omega),\ \frac{\partial\gamma}{\partial\nu} = 0\Big\}.$

Then

$$A^* = \begin{pmatrix} \frac{1}{\rho c}\Big(k\Delta - \frac{\ell}{\tau}\Big) & \frac{2}{\tau} \\ -\frac{\ell}{2\tau\rho c}\Big(\xi^2\Delta + \frac{1}{2a}\Big) & \frac{1}{\tau}\Big(\xi^2\Delta + \frac{1}{2a}\Big) \end{pmatrix}$$

$D(A^*) = \Big\{(\psi,\gamma) \in H^2(\Omega)\times H^2(\Omega);\ \frac{\partial\psi}{\partial\nu} + h\psi = 0,\ \frac{\partial\gamma}{\partial\nu} = \frac{2\rho c}{\ell}\frac{\partial\psi}{\partial\nu}\Big\}$, and thus the adjoint state system is

$$\begin{pmatrix} p^\varepsilon \\ q^\varepsilon \end{pmatrix}_t = -A^*\begin{pmatrix} p^\varepsilon \\ q^\varepsilon \end{pmatrix} + \begin{pmatrix} \frac{\partial}{\partial u^\varepsilon} L_0^\varepsilon(w) \\ \frac{\partial}{\partial \varphi^\varepsilon} L_0^\varepsilon(w) \end{pmatrix}$$

i.e.

$$\begin{cases} p_t^\varepsilon + \dfrac{k}{\rho c}p_{xx}^\varepsilon - \dfrac{\ell}{\tau\rho c}p^\varepsilon + \dfrac{2}{\tau}q^\varepsilon = \beta(u^\varepsilon - \delta_2)^+ \cdot \chi_{Q_0} & \text{in } Q_0^\varepsilon, \\ p_x^\varepsilon + hp^\varepsilon = 0 & \text{on } \Sigma^\varepsilon, \\ p_-^\varepsilon(\varepsilon, x) = 0, \quad p_-^\varepsilon(T,x) = 0 & x \in \Omega_T, \end{cases} \tag{3.130}$$

$$\begin{cases} q_t^\varepsilon - \frac{\ell\xi^2}{2\tau\rho c}p_{xx}^\varepsilon - \frac{\ell}{4a\tau\rho c}p^\varepsilon + \frac{\xi^2}{\tau}q_{xx}^\varepsilon + \frac{1}{2a\tau}q^\varepsilon = (\varphi^\varepsilon - 1 - \delta_1)\cdot\chi_{Q_0} & \text{in } Q_0^\varepsilon, \\ q_x^\varepsilon = \dfrac{\ell}{2\rho c}p_x^\varepsilon & \text{on } \Sigma^\varepsilon, \\ q_x^\varepsilon = 0 & \text{on } \Sigma_0^\varepsilon, \\ q_-^\varepsilon(\varepsilon,x) = \exp\Big(\int_0^\varepsilon \frac{3}{2a}z^2(t,\cdot)dt\Big)q_+^\varepsilon(\varepsilon,x) \quad q_-^\varepsilon(T,x) = 0, & x \in \Omega_T, \end{cases} \tag{3.131}$$

Let us introduce the cost functional

$$L_1^\varepsilon(w) = L_0^\varepsilon(w) + \frac{1}{2} I_{\mathcal{U}}(w)$$

where, as usually, $I_{\mathcal{U}}(w)$ is the indicator function of the set $\mathcal{U}$.

If w^* is an optimal controller of problem (P_{inv}^ε), then

$$\frac{L_1^\varepsilon(w^* + \lambda\tilde{w}) - L_1^\varepsilon(w^*)}{\lambda} \geq 0 \quad \forall\, \lambda > 0.$$

that leads to (letting λ tend to zero)

$$\beta \int_Q \tilde{u}^\varepsilon\ (u^\varepsilon - \delta_2)^+ \cdot \chi_{Q_0}\ dtdx + \int_Q \tilde{\varphi}^\varepsilon\ (\varphi^\varepsilon - 1 - \delta_1) \cdot \chi_{Q_0}\ dtdx \qquad (3.132)$$

$$+ \int_0^T w^*\ \tilde{w}\ dt + I'_{\mathcal{U}}(w^*, \tilde{w}) \geq 0 \quad \forall\, \tilde{w} \in T_{\mathcal{U}}(w^*).$$

Multiplying $(3.130)_1$ by $\tilde{u}^\varepsilon$ and $(3.131)_1$ by $\tilde{\varphi}^\varepsilon$, using integration by parts and Green's formula, we get

$$\int_{Q_0^\varepsilon} p_t^\varepsilon \tilde{u}^\varepsilon dt\ dx + \frac{k}{\rho c} \int_{Q_0^\varepsilon} p^\varepsilon \tilde{u}_{xx}^\varepsilon dt\ dx - \frac{\ell}{\tau \rho c} \int_{Q_0^\varepsilon} p^\varepsilon \tilde{u}^\varepsilon dt\ dx \qquad (3.133)$$

$$+ \frac{2}{\tau} \int_{Q_0^\varepsilon} q^\varepsilon \tilde{u}^\varepsilon dt\ dx + \frac{k}{\rho c} \int_{\Sigma^\varepsilon} \left(p^\varepsilon \tilde{u}_x^\varepsilon - p_x^\varepsilon \tilde{u}^\varepsilon \right) dt\ d\gamma = \beta \int_{Q_0^\varepsilon} (u^\varepsilon - \delta_2)^+ \cdot \chi_{Q_0}\ \tilde{u}^\varepsilon\ dtdx,$$

$$\int_{Q_0^\varepsilon} q_t^\varepsilon \tilde{\varphi}^\varepsilon dt\ dx - \frac{\ell \xi^2}{2\tau \rho c} \int_{Q_0^\varepsilon} p^\varepsilon \tilde{\varphi}_{xx}^\varepsilon dt\ dx - \frac{\ell}{4a\tau \rho c} \int_{Q_0^\varepsilon} p^\varepsilon \tilde{\varphi}^\varepsilon dt\ dx \qquad (3.134)$$

$$+ \frac{\ell \xi^2}{2\tau \rho c} \int_{\Sigma^\varepsilon} \left(p_x^\varepsilon \tilde{\varphi}^\varepsilon - p^\varepsilon \tilde{\varphi}_x^\varepsilon \right) dt\ d\gamma + \frac{\xi^2}{\tau} \int_{\Sigma^\varepsilon} \left(q^\varepsilon \tilde{\varphi}_x^\varepsilon - \tilde{\varphi}^\varepsilon q_x^\varepsilon \right)\ dtd\gamma$$

$$+ \frac{\xi^2}{\tau} \int_{Q_0^\varepsilon} q^\varepsilon \tilde{\varphi}_{xx}^\varepsilon dt\ dx + \frac{1}{2a\tau} \int_{Q_0^\varepsilon} q^\varepsilon \tilde{\varphi}^\varepsilon dt\ dx = \int_{Q_0^\varepsilon} (\varphi^\varepsilon - 1 - \delta_1) \cdot \chi_{Q_0}\ \tilde{\varphi}^\varepsilon\ dtdx.$$

Now we multiply (3.125) by p^ε, $(3.130)_2$ by $\tilde{u}^\varepsilon$, by subtraction we get

$$p_x^\varepsilon \tilde{u}^\varepsilon - \tilde{u}_x^\varepsilon p^\varepsilon = -p^\varepsilon \tilde{w}. \tag{3.135}$$

Adding (3.133)-(3.134) and using $(3.125)_2$, $(3.131)_2$, (3.135), we obtain

$$\int\limits_{Q_0^\varepsilon} p_t^\varepsilon \tilde{u}^\varepsilon dt\ dx + \int\limits_{Q_0^\varepsilon} q_t^\varepsilon \tilde{\varphi}^\varepsilon dt\ dx + \frac{k}{\rho c}\int\limits_{\Sigma^\varepsilon} p^\varepsilon \tilde{w} dt\ d\gamma +$$

$$+ \int\limits_{Q_0^\varepsilon} p^\varepsilon \Big[\frac{k}{\rho c}\tilde{u}_{xx}^\varepsilon - \frac{\ell\xi^2}{2\tau\rho c}\tilde{\varphi}_{xx}^\varepsilon - \frac{\ell}{4a\tau\rho c}\tilde{\varphi}^\varepsilon - \frac{\ell}{\tau\rho c}\tilde{u}^\varepsilon\Big] dt\ dx +$$

$$+ \int\limits_{Q_0^\varepsilon} q^\varepsilon \Big[\frac{\xi^2}{\tau}\tilde{\varphi}_{xx}^\varepsilon + \frac{1}{2a\tau}\tilde{\varphi}^\varepsilon + \frac{2}{\tau}\tilde{u}^\varepsilon\Big] dt\ dx \ =$$

$$= \beta \int\limits_{Q_0^\varepsilon} (u^\varepsilon - \delta_2)^+ \cdot \chi_{Q_0}\ \tilde{u}^\varepsilon\ dtdx + \int\limits_{Q_0^\varepsilon} (\varphi^\varepsilon - 1 - \delta_1) \cdot \chi_{Q_0}\ \tilde{\varphi}^\varepsilon\ dtdx,$$

i.e., making use of (3.124), the last relation leads to

$$\int\limits_{Q_0^\varepsilon} \left(p_t^\varepsilon \tilde{u}^\varepsilon + p^\varepsilon \tilde{u}_t^\varepsilon + q_t^\varepsilon \tilde{\varphi}^\varepsilon + q^\varepsilon \tilde{\varphi}_t^\varepsilon\right) dt\ dx + \frac{k}{\rho c}\int\limits_{\Sigma^\varepsilon} p^\varepsilon \tilde{w} dt\ d\gamma =$$

$$= \beta \int\limits_{Q_0^\varepsilon} (u^\varepsilon - \delta_2)^+ \cdot \chi_{Q_0}\ \tilde{u}^\varepsilon\ dtdx + \int\limits_{Q_0^\varepsilon} (\varphi^\varepsilon - 1 - \delta_1) \cdot \chi_{Q_0}\ \tilde{\varphi}^\varepsilon\ dtdx,$$

By Fubini's theorem and definition of distributional derivative, the latter relation give us

$$\frac{k}{\rho c}\int\limits_{\Sigma^\varepsilon} p^\varepsilon \tilde{w} dt\ d\gamma = \beta \int\limits_{Q_0^\varepsilon} (u^\varepsilon - \delta_2)^+ \cdot \chi_{Q_0}\ \tilde{u}^\varepsilon\ dtdx + \int\limits_{Q_0^\varepsilon} (\varphi^\varepsilon - 1 - \delta_1) \cdot \chi_{Q_0}\ \tilde{\varphi}^\varepsilon\ dtdx,$$

and then (3.132) becomes

$$\frac{k}{\rho c}\int\limits_{\Sigma^\varepsilon} p^\varepsilon \tilde{w}\; dtd\gamma + \int\limits_0^T w^* \tilde{w}\; dt + I'_{\mathcal{U}}(w^*, \tilde{w}) \geq 0 \qquad \forall\, \tilde{w} \in T_{\mathcal{U}}(w^*)$$

or

$$\int\limits_0^T \Big[\frac{k}{\rho c} p^\varepsilon(s, b_1) + w^*(s)\Big] \tilde{w}(s)\; ds + I'_{\mathcal{U}}(w^*, \tilde{w}) \geq 0 \qquad \forall\, \tilde{w} \in T_{\mathcal{U}}(w^*).$$

The last inequality is equivalent with

$$-r(t) \in \partial I_{\mathcal{U}}(w^*) \qquad \text{a.p.t. } (t, x) \in [0, T],$$

where $r(t) = \frac{k}{\rho c} p^\varepsilon(t, b_1) + w(t)$, and thus we can conclude that

$$w^*(t) = \begin{cases} 0, & \text{if } r(t) > 0 \\ -R, & \text{if } r(t) < 0. \end{cases} \tag{3.136}$$

Summing up, we have proved the following maximum principle for problem (P^ε_{inv})

Theorem 3.4 *Let $(u^{*,\varepsilon}, \varphi^{*,\varepsilon}, w^*)$ be optimal in problem (P^ε_{inv}). Then the optimal control is given by* (3.136) *where $(p^\varepsilon, q^\varepsilon)$ satisfy along with $u^{*,\varepsilon}, \varphi^{*,\varepsilon}$ the dual system* (3.130)-(3.131).

3.6 A state constraint optimal control problem: $(\mathbf{P}_S)$

Let $\Omega \subset I\!R^N$, $N = 1, 2, 3$, be a bounded domain with smooth boundary $\partial\Omega$ and let $T > 0$. Consider the controlled phase-field transition system

$$\begin{cases} u_t + \frac{\ell}{2}\varphi_t - k\Delta u = mv & \text{in } \; Q, \\ \frac{\partial}{\partial \nu} u = 0 & \text{on } \; \Sigma, \\ u(0, x) = u_0(x) & \text{on } \; \Omega, \end{cases} \tag{3.137}$$

$$\begin{cases} \tau\varphi_t - \xi^2\Delta\varphi + \frac{1}{2a}(\varphi^3 - \varphi) - 2u = mw & \text{in} \quad Q, \\ \varphi = 0 & \text{on} \quad \Sigma, \\ \varphi(0,x) = \varphi_0(x) & \text{on} \quad \Omega, \end{cases} \tag{3.138}$$

where $u_0, \varphi_0 \in H^1(\Omega)$ with $\frac{\partial}{\partial\nu}u_0 = 0$ and $\varphi_0 = 0$ on $\partial\Omega$, $m(x)$ is the characteristic function of an open subset ω of Ω, and $v(t,x)$, $w(t,x)$ are given functions.

It is known (see Theorem 2.1 in Chapter 2) that for each $v \in L^2(Q)$, $w \in W^{2,1}([0,T];L^2(\Omega))$, $u_0 \in H^1(\Omega)$, $\varphi_0 \in L^4(\Omega) \cap H^1(\Omega)$, the system (3.137)-(3.138) has a unique solution $u, \varphi \in W$, where

$$W = W^{2,1}([0,T];L^2(\Omega)) \cap L^2([0,T];H^2(\Omega)).$$

Let $S \subset L^2(\Omega)$ be a closed convex subset. We consider the optimal control problem

(P_S) *Minimize* $L(u,\varphi,v,w)$ *over all* $(u,\varphi,v,w) \in A_{ad}$

where

$$L(u,\varphi,v,w) = \int\limits_Q \left[\alpha u^2(t,x) + v^2(t,x) + w^2(t,x)\right] dtdx$$

with $\alpha > 0$, and

$A_{ad} = \{(u,\varphi,v,w) \in \left(W\right)^2 \times \left(L^2(Q)\right)^2$: (u,φ) *is the solution of* (3.137)-(3.138) *corresponding to* (v,w), *satisfying* $\varphi(T,x) \in S\}$.

The first question regarding the problem (P_S) is if there is an admissible solution, i.e., if the set A_{ad} is nonempty. Taking into account Theorem 1 in Barbu [77], we may assume in the sequel that for $v \in L^2(Q)$, $w \in W^{2,1}([0,T];L^2(\Omega))$, the problem ($P_S$) admit at least one admissible solution.

We point out that the problem (P_S) is an *optimal control problem with the distributed control acting on a subset* $\omega \subset \Omega$ and with the state constraint. For other researches in this context see V. Barbu & G. Wang [78], C. Moroşanu & G. Wang [37], G. Wang [79]. Works devoted to the distributed optimal control problem governed by the phase field transition system with the distributed control acting on Ω and without state constraint, are due to Z. Chen & K.-H. Hoffmann [41], M. Heinkenschloss & E.W. Sachs [42], M. Heinkenschloss & F. Tröltzsch [43] and K.-H. Hoffmann & L. Jiang [44].

Let $(u^*, \varphi^*, v^*, w^*)$ be optimal for (P_S). The *variational system* of (3.137)-(3.138) at $(u^*, \varphi^*, v^*, w^*)$ is given by

$$\begin{cases} y_t = k\Delta y - \dfrac{\ell}{\tau} y - \dfrac{\ell}{2\tau} Dz + m\bar{v} - \dfrac{\ell}{2\tau} m\bar{w} & \text{in} \quad Q, \\ z_t = \dfrac{1}{\tau} Dz + \dfrac{2}{\tau} y + m\bar{w} & \text{in} \quad Q, \\ \dfrac{\partial}{\partial \nu} y = z = 0 & \text{on} \quad \Sigma, \\ y(0,x) = z(0,x) = 0 & \text{in} \quad \Omega, \end{cases} \tag{3.139}$$

where $D = \xi^2\Delta + \frac{1}{2a}(1 - 3(\varphi^*)^2)$. We denote the solution of (3.139) corresponding to $(\bar{v}, \bar{w})$ by $(y(t, x; \bar{v}, \bar{w}), z(t, x; \bar{v}, \bar{w}))$. Define

$$W_{\delta_1} = \{z_T - x_T \in L^2(\Omega) : \ x_T \in S, \ \ z_T(x) = z(T, x; \bar{v}, \bar{w}) \ \textit{for some } \bar{v}, \bar{w} \in L^2(Q) \textit{ satisfying } \|\bar{v}\|_{L^2(Q)} + \|\bar{w}\|_{L^2(Q)} \le \delta_1\}.$$

We suppose

($\mathbf{H}_S$) *The set* W_{δ_1} *has finite codimensionality*[2] *in* $L^2(\Omega)$ *for some* $\delta_1 > 0$.

Our main result in studying problem (P_S) is the following

Theorem 3.5. *Let* $(u^*, \varphi^*, v^*, w^*)$ *be optimal for* (P_S). *Suppose that* (H_S) *holds. Then there exist* $(p, q) \in \left[W^{2,1}([0,T]; (H^1(\Omega))') \cap L^2([0,T]; H^1(\Omega))\right]^2$ *and* $\mu_0 \neq 0$, *and* $b_0 \in L^2(\Omega)$, *such that*

$$\begin{cases} p_t + k\Delta p - \frac{\ell}{\tau} p + \frac{2}{\tau} q = 2\alpha\mu_0 u^* & \text{in} \quad Q, \\ \frac{\partial}{\partial \nu} p = 0 & \text{on} \quad \Sigma, \\ p(T,x) = 0 & \text{on} \quad \Omega, \end{cases}$$

[2] **Definition.** Let Z be a Banach space. A subset S of Z is said to be `finite codimensional in Z` (denoted by $codim_Z < \infty$), if there exists a point $z_0 \in \bar{co}S$ such that $span\{S - z_0\}$ - the closed subspace spanned by $\{z - z_0 : z \in S\}$, is a finite codimensional subspace of Z and $\bar{co}(S - z_0)$ has a nonempty interior in this subspace.

$$\begin{cases} q_t - \frac{\ell}{2\tau} Dp + \frac{1}{\tau} Dq = 0 & \text{in} \quad Q, \\ q = 0 & \text{on} \quad \Sigma, \\ q(T,x) = -b_0 & \text{on} \quad \Omega, \end{cases}$$

$$mp = 2\mu_0 v^*,$$

$$mq = \frac{2\mu_0}{\tau} w^* + \mu_0 \ell v^*,$$

$$\int_\Omega b_0(\varphi^*(T,x) - x_T)\, dx \geq 0 \qquad \forall x_T \in S.$$

If $S = \{x_T \in L^2(\Omega) : \ \|x_T - 1\|_{L^2(\Omega)} \leq \delta, \delta > 0\}$, then (H_S) holds automatically.

3.6a The approximation problem $(\mathbf{P}_S^\varepsilon)$

Firstly we shall prove the following technical lemmas

Lemma 3.3. *Let $u_0, \varphi_0 \in H^1(\Omega)$. The solution (u, φ) of system (3.137)-(3.138) corresponding to $v, w \in L^2(Q)$ satisfies the following estimates*

$$\|\varphi(t)\|^2_{H^1(\Omega)} + \|u(t)\|^2_{H^1(\Omega)} \leq C(1 + \|mv\|^2_{L^2(Q)} + \|mw\|^2_{L^2(Q)}) \quad \forall t \in [0,T], \tag{3.140}$$

$$\|\varphi_t\|^2_{L^2([0,T];L^2(\Omega))} + \|u_t\|^2_{L^2([0,T];L^2(\Omega))} \leq C(1 + \|mv\|^2_{L^2(Q)} + \|mw\|^2_{L^2(Q)}), \tag{3.141}$$

and

$$\|\varphi\|^2_{L^2([0,T];H^2(\Omega))} + \|u\|^2_{L^2([0,T];H^2(\Omega))} \leq C[1 + \|mv\|^2_{L^2(Q)} + \|mw\|^2_{L^2(Q)}]^3 \tag{3.142}$$

where $C > 0$ is a constant independent of v and w.

Proof. Multiplying $(3.137)_1$ by $\frac{4}{\ell} u$ and u_t, then integrating them respectively, we obtain

$$\frac{2}{\ell} \int_\Omega u^2(x,t) dx + 2 \int_0^t \int_\Omega u \varphi_t dx ds + \frac{4k}{\ell} \int_0^t \int_\Omega |\nabla u|^2 dx ds \tag{3.143}$$

$$= \frac{2}{\ell}\int_{\Omega} u_0^2 dx + \frac{4}{\ell}\int_0^t\int_{\Omega}(mv)u dx ds,$$

and

$$\int_0^t\int_{\Omega} u_t^2 dx ds + \frac{2}{\ell}\int_0^t\int_{\Omega} u_t\varphi_t dx ds + \frac{k}{2}\int_{\Omega}|\nabla u(x,t)|^2 dx \tag{3.144}$$

$$= \frac{k}{2}\int_{\Omega}|\nabla u_0|^2 dx + \int_0^t\int_{\Omega}(mv)u_t dx ds.$$

Multiplying $(3.138)_1$ by φ_t, integrating it on $(0,t)\times\Omega$, we obtain

$$\tau\int_0^t\int_{\Omega}\varphi_t^2 dx ds + \frac{\xi^2}{2}\int_{\Omega}|\nabla\varphi(x,t)|^2 dx + \frac{1}{8a}\int_{\Omega}\varphi^4 dx \tag{3.145}$$

$$= \frac{\xi^2}{2}\int_{\Omega}|\nabla\varphi_0|^2 dx + \frac{1}{8a}\int_{\Omega}\varphi_0^4 dx + 2\int_0^t\int_{\Omega}u\varphi_t dx ds$$

$$+\int_0^t\int_{\Omega}(mw)\varphi_t dx ds + \frac{1}{4a}\int_{\Omega}\varphi^2 dx - \frac{1}{4a}\int_{\Omega}\varphi_0^2 dx.$$

It follows from (3.143) and (3.145) that (after some simple manipulations)

$$\int_{\Omega}u^2(x,t)dx + \int_0^t\int_{\Omega}|\nabla u|^2 dx ds \tag{3.146}$$

$$+\int_0^t\int_{\Omega}\varphi_t^2 dx ds + \int_{\Omega}|\nabla\varphi(x,t)|^2 dx + \int_{\Omega}\varphi^4 dx$$

$$\leq C\Big[1 + \|mv\|^2_{L^2(Q)} + \|mw\|^2_{L^2(Q)} + \int_0^t\int_{\Omega}u^2 dx ds + \int_0^t\int_{\Omega}\varphi^2 dx ds\Big]$$

where $C > 0$ is independent of v and w. By (3.146) we get (we have used Poincaré inequality)

$$\int_{\Omega} \left[u^2(x,t) + |\nabla\varphi(x,t)|^2\right] dx \tag{3.147}$$

$$\leq C\left[1 + \|mv\|^2_{L^2(Q)} + \|mw\|^2_{L^2(Q)} + \int_0^t \int_{\Omega} (u^2 + |\nabla\varphi|^2) dx ds\right] \quad \forall t \in [0,T].$$

Using Gronwall's inequality, we derive from (3.147) the estimate

$$\int_{\Omega} \left[u^2(x,t) + |\nabla\varphi(x,t)|^2\right] dx \tag{3.148}$$

$$\leq C(T)\left[1 + \|mv\|^2_{L^2(Q)} + \|mw\|^2_{L^2(Q)}\right] \quad \forall t \in [0,T]$$

which yields

$$\|u\|^2_{C([0,T];L^2(\Omega))} + \|\varphi\|^2_{C([0,T];H^1(\Omega))} \tag{3.149}$$

$$\leq C[1 + \|mv\|^2_{L^2(Q)} + \|mw\|^2_{L^2(Q)}].$$

Together with (3.146), (3.149) implies

$$\|\varphi_t\|^2_{L^2(0,T;L^2(\Omega))} \leq C[1 + \|mv\|^2_{L^2(Q)} + \|mw\|^2_{L^2(Q)}]. \tag{3.150}$$

Now, from relation (3.144) and the inequality above, we conclude

$$\|u_t\|^2_{L^2([0,T];L^2(\Omega))} \leq C[1 + \|mv\|^2_{L^2(Q)} + \|mw\|^2_{L^2(Q)}]. \tag{3.151}$$

The relations (3.150)-(3.151) implies estimate (3.141). Now, the estimate (3.140) may be easily deduced on the basis of estimates (3.141), (3.144) and (3.148).

By Sobolev's embedding theorem ($H^1(\Omega) \subset L^6(\Omega)$, $\Omega \subset I\!R^n, n \leq 1,2,3$) we have from (3.149) that

$$\|\varphi^3\|^2_{L^2(0,T;L^2(\Omega))} \leq C \int_0^T \|\varphi\|^6_{H^1(\Omega)} dt$$

$$\leq C[1 + \|mv\|^2_{L^2(Q)} + \|mw\|^2_{L^2(Q)}]^3.$$

This, together with (3.137)-(3.139) and (3.150)-(3.151), implies

$$\|\Delta\varphi\|^2_{L^2(0,T;L^2(\Omega))} + \|\Delta u\|^2_{L^2(0,T;L^2(\Omega))}$$

$$\leq C[1 + \|mv\|^2_{L^2(Q)} + \|mw\|^2_{L^2(Q)}]^3.$$

Then by the same argument as in [80], we obtain (3.142). This completes the proof. **q.e.d.**

Lemma 3.4. *Let $v_n, w_n \in L^2(Q)$ be such that $v_n \to \tilde{v}$ strongly in $L^2(Q)$ and $w_n \to \tilde{w}$ strongly in $L^2(Q)$. Let (u_n, φ_n) and $(\tilde{u}, \tilde{\varphi})$ be the solution of (3.137) and (3.138) corresponding to (v_n, w_n) and $(\tilde{v}, \tilde{w})$, respectively. Then there exists subsequences of (u_n, φ_n), still denoted by itself, such that*

$$u_n \to \tilde{u} \quad \text{strongly in} \quad W \cap C([0,T]; H^1(\Omega)),$$
$$\varphi_n \to \tilde{\varphi} \quad \text{strongly in} \quad W \cap C([0,T]; H^1(\Omega)).$$

Proof. By Lemma 3.3 we have

$$\|\varphi_n\|^2_{C([0,T];H^1(\Omega))} + \|u_n\|^2_{C([0,T];H^1(\Omega))} \leq C, \tag{3.152}$$

$$\|\varphi_n\|^2_{L^2([0,T];H^2(\Omega))} + \|u_n\|^2_{L^2([0,T];H^2(\Omega))} \leq C, \tag{3.153}$$

and

$$\|\varphi_n'\|^2_{L^2([0,T];L^2(\Omega))} + \|u_n'\|^2_{L^2([0,T];L^2(\Omega))} \leq C \tag{3.154}$$

where u_n' and φ_n' denote $\frac{\partial}{\partial t}u_n$ and $\frac{\partial}{\partial t}\varphi_n$, respectively. Then relations (3.152)-(3.154), Ascoli-Arezala Theorem and Aubin Compactness Theorem (see [80]) enable us to conclude that a subsequence of (u_n, φ_n) can be found such that

$$(\varphi_n, u_n) \to (\bar{\varphi}, \bar{u}) \qquad \begin{array}{l}\text{strongly in } W1,\\[1ex] \text{weakly in } W.\end{array} \tag{3.155}$$

where $W1 = L^2([0,T]; H^1(\Omega)) \cap C([0,T]; L^2(\Omega))$. By (3.152) and by Sobolev's Embedding Theorem, we obtain

$$\|\varphi_n^2 + \varphi_n\bar{\varphi} + \bar{\varphi}^2\|_{C([0,T];L^3(\Omega))} \leq C,$$

which implies

$$\begin{aligned}\int_\Omega (\varphi_n^3-\bar{\varphi}^3)^2dx &\leq \Big[\int_\Omega (\varphi_n-\bar{\varphi})^6dx\Big]^{1/3}\Big[\int_\Omega |\varphi_n^2+\varphi_n\bar{\varphi}+\bar{\varphi}^2|^2dx\Big]^{2/3}\\ &\leq C\|\varphi_n-\bar{\varphi}\|^2_{L^6(\Omega)}.\end{aligned}$$

This, combined with (3.155) and Sobolev's Embedding Theorem, indicates that

$$\varphi_n^3 \to \bar{\varphi}^3 \quad \text{strongly in} \quad L^2(\Omega). \tag{3.156}$$

By (3.155) and (3.156) we may pass to the limit for $n \to \infty$ in the system satisfied by (u_n, φ_n) and (v_n, w_v) to get $(\bar{u}, \bar{\varphi}) = (\tilde{u}, \tilde{\varphi})$. On the other hand we have

$$\begin{cases} (u_n-\tilde{u})_t+\dfrac{\ell}{2}(\varphi_n-\tilde{\varphi})_t-k\Delta(u_n-\tilde{u}) & \\ \qquad = m(v_n-\tilde{v}) & \text{in} \quad Q,\\ \frac{\partial}{\partial\nu}(u_n-\tilde{u})=0 & \text{on} \quad \Sigma,\\ (u_n-\tilde{u})(0,x)=0 & \text{on} \quad \Omega, \end{cases} \tag{3.157}$$

$$\begin{cases} \tau(\varphi_n-\tilde{\varphi})_t-\xi^2\Delta(\varphi_n-\tilde{\varphi})+\frac{1}{2a}[(\varphi_n^3-\tilde{\varphi}^3)] & \\ \qquad = \frac{1}{2a}(\varphi_n-\tilde{\varphi})]+2(u_n-\tilde{u})+m(w_n-\tilde{w}) & \text{in } Q,\\ (\varphi_n-\tilde{\varphi})=0 & \text{on } \Sigma,\\ (\varphi_n-\tilde{\varphi})(0,x)=0 & \text{on } \Omega. \end{cases} \tag{3.158}$$

Multiplying $(3.158)_1$ by $(\varphi_n-\tilde{\varphi})_t$, then integrating on $(0,t)\times\Omega$, we get

$$\tau\int_0^t\int_\Omega(\varphi_n-\tilde{\varphi})_t^2dxds+\frac{\xi^2}{2}\int_\Omega|\nabla(\varphi_n-\tilde{\varphi})(x,t)|^2dx$$

$$\leq \frac{\tau}{2}\int_0^t\int_\Omega(\varphi_n-\tilde{\varphi})_t^2dxds+C\Bigg[\int_0^t\int_\Omega(\varphi_n^3-\tilde{\varphi}^3)^2dxds$$

$$+\int_0^t\int_\Omega (\varphi_n - \tilde{\varphi})^2 dxds + \int_0^t\int_\Omega |m(w_n - \tilde{w})|^2 dxds + 2\int_0^t\int_\Omega (u_n - \tilde{u})^2 dxds\Bigg],$$

which, combined with (3.155) and (3.156), implies

$$\varphi_n' \to \tilde{\varphi}' \qquad \text{strongly in } L^2([0,T];L^2(\Omega)) \tag{3.159}$$

$$\varphi_n \to \tilde{\varphi} \qquad \text{strongly in } C([0,T];H^1(\Omega)). \tag{3.160}$$

By (3.155), (3.158), (3.159) and (3.160), we get

$$\|\Delta(\varphi_n - \tilde{\varphi})\|_{L^2([0,T];L^2(\Omega))} \to 0$$

which shows

$$\varphi_n \to \tilde{\varphi} \qquad \text{weakly in } L^2([0,T];H^2(\Omega)). \tag{3.161}$$

Next, by multiplying $(3.157)_1$ by $(u_n - \tilde{u})_t$ and then integrating on $(0,t)\times\Omega$, one can show that

$$u_n' \to \tilde{u}' \qquad \text{strongly in} \quad L^2([0,T];L^2(\Omega))$$

and

$$u_n \to \tilde{u} \qquad \text{strongly in} \quad C([0,T];H^1(\Omega)),$$

which together with (3.157) and (3.160) implies

$$u_n \to \tilde{u} \qquad \text{strongly in} \quad L^2([0,T];H^2(\Omega)).$$

This complete the proof of Lemma 3.4. **q.e.d.**

Now we let $(u^*, \varphi^*, v^*, w^*)$ be optimal for problem (P_S). For each $\varepsilon > 0$, we define the penalty problem as follows

$$(P_S^\varepsilon) \qquad \text{Min } \; L^\varepsilon(u,\varphi,v,w) = \text{Min } \left\{\int_Q [\alpha u^2 + v^2 + w^2] dxdt\right.$$

$$+\frac{1}{2}\|v - v^*\|^2_{L^2(Q)} + \frac{1}{2}\|w - w^*\|^2_{L^2(Q)} + \frac{1}{2\varepsilon}[d_S(\varphi(T,x)) + \varepsilon]^2$$

$$\text{over all } (u, \varphi, v, w) \in W^2 \times \left(L^2(Q)\right)^2 \Big\},$$

such that (u, φ) is the solution of (3.137)-(3.138) corresponding to (v, w). Here $d_S(\varphi(T, x))$ denotes the distance of $\varphi(T, x)$ to S in $L^2(\Omega)$.

Lemma 3.5. *Problem* (P_S^ε) *has at least one optimal solution.*

Proof. Let $(u^n, \varphi^n, v^n, w^n) \in W$ be such that

$$inf \ \ (P_S^\varepsilon) \leq L^\varepsilon(u^n, \varphi^n, v^n, w^n) \leq inf \ \ (P_S^\varepsilon) + \frac{1}{n}. \tag{3.162}$$

It is obvious that $\{v^n\}$ and $\{w^n\}$ are bounded in $L^2(Q)$. Therefore we can select subsequences of $\{v^n\}$ and of $\{w^n\}$, still denoted by themselves, such that

$$v^n \to \tilde{v} \quad \text{and} \quad w^n \to \tilde{w} \quad \text{weakly in } L^2(Q). \tag{3.163}$$

Let (u^n, φ^n) and $(\tilde{u}, \tilde{\varphi})$ be the solution of (3.137)-(3.138) corresponding to (v^n, w^n) and (v, w), respectively. By Lemma 3.3 and by the same arguments as in the proof of Lemma 3.4, we have that there exists a subsequence of $\{u^n\}$ and $\{\varphi^n\}$, still denoted by itself, such that

$$u^n \to \tilde{u}, \quad \varphi^n \to \tilde{\varphi} \qquad \begin{array}{l} \text{strongly in } \ W1, \\ \text{weakly in } \ W, \end{array} \tag{3.164}$$

where $W1 = L^2([0, T]; H^1(\Omega)) \cap C([0, T]; L^2(\Omega))$ and

$$(\varphi^n)^3 \to (\tilde{\varphi})^3 \qquad \text{strongly in} \quad L^2([0, T]; L^2(\Omega)). \tag{3.165}$$

By (3.164)-(3.165), we may pass to the limit for $n \to \infty$ in the system satisfied by (u^n, φ^n) and (v^n, w^n) to get that $(\tilde{u}, \tilde{\varphi})$ is the solution of (3.137)-(3.138) corresponding to $(\tilde{v}, \tilde{w})$. By (3.164) again, we implies

$$d_S(\varphi_n(T, x)) \to d_S(\tilde{\varphi}(T, x)) \tag{3.166}$$

because d_S is continuous in $L^2(\Omega)$. Thus we obtain

$$\underline{\lim}_{n\to\infty} L^\varepsilon(u^n, \varphi^n, v^n, w^n) \geq \int_Q [\alpha\tilde{u}^2 + \tilde{v}^2 + \tilde{w}^2] dxdt$$

$$+\frac{1}{2}\|\tilde{v}-v^*\|^2_{L^2(Q)}+\frac{1}{2}\|\tilde{w}-w^*\|^2_{L^2(Q)}+\frac{1}{2\varepsilon}[d_S(\tilde{\varphi}(T,x))+\varepsilon]^2$$
$$= L^\varepsilon(\tilde{u},\tilde{\varphi},\tilde{v},\tilde{w}).$$

So $(\tilde{u},\tilde{\varphi},\tilde{v},\tilde{w})$ is a optimal solution to $(\mathrm{P}^\varepsilon_S)$. **q.e.d.**

Lemma 3.6. *Let $(u^\varepsilon,\varphi^\varepsilon,v^\varepsilon,w^\varepsilon)$ be optimal for problem (P^ε_S). Then there exists a generalized subsequence of ε, still denoted by itself, such that*

$$u^\varepsilon\to u^* \quad \varphi^\varepsilon\to\varphi^* \quad \text{strongly in} \quad W\cap C([0,T];H^1(\Omega)).$$

Proof. It is obvious that

$$L^\varepsilon(u^\varepsilon,\varphi^\varepsilon,v^\varepsilon,w^\varepsilon)\le L^\varepsilon(u^*,\varphi^*,v^*,w^*)=\int_Q[\alpha u^*+v^*+w^*]dxdt+\frac{\varepsilon}{2}.$$

which implies

$$\overline{\lim}_{\varepsilon\to 0}L^\varepsilon(u^\varepsilon,\varphi^\varepsilon,v^\varepsilon,w^\varepsilon)\le\int_Q[\alpha u^*+v^*+w^*]dxdt. \tag{3.167}$$

From (3.167), $\{v^\varepsilon\}$ and $\{w^\varepsilon\}$ are bounded in $L^2(Q)$. Thus there exists a generalized subsequence of ε, still denoted by itself, such that

$$v^\varepsilon\to\tilde{v}, \quad w^\varepsilon\to\tilde{w} \quad \text{weakly in} \quad L^2(Q) \quad \text{as } \varepsilon\to 0. \tag{3.168}$$

By (3.168) and Lemma 3.3 (following the same arguments as in the proof of Lemma 3.4) it results that

$$u^\varepsilon\to\tilde{u}, \quad \varphi^\varepsilon\to\tilde{\varphi} \qquad \begin{array}{l}\text{strongly in} \quad W1,\\ \text{weakly in} \quad W,\end{array}$$

where $W1=L^2([0,T];H^1(\Omega))\cap C([0,T];L^2(\Omega))$, and

$$(\varphi^\varepsilon)^3\to(\tilde{\varphi})^3 \quad \text{strongly in} \quad L^2([0,T];L^2(\Omega)),$$

where $(\tilde{u},\tilde{\varphi})$ is the solution of (3.137)-(3.138) corresponding to $(\tilde{v},\tilde{w})$. In particular, we have

$$\varphi^\varepsilon(T,x)\to\tilde{\varphi}(T,x) \quad \text{strongly in} \quad L^2(\Omega).$$

This implies

$$d_S(\varphi^\varepsilon(T,x)) \to d_S(\tilde{\varphi}(T,x)), \tag{3.169}$$

because S is closed and convex in $L^2(\Omega)$.

On the other hand, we have from (3.167) that

$$\frac{1}{2\varepsilon}(d_S(\varphi^\varepsilon(T,x)) + \varepsilon)^2 \le C$$

which implies $d_S(\varphi^\varepsilon(T,x)) \to 0$ as $\varepsilon \to 0$. This, combined with (3.169) indicates that $\tilde{\varphi}(T,x) \in S$ and thus $(\tilde{u},\tilde{\varphi},\tilde{v},\tilde{w}) \in A_{ad}$.

Since $L^\varepsilon(u^\varepsilon,\varphi^\varepsilon,v^\varepsilon,w^\varepsilon)$ is weakly lower semicontinuous, it follows that

$$\underline{\lim}_{\varepsilon\to 0} L^\varepsilon(u^\varepsilon,\varphi^\varepsilon,v^\varepsilon,w^\varepsilon) \ge L^\varepsilon(\tilde{u},\tilde{\varphi},\tilde{v},\tilde{w}) \ge L^\varepsilon(u^*,\varphi^*,v^*,w^*)$$

which together with (3.167) shows that $v^* = \tilde{v}$, $w^* = \tilde{w}$ and $v^\varepsilon \to v^*$, $w^\varepsilon \to w^*$, strongly in $L^2(Q)$. Then by Lemma 3.5 the proof of Lemma 3.6 is complete. **q.e.d.**

3.6b The maximum principle for problems ($\mathbf{P}_S^\varepsilon$) and ($\mathbf{P}_S$)

We start this subsection with the necessary conditions for $(u^\varepsilon,\varphi^\varepsilon,v^\varepsilon,w^\varepsilon)$ to be optimal for (P_S^ε). Let $(\bar{v},\bar{w}) \in (L^2(Q))^2$ be arbitrary but fixed and $\lambda > 0$. Set $v_\varepsilon^\lambda = v_\varepsilon + \lambda\bar{v}$, $w_\varepsilon^\lambda = w_\varepsilon + \lambda\bar{w}$. Let $(u_\varepsilon^\lambda,\varphi_\varepsilon^\lambda)$ the solution of (3.137)-(3.138) corresponding to $(v_\varepsilon^\lambda,w_\varepsilon^\lambda)$. It is clear that $v_\varepsilon^\lambda \to v_\varepsilon$ and $w_\varepsilon^\lambda \to w_\varepsilon$ strongly in $L^2(Q)$ as $\lambda \to 0$. Thus it follows from Lemma 3.4 that, as $\lambda \to 0$,

$$u_\varepsilon^\lambda \to u_\varepsilon,\quad \varphi_\varepsilon^\lambda \to \varphi_\varepsilon \quad \text{strongly in } W \cap C([0,T];H^1(\Omega)), \tag{3.170}$$

and

$$(\varphi_\varepsilon^\lambda)^3 \to \varphi_\varepsilon^3 \quad \text{strongly in } L^2(Q). \tag{3.171}$$

Let $y_\varepsilon^\lambda = \frac{u_\varepsilon^\lambda - u_\varepsilon}{\lambda}$, $z_\varepsilon^\lambda = \frac{\varphi_\varepsilon^\lambda - \varphi_\varepsilon}{\lambda}$. Then

$$\begin{cases} (y_\varepsilon^\lambda)_t + \frac{\ell}{2}(z_\varepsilon^\lambda)_t - k\Delta y_\varepsilon^\lambda = m\bar{v} & \text{in } Q, \\ \frac{\partial}{\partial\nu} y_\varepsilon^\lambda = 0 & \text{on } \Sigma, \\ y_\varepsilon^\lambda(0,x) = u_0(x) & \text{on } \Omega, \end{cases} \tag{3.172}$$

$$\begin{cases} \tau(z_\varepsilon^\lambda)_t - \xi^2\Delta z_\varepsilon^\lambda + \frac{1}{2a}\left[\frac{(\varphi_\varepsilon^\lambda)^3-\varphi_\varepsilon^3}{\lambda} - z_\varepsilon^\lambda\right] - 2y_\varepsilon^\lambda = m\bar{w} & \text{in} \quad Q, \\ z_\varepsilon^\lambda = 0 & \text{on} \quad \Sigma, \\ z_\varepsilon^\lambda(0,x) = \varphi_0(x) & \text{on} \quad \Omega. \end{cases} \tag{3.173}$$

We check now the convergences of $\{y_\varepsilon^\lambda\}_{\lambda>0}$ and $\{z_\varepsilon^\lambda\}_{\lambda>0}$ as $\lambda \to 0$. To this end, we rewrite (3.173) as

$$\begin{cases} \tau(z_\varepsilon^\lambda)_t - \xi^2\Delta z_\varepsilon^\lambda + \frac{1}{2a}(\psi_\varepsilon^\lambda - 1)z_\varepsilon^\lambda - 2y_\varepsilon^\lambda = m\bar{w} & \text{in} \quad Q, \\ z_\varepsilon^\lambda = 0 & \text{on} \quad \Sigma, \\ z_\varepsilon^\lambda(0,x) = \varphi_0(x) & \text{on} \quad \Omega. \end{cases} \tag{3.174}$$

where

$$\psi_\varepsilon^\lambda(t,x) = \int_0^1 3[\varphi_\varepsilon(t,x) + \theta(\varphi_\varepsilon^\lambda(t,x) - \varphi_\varepsilon(t,x))]^2 d\theta.$$

It is obvious that $\psi_\varepsilon^\lambda \to 3\varphi_\varepsilon^2$ strongly in $C([0,T];L^3(\Omega))$ as $\lambda \to 0$. In particular this implies

$$\|\psi_\varepsilon^\lambda\|_{C([0,T];L^3(\Omega))} \leq C. \tag{3.175}$$

Multiplying $(3.172)_1$ by $\frac{4}{\ell}y_\varepsilon^\lambda$, integrating on $(0,t)\times\Omega$, we get

$$\frac{2}{\ell}\|y_\varepsilon^\lambda(t)\|^2_{L^2(\Omega)} + 2\int_0^t\int_\Omega (z_\varepsilon^\lambda)_t y_\varepsilon^\lambda dxds + \frac{4k}{\ell}\int_0^t\int_\Omega |\nabla y_\varepsilon^\lambda|^2 dxds \tag{3.176}$$

$$= \int_0^t\int_\Omega m\bar{v}y_\varepsilon^\lambda dxds.$$

Multiplying $(3.174)_1$ by $(z_\varepsilon^\lambda)_t$, integrating on $(0,t)\times\Omega$, we obtain

$$\tau\int_0^t\int_\Omega (z_\varepsilon^\lambda)_t^2 dxds + \xi^2\int_\Omega |\nabla z_\varepsilon^\lambda|^2 dx \tag{3.177}$$

$$-2\int_0^t\int_\Omega y_\varepsilon^\lambda (z_\varepsilon^\lambda)_t dxds + \frac{1}{2a}\int_0^t\int_\Omega \psi_\varepsilon^\lambda z_\varepsilon^\lambda (z_\varepsilon^\lambda)_t dxds$$

$$= \frac{1}{2a}\int_0^t\int_\Omega z_\varepsilon^\lambda (z_\varepsilon^\lambda)_t dxds + \int_0^t\int_\Omega m\bar{w}(z_\varepsilon^\lambda)_t dxds.$$

By (3.176) and (3.177), after some calculations involving Hölder's inequality, Young's inequality and Sobolev's Embedding Theorem, we obtain

$$\|y_\varepsilon^\lambda(t)\|^2_{L^2(\Omega)} + \int_0^t\int_\Omega |\nabla y_\varepsilon^\lambda|^2 dxds + \int_0^t\int_\Omega (z_\varepsilon^\lambda)_t^2 dxds + \|z_\varepsilon^\lambda(t)\|^2_{H^1(\Omega)}$$

$$\leq C\left[\|m\bar{v}\|^2_{L^2(Q)} + \|m\bar{w}\|^2_{L^2(Q)} + \int_0^t\int_\Omega [(y_\varepsilon^\lambda)^2 + (z_\varepsilon^\lambda)^2] dxds\right]$$

$$+\frac{1}{2}\int_0^t\int_\Omega (z_\varepsilon^\lambda)_t^2 dxds$$

$$+\int_0^t \left[\int_\Omega |\psi_\varepsilon^\lambda|^3 dx\right]^{1/3}\left[\int_\Omega (z_\varepsilon^\lambda)_t^2 dx\right]^{1/2}\left[\int_\Omega (z_\varepsilon^\lambda)^6 dx\right]^{1/6} ds$$

$$\leq \frac{3}{4}\int_0^t\int_\Omega (z_\varepsilon^\lambda)_t^2 dxds$$

$$+C\left[1+\int_0^t \left(\|y_\varepsilon^\lambda\|^2_{L^2(\Omega)} + \|z_\varepsilon^\lambda\|^2_{L^2(\Omega)} + \|z_\varepsilon^\lambda\|^2_{L^6(\Omega)}\right) ds\right]$$

$$\leq \frac{3}{4}\int_0^t\int_\Omega (z_\varepsilon^\lambda)_t^2 dxds + C\left[1+\int_0^t \left(\|y_\varepsilon^\lambda\|^2_{L^2(\Omega)} + \|z_\varepsilon^\lambda\|^2_{H^1(\Omega)}\right) ds\right]$$

which yields

$$\|y_\varepsilon^\lambda(t)\|_{L^2(\Omega)}^2 + \int_0^t\int_\Omega |\nabla y_\varepsilon^\lambda|^2 dxds + \int_0^t\int_\Omega (z_\varepsilon^\lambda)_t^2 dxds + \|z_\varepsilon^\lambda(t)\|_{H^1(\Omega)}^2 \quad (3.178)$$

$$\leq C\left(1 + \int_0^t \left(\|y_\varepsilon^\lambda\|_{L^2(\Omega)}^2 + \|z_\varepsilon^\lambda\|_{H^1(\Omega)}^2\right) ds\right) \qquad \forall t \in [0,T].$$

By Gronwall's inequality, we get

$$\|y_\varepsilon^\lambda(t)\|_{C([0,T];L^2(\Omega))} + \|z_\varepsilon^\lambda\|_{C([0,T];H^1(\Omega))} \leq C. \tag{3.179}$$

By (3.178) and (3.179) we imply

$$\|z_\varepsilon^\lambda\|_{L^2([0,T];L^2(\Omega))} \leq C. \tag{3.180}$$

Now it follow from (3.174), (3.175), (3.179) and (3.180) that

$$\|z_\varepsilon^\lambda\|_{L^2([0,T];H^2(\Omega))} \leq C.$$

By the Aubin Compactness Theorem (see [80]) and the Ascoli-Arezala Theorem, there exist a generalized subsequence λ, still denoted by itself, and $z_\varepsilon \in W \cap C([0,T];H^1(\Omega))$ such that

$$z_\varepsilon^\lambda \to z_\varepsilon \qquad \begin{array}{l}\text{strongly in } L^2([0,T];H^1(\Omega)) \cap C([0,T];L^2(\Omega)),\\ \text{weakly in } W, \quad \text{as } \lambda \to 0.\end{array} \tag{3.181}$$

Since $\psi_\varepsilon^\lambda \to 3\varphi_\varepsilon^2$ strongly in $C([0,T];L^3(\Omega))$ as $\lambda \to 0$, it follows from (3.181) and Sobolev's Embedding Theorem that

$$z_\varepsilon^\lambda \psi_\varepsilon^\lambda \to 3\varphi_\varepsilon^2 z_\varepsilon \quad \text{strongly in } L^2(\Omega) \ \text{ as } \lambda \to 0. \tag{3.182}$$

Similarly, we can get that there exists a generalized subsequence of λ, still denoted by itself, and $y_\varepsilon \in W \cap C([0,T];H^1(\Omega))$ such that

$$y_\varepsilon^\lambda \to y_\varepsilon \qquad \begin{array}{l}\text{strongly in } L^2([0,T];H^1(\Omega)) \cap C([0,T];L^2(\Omega)),\\ \text{weakly in } W \quad \text{as } \lambda \to 0.\end{array} \tag{3.183}$$

By (3.181)-(3.183) we may pass to the limit for $\lambda \to 0$ in (3.172) and (3.174) to derive the variational system for problem $(\mathrm{P}_S^\varepsilon)$

$$\begin{cases} (y_\varepsilon)_t + \dfrac{\ell}{2}(z_\varepsilon)_t - k\Delta y_\varepsilon = m\bar{v} & \text{in} \quad Q, \\ \dfrac{\partial}{\partial\nu} y_\varepsilon = 0 & \text{on} \quad \Sigma, \\ y_\varepsilon(0,x) = u_0(x) & \text{on} \quad \Omega, \end{cases} \tag{3.184}$$

$$\begin{cases} \tau(z_\varepsilon)_t - \xi^2\Delta z_\varepsilon + \frac{1}{2a}(3\varphi_\varepsilon^2 - 1)z_\varepsilon - 2y_\varepsilon = m\bar{w} & \text{in} \quad Q, \\ z_\varepsilon = 0 & \text{on} \quad \Sigma, \\ z_\varepsilon(0,x) = \varphi_0(x) & \text{on} \quad \Omega. \end{cases} \tag{3.185}$$

By (3.181), we have in particular,

$$z_\varepsilon^\lambda(T,x) \to z_\varepsilon(T,x) \quad \text{strongly in } L^2(\Omega) \quad \text{as } \lambda \to 0$$

which implies (see Wang [79])

$$\frac{1}{\lambda}\frac{1}{2\varepsilon}\left\{\left[d_S(\varphi_\varepsilon^\lambda(T,x)) + \varepsilon\right]^2 - \left[d_S(\varphi_\varepsilon(T,x)) + \varepsilon\right]^2\right\} \to \tag{3.186}$$

$$\frac{d_S(\varphi_\varepsilon(T,x)) + \varepsilon}{\varepsilon} \int_\Omega b_\varepsilon(x) z_\varepsilon(T,x)dx \quad \text{as } \lambda \to 0$$

where $b_\varepsilon \in \partial d_S(\varphi_\varepsilon(T,x))$ satisfying

$$\|b_\varepsilon\|_{L^2(\Omega)} = \begin{cases} 1 & \text{if } \varphi_\varepsilon(T,x) \notin S \\ 0 & \text{if } \varphi_\varepsilon(T,x) \in S. \end{cases} \tag{3.187}$$

Let $\mu_\varepsilon = \frac{\varepsilon}{d_S(\varphi_\varepsilon(T,x))+\varepsilon}$. Then it is clear that

$$1 \le \mu_\varepsilon + \|b_\varepsilon\|_{L^2(\Omega)} \le 2. \tag{3.188}$$

Since $(u_\varepsilon, \varphi_\varepsilon, v_\varepsilon, w_\varepsilon)$ is optimal controller for problem $(\mathrm{P}_S^\varepsilon)$, we have

$$\frac{L^\varepsilon(u_\varepsilon^\lambda, \varphi_\varepsilon^\lambda, v_\varepsilon^\lambda, w_\varepsilon^\lambda) - L^\varepsilon(u_\varepsilon, \varphi_\varepsilon, v_\varepsilon, w_\varepsilon)}{\lambda} \ge 0 \quad \forall \lambda > 0,$$

which together with (3.181), (3.183) and (3.186) yields

$$\int_0^T \int_\Omega \left[2\alpha\mu_\varepsilon u_\varepsilon y_\varepsilon + 2\mu_\varepsilon v_\varepsilon \bar{v} + 2\mu_\varepsilon w_\varepsilon \bar{w} + \mu_\varepsilon(v_\varepsilon - v^*)\bar{v}\right. \tag{3.189}$$

$$\left.+\mu_\varepsilon(w_\varepsilon - w^*)\bar{w}\right] dxdt + \int_\Omega b_\varepsilon z_\varepsilon(T,x)dx \geq 0 \quad \forall \bar{v}, \bar{w} \in L^2(Q).$$

Since $b_\varepsilon \in \partial d_S(\varphi_\varepsilon(T,x))$, we have

$$\int_\Omega b_\varepsilon(\varphi_\varepsilon(T,x) - x_T)dx \geq 0 \quad \forall x_T \in S. \tag{3.190}$$

We may regard (3.189) and (3.190) as necessary conditions of $(u_\varepsilon, \varphi_\varepsilon, v_\varepsilon, w_\varepsilon)$.

Next we shall pass to the limit for $\varepsilon \to 0$ in (3.189) and (3.190) to derive the **Maximum Principle for ($\mathbf{P}_S$)**. To this end we first note that by (3.188), we may have a subsequence of ε, still denoted by itself, such that

$$\mu_\varepsilon \to \mu_0, \quad b_\varepsilon \to b_0 \quad \text{weakly in } L^2(\Omega). \tag{3.191}$$

By Lemma 3.6 and by the same argument as before, there exists a subsequence of ε, still denoted by itself, such that

$$y_\varepsilon \to y, \quad z_\varepsilon \to z \quad \begin{array}{l}\text{strongly in } W1,\\ \text{weakly in } W,\end{array} \tag{3.192}$$

where $W1 = L^2([0,T]; H^1(\Omega)) \cap C([0,T]; L^2(\Omega))$ and (y,z) satisfies (the *variational system* for promlem (P_S))

$$\begin{cases} y_t = k\Delta y - \dfrac{\ell}{\tau}y - \dfrac{\ell}{2\tau}Dz + m\bar{v} - \dfrac{\ell}{2\tau}m\bar{w} & \text{in } Q, \\ \dfrac{\partial}{\partial \nu}y = 0 & \text{on } \Sigma, \\ y(0,x) = 0 & \text{in } \Omega, \end{cases} \tag{3.193}$$

$$\begin{cases} z_t = \dfrac{1}{\tau}Dz + \dfrac{2}{\tau}y + \dfrac{1}{\tau}m\bar{w} & \text{in} \quad Q, \\ z = 0 & \text{on} \quad \Sigma, \\ z(x,0) = 0 & \text{in} \quad \Omega, \end{cases} \tag{3.194}$$

which $D = \xi^2\Delta + \frac{1}{2a}(1 - 3(\varphi^*)^2)$ (see also relations (3.184), (3.185)).

By (3.191)-(3.192) we may pass to the limit in (3.189) and (3.190) to derive respectively that

$$\int_0^T \int_\Omega (2\alpha\mu_0 u^* y + 2\mu_0 v^* \bar{v} + 2\mu_0 w^* \bar{w})dxdt \tag{3.195}$$

$$+ \int_\Omega b_0 z(T,x)dx \geq 0 \qquad \forall \bar{v}, \bar{w} \in L^2(Q)$$

and

$$\int_\Omega b_0(\varphi^*(T,x) - x_T)dx \geq 0 \qquad \forall x_T \in S. \tag{3.196}$$

Let $(p,q) \in \left[W^{1,2}([0,T];(H^1(\Omega))') \cap W1\right]^2$ be the solution to[3]

$$\begin{cases} p_t + k\Delta p - \frac{\ell}{\tau}p + \frac{2}{\tau}q = 2\alpha\mu_0 u^* & \text{in} \quad Q, \\ \frac{\partial}{\partial\nu}p = 0 & \text{on} \quad \Sigma, \\ p(T,x) = 0 & \text{on} \quad \Omega, \end{cases} \tag{3.197}$$

$$\begin{cases} q_t - \frac{\ell}{2\tau}Dp + \frac{1}{\tau}Dq = 0 & \text{in} \quad Q, \\ q = 0 & \text{on} \quad \Sigma, \\ q(T,x) = -b_0 & \text{on} \quad \Omega. \end{cases} \tag{3.198}$$

[3] we remember that $W1 = L^2([0,T];H^1(\Omega)) \cap C([0,T];L^2(\Omega))$

By (3.193) and (3.197), we obtain

$$\int_0^T\int_\Omega \left(2\alpha\mu_0 u^* - \frac{2}{\tau}q\right) y \, dxdt \tag{3.199}$$

$$= \int_0^T\int_\Omega \frac{\ell}{2\tau} Dzpdxdt - \int_0^T\int_\Omega \left(m\bar{v} + \frac{2}{2\tau}m\bar{w}\right) p \, dxdt.$$

By (3.194) and (3.198), we get

$$-\int_0^T\int_\Omega \frac{\ell}{2\tau} Dzpdxdt - \int_0^T\int_\Omega \frac{2}{\tau} yqdxdt \tag{3.200}$$

$$-\int_0^T\int_\Omega \frac{1}{\tau} m\bar{w}qdxdt = \int_\Omega z(T,x)b_0 \, dx.$$

It follows from (3.195), (3.199) and (3.200) that

$$\int_0^T\int_\Omega (2\mu_0 v^* - mp)\,\bar{v} \, dxdt \tag{3.201}$$

$$+\int_0^T\int_\Omega \left(2\mu_0 w^* - \frac{1}{\tau}mq + \frac{\ell}{2\tau}mp\right)\bar{w}dxdt \geq 0 \qquad \forall \bar{v}, \bar{w} \in L^2(Q),$$

which implies

$$mp = 2\mu_0 v^*, \qquad mq = \frac{2\mu_0}{\tau} w^* + \mu_0 \ell v^*. \tag{3.202}$$

Now (3.202) together with (3.196)-(3.198) gives the **maximum principle for problem ($\mathbf{P}_S$)**.

Finally we shall show $\mu_0 \neq 0$. To this end we show first $(\mu_0, p, q) \neq 0$. By (3.189) and (3.190), we have

$$\int_\Omega b_\varepsilon(\varphi^*(T,x) + z(T,x) - x_T)dx \geq -\zeta_\varepsilon \quad \forall x_T \in S, \tag{3.203}$$

where

$$\zeta_\varepsilon = \int_0^T \int_\Omega \big[2\alpha\mu_\varepsilon u_\varepsilon y_\varepsilon + 2\mu_\varepsilon v_\varepsilon \bar{v} + 2\mu_\varepsilon w_\varepsilon \bar{w} + \mu_\varepsilon(v_\varepsilon - v^*)\bar{v}$$

$$+\mu_\varepsilon(w_\varepsilon - w^*)\bar{w}\big]dxdt + \int_\Omega b_\varepsilon\big[z_\varepsilon(T,x) - z(T,x) + \varphi_\varepsilon(T,x) - \varphi^*(T,x)\big]dx.$$

Suppose $\mu_0 = 0$. Then it follows from (3.188) that $\mu_\varepsilon \to 0$ and

$$2 \geq \|b_\varepsilon\|_{L^2(\Omega)} \geq \bar{\delta} > 0 \qquad \forall \varepsilon < \varepsilon_1 \tag{3.204}$$

for some $\bar{\delta} > 0$ and $\varepsilon_1 > 0$.

One can check that $\zeta_\varepsilon \to 0$ uniformly for $(\bar{v}, \bar{w}) \in (L^2(Q))^2$ with $\|\bar{v}\|_{L^2(Q)} + \|\bar{w}\|_{L^2(Q)} \leq \delta_1$ where δ_1 is an arbitrary but fixed positive number. By (H_S), W_{δ_1} has finite codimensionality in $L^2(\Omega)$ for some $\delta_1 > 0$, so does $W_{\delta_1} + \{\varphi^*(T,x)\}$. Thanks to the Lemma 3.6 in Chapter 4 of [81], we conclude from (3.203) and (3.204) that $b_0 \neq 0$, which combined with (3.197)-(3.198) implies that $(p,q) \neq 0$. Thus we have proved

$$(b_0, p, q) \neq 0. \tag{3.205}$$

Suppose $\mu_0 = 0$, by taking the limit for $\varepsilon \to 0$ in (3.191) we get

$$\int_\Omega b_0(\varphi^*(T,x) + z(T,x) - x_T)dx \geq 0 \quad \forall x_T \in S$$

which implies

$$\int_\Omega b_0 z(T,x)dx \geq 0 \quad \forall \bar{v}, \bar{w} \in L^2(Q). \tag{3.206}$$

On the other hand, by the same arguments as in Barbu [77], we get that for each $z_T \in H^2(\Omega)$ with $\|z_T\|_{H^2(\Omega)} \leq \tilde{\delta}$ ($\tilde{\delta}$ is a positive constant), there exists $(\tilde{v}, \tilde{w}) \in (L^2(Q))^2$, such that $z(T, x; \tilde{v}, \tilde{w}) = z_T$, where $(y(t, x; \tilde{v}, \tilde{w}), z(t, x; \tilde{v}, \tilde{w}))$ is the solution of (3.139) corresponding to $(\tilde{v}, \tilde{w})$. Thus it follows from (3.206) that

$$\int_\Omega b_0 z_T dx \geq 0 \quad \forall z_T \in H^2(\Omega) \text{ with } \quad \|z_T\|_{H^2(\Omega)} \leq \tilde{\delta}$$

which implies $b_0 = 0$. This contradicts (3.205). So $\mu_0 \neq 0$. This completes the proof of Theorem 3.5. **q.e.d.**

3.7 A product formula approach to a non-homogeneous boundary optimal control problem governed by nonlinear phase-field transition system

In this Section, we prove the convergence of an iterative scheme of fractional steps type for a non-homogeneous Cauchy-Neumann boundary optimal control problem governed by nonlinear phase-field system. Consequently, the necessary optimality conditions were established for approximating proces.

Consider the following nonlinear optimal control problem

$$\text{(P)} \quad \textit{Minimize } j(w) = \frac{1}{2}\int_Q (\varphi(t,x) + 1 + \delta_1)^2 \cdot \chi_{Q_0}\, dtdx$$

$$+\frac{1}{2}\int_Q \left((u(t,x) + \delta_2 - \delta_1)^+\right)^2 \cdot \chi_{Q_0}\, dtdx + \frac{1}{2}\int_\Sigma w^2(t,x)\, dtd\gamma,$$

on all (u, φ) *satisfying* (3.1)-(3.3),

where χ_{Q_0} is the characteristic function of Q_0, $Q_0 \subset Q$.

We define as *interface at the moment* t the set Ω_t given by

$$\Omega_t = \{x \in \Omega;\ |u(t,x)| \leq -\delta_2,\ |\varphi(t,x)| \leq 1 + \delta_1, \quad \delta_1, \delta_2 \geq 0\}.$$

It is known (see Theorem 2.1 in Chapter 2) that for each $u_0, \varphi_0 \in W^{1,\infty}(\Omega)$ satisfying $\frac{\partial u_0}{\partial \nu} + hu_0 = w(t,x)$, $\frac{\partial \varphi_0}{\partial \nu} = 0$ and $w \in W^{1,\infty}(\Sigma)$, the system (3.1)-(3.3) has a unique solution $u, \varphi \in W$, where

$$W = W^{2,1}([0,T]; L^2(\Omega)) \cap L^2([0,T]; H^2(\Omega)).$$

For $\forall \varepsilon > 0$, we associate to the nonlinear system (3.1)-(3.3) the following approximating scheme:[4]

$$\begin{cases} \rho V \frac{\partial}{\partial t} u^\varepsilon + \frac{\ell}{2}\frac{\partial}{\partial t} \varphi^\varepsilon = k\Delta u^\varepsilon \\ \tau \frac{\partial}{\partial t} \varphi^\varepsilon = \xi^2 \Delta\varphi^\varepsilon + \frac{1}{2a}\varphi^\varepsilon + 2u^\varepsilon \end{cases} \quad \text{in } Q_i^\varepsilon, \tag{3.207}$$

$$\frac{\partial}{\partial \nu} u^\varepsilon + h u^\varepsilon = w(t,x), \quad \frac{\partial}{\partial \nu}\varphi^\varepsilon = 0 \quad \text{on } \Sigma_i^\varepsilon, \tag{3.208}$$

$$\begin{cases} u_+^\varepsilon(i\varepsilon, x) = u_-^\varepsilon(i\varepsilon, x), \quad u_-^\varepsilon(0,x) = u_0(x) \\ \varphi_+^\varepsilon(i\varepsilon, x) = z((i+1)\varepsilon, \varphi_-^\varepsilon(i\varepsilon, x)) \end{cases} \quad \text{on } \Omega, \tag{3.209}$$

where $z(\cdot, \varphi_-^\varepsilon(i\varepsilon, x))$ is the solution of the Cauchy problem:

$$\begin{cases} z'(s) + \frac{1}{2a} z^3(s) = 0 \quad s \in (i\varepsilon, (i+1)\varepsilon), \\ z(i\varepsilon) = \varphi_-^\varepsilon(i\varepsilon, x), \quad \varphi_-^\varepsilon(0,x) = \varphi_0(x), \end{cases} \tag{3.210}$$

for $i = 0, 1, \cdots, M_\varepsilon - 1$, with $M_\varepsilon = \left[\frac{T}{\varepsilon}\right]$, $Q_{M_\varepsilon - 1}^\varepsilon = [(M_\varepsilon - 1)\varepsilon, T] \times \Omega$, $\varphi_+^\varepsilon(i\varepsilon, x) = \lim\limits_{t \downarrow i\varepsilon} \varphi^\varepsilon(t,x)$, $\varphi_-^\varepsilon(i\varepsilon, x) = \lim\limits_{t \uparrow i\varepsilon} \varphi^\varepsilon(t,x)$.

Than, corresponding to problem (P), we consider the approximating optimal control problem:

$$(\mathrm{P}^\varepsilon) \quad \textit{Minimize } j^\varepsilon(w) = \frac{1}{2}\int_Q (\varphi^\varepsilon(t,x) + 1 + \delta_1)^2 \cdot \chi_{Q_0} dt dx$$
$$+ \frac{1}{2}\int_Q \left((u^\varepsilon(t,x) + \delta_2 - \delta_1)^+\right)^2 \cdot \chi_{Q_0} dt dx + \frac{1}{2}\int_\Sigma w^2(t,x) dt d\gamma,$$

on all $(u^\varepsilon, \varphi^\varepsilon)$ *solution of* (3.207)-(3.210) *corresponding to* w.

3.7a The existence in probems (P) and (P$^\varepsilon$)

The existence of the optimalities in problem (P) and (P^ε) are proved in the present Section. The definition of weak solution for (3.1)-(3.2) and (3.207)-(3.210) was introduced in Chapter 2 (see Definition 2.5 and Definition 2.6, respectively).

[4] $Q_i^\varepsilon = (i\varepsilon, (i+1)\varepsilon) \times \Omega$, $\Sigma_i^\varepsilon = (i\varepsilon, (i+1)\varepsilon) \times \partial\Omega$.

Proposition 3.5. *For $u_0, \varphi_0 \in W^{1,\infty}(\Omega) = \{y : \ y, \nabla y \in L^\infty(\Omega)\}$ satisfying $\frac{\partial}{\partial \nu} u_0 + hu_0 = w(t,x)$, $\frac{\partial \varphi_0}{\partial \nu} = 0$ and $w \in W^1([0,T]; L^2(\partial\Omega))$, problem (P) has at least one solution (u^*, φ^*, w^*).*

Proof. Consider $j(w) : \widehat{W} \to I\!R$ given by

$$j(w) = \frac{1}{2} \int_Q (\varphi^w(t,x) + 1 + \delta_1)^2 \cdot \chi_{Q_0} \, dtdx$$

$$+ \frac{1}{2} \int_Q \left((u^w(t,x) + \delta_2 - \delta_1)^+\right)^2 \cdot \chi_{Q_0} \, dtdx + \frac{1}{2} \int_\Sigma w^2(t,x) \, dtd\gamma, \quad (3.211)$$

where (u^w, φ^w) is the solution of (3.1)-(3.3) corresponding to $w \in \widehat{W}$,

$$\widehat{W} = W^1([0,T]; L^2(\partial\Omega)) \cap L^\infty(\Sigma).$$

Let $0 \le d = \inf \{j(w), \ w \in \widehat{W}\}$ and let $\{w_n\} \subset \widehat{W}$ be such that

$$d \le j(w_n) \le d + \frac{1}{n}. \quad (3.212)$$

It is obvious that the sequence $\{w_n\}$ is bounded in $\widehat{W}$. Therefore, we can select subsequences of $\{w_n\}$, still denoted by itself, such that

$$w_n \to w^* \ \text{ weak star in } \ \widehat{W}.$$

Let (u_n, φ_n) be the solution of problem (P) corresponding to w_n, i.e.,

$$\begin{cases} \rho V \frac{\partial}{\partial t} u_n + \frac{\ell}{2} \frac{\partial}{\partial t} \varphi_n = k \Delta u_n & \\ \tau \frac{\partial}{\partial t} \varphi_n = \xi^2 \Delta \varphi_n + \dfrac{1}{2a}(\varphi_n - \varphi_n^3) + 2u_n & \text{in } Q, \\ \frac{\partial}{\partial \nu} u_n + hu_n = w_n(t,x) \quad \frac{\partial}{\partial \nu} \varphi_n = 0 & \text{on } \Sigma, \\ u_n(0,x) = u_0(x) \quad \varphi_n(0,x) = \varphi_0(x) & \text{on } \Omega. \end{cases} \quad (3.213)$$

Taking into account the estimates (3.4)-(3.6) and the Sobolev embedding theorem, we have (on a subsequence, again denoted $\{n\}$)

$$u_n \to u^* \quad \text{weakly in } L^2([0,T]; H^2(\Omega)) \text{ and strongly in } L^2([0,T]; H^1(\Omega)),$$

$$
\begin{array}{lll}
\varphi_n \to \varphi^* & \text{weakly in} & L^2([0,T];H^2(\Omega)) \\
 & \text{and strongly in} & L^2([0,T];H^1(\Omega)), \\
\frac{\partial u_n}{\partial t} \to \frac{\partial u^*}{\partial t} & \text{weakly in} & L^2([0,T];L^2(\Omega)), \\
\frac{\partial \varphi_n}{\partial t} \to \frac{\partial \varphi^*}{\partial t} & \text{weakly in} & L^2([0,T];L^2(\Omega)), \\
\Delta u_n \to \Delta u^* & \text{weakly in} & L^2([0,T];L^2(\Omega)), \\
\Delta \varphi_n \to \Delta \varphi^* & \text{weakly in} & L^2([0,T];L^2(\Omega)),
\end{array}
$$

and, by the Ascoli-Arzéla theorem

$$
\begin{array}{l}
u_n \to u^* \text{ strongly in } C([0,T];L^2(\Omega)), \\
\varphi_n \to \varphi^* \text{ strongly in } C([0,T];L^2(\Omega)).
\end{array}
$$

By Sobolev's embedding theorem ($H^1(\Omega) \subset L^6(\Omega)$, $\Omega \subset \mathbb{R}^n$, $n \leq 3$) and thanks to the relations of convergence established above, we obtain

$$
\|\varphi_n^2 + \varphi_n\varphi^* + (\varphi^*)^2\|_{L^2([0,T];L^3(\Omega))} \leq C,
$$

which implies

$$
\begin{aligned}
&\int_Q (\varphi_n^3 - (\varphi^*)^3)^2 dxdt \\
&\leq \left[\int_Q (\varphi_n - \varphi^*)^6 dxdt\right]^{1/3}\left[\int_Q |\varphi_n^2 + \varphi_n\varphi^* + (\varphi^*)^2|^2 dxdt\right]^{2/3} \\
&\leq C\int_0^T \|\varphi_n - \varphi^*\|_{L^6(\Omega)}^2 dt.
\end{aligned}
$$

This indicates that

$$
\varphi_n^3 \to (\varphi^*)^3 \quad \text{strongly in } L^2([0,T];L^2(\Omega)).
$$

So, letting n tend to $+\infty$ in (3.213), we get

$$
\begin{cases}
\rho V \frac{\partial}{\partial t}\, u^* + \frac{\ell}{2}\frac{\partial}{\partial t}\, \varphi^* = k\Delta u^* & \\
\tau \frac{\partial}{\partial t}\, \varphi^* = \xi^2 \Delta\varphi^* + \dfrac{1}{2a}\big(\varphi^* - (\varphi^*)^3\big) + 2u^* & \text{in } Q, \\
\frac{\partial}{\partial \nu}\, u^* + hu^* = w^*(t,x) \quad \frac{\partial}{\partial \nu}\, \varphi^* = 0 & \text{on } \Sigma, \\
u^*(0,x) = u_0(x) \quad \varphi^*(0,x) = \varphi_0(x) & \text{on } \Omega.
\end{cases}
\tag{3.214}
$$

Thus, the uniqueness of solution for (3.214) implies that (u^*, φ^*) is the solution of problem (3.1)-(3.3) corresponding to $w^* \in \widehat{W}$.

Since j is continuous, then by (3.212), we see that $d = j(w^*)$ and the proof of Proposition 3.5 is completed. **q.e.d.**

As regards the existence in problem (P^ε), we have

Proposition 3.6. *For* $u_0, \varphi_0 \in W^{1,\infty}(\Omega) = \{y : \; y, \nabla y \in L^\infty(\Omega)\}$ *satisfying* $\frac{\partial}{\partial \nu} u_0 + h u_0 = w(t,x)$, $\frac{\partial \varphi_0}{\partial \nu} = 0$ *and* $w \in W^1([0,T]; L^2(\partial\Omega))$, *problem* (P^ε) *has at least one solution* $(u_\varepsilon^*, \varphi_\varepsilon^*, w_\varepsilon^*)$.

Proof. In the proof of Proposition 3.6 (see C. Moroşanu [19]) we have used the estimates established in Section 3.1, the Sobolev imbedding theorem, the Ascoli-Arzéla theorem and the continuity of $j^\varepsilon(w)$.

3.7b The convergence of problem (P^ε): Theorem 3.6

The main result of this Section is

Theorem 3.6 *Let* $\{w_\varepsilon^*\}$ *be a sequence of optimal controllers for problems* (P^ε). *Then*

$$\lim_{\varepsilon \to 0} (\inf \; j^\varepsilon(w)) = \inf \; \{j(w); \quad w \in \mathcal{U}\} \quad \text{and} \tag{3.215}$$

$$\lim_{\varepsilon \to 0} j(w_\varepsilon^*) = \inf \; \{j(w); \quad w \in \mathcal{U}\}. \tag{3.216}$$

Moreover, every weak limit point of $\{w_\varepsilon^*\}$ *is an optimal controller for problem* (P).

As already know, Theorem 3.6 amounts to saying that (P^ε) approximates problem (P) and an optimal control $\{w_\varepsilon^*\}$ of (P^ε) is a suboptimal control for problem (P).

The main ingredient of the proof of Theorem 3.6 is the following lemma

Lemma 3.7. *If* $\{w_\varepsilon^*\}$ *is a sequence of optimal controllers for problems* (P^ε) *then there exists* $\{\varepsilon_n\} \to 0$ *such that*

$$\begin{array}{lll} w_{\varepsilon_n}^* \to w^* & \text{weak star in} & L^\infty(\Sigma), \\ \varphi_{\varepsilon_n}^* \to \varphi^* & \text{strongly in} & L^2([0,T]; H^1(\Omega)), \\ u_{\varepsilon_n}^* \to u^* & \text{strongly in} & L^2([0,T]; H^1(\Omega)), \end{array}$$

where $(u_{\varepsilon_n}^*, \varphi_{\varepsilon_n}^*, w_{\varepsilon_n}^*) = (u_{\varepsilon_n}^{w_{\varepsilon_n}^*}, \varphi_{\varepsilon_n}^{w_{\varepsilon_n}^*}, w_{\varepsilon_n}^*)$ *is the solution to* (3.207)-(3.210) *corresponding to* $w = w_{\varepsilon_n}^*$ *and* $(u^*, \varphi^*, w^*) = (u^{w^*}, \varphi^{w^*}, w^*)$ *is the solution to* (3.1)-(3.3) *corresponding to* $w = w^*$.

Proof. For $\{w_\varepsilon^*\}$ independent of ε this lemma was proved in Section 3.3, Lemma 3.1. We omit them.

Proof of Theorem 3.6. Let $\{w_\varepsilon^*\}$ be an optimal controller for problem (P^ε) and let $(u_\varepsilon^*, \varphi_\varepsilon^*, w_\varepsilon^*)$ be the corresponding solution of (3.207)-(3.210) with $w = w_\varepsilon^*$. By virtue of Lemma 3.7 it results that there exist $w^* \in L^\infty(\Sigma)$ and $\{\varepsilon_n\}$ such that

$$\begin{array}{lll} w_{\varepsilon_n}^* \to w^* & \text{weak star in} & L^\infty(\Sigma), \\ \varphi_{\varepsilon_n}^{w_{\varepsilon_n}^*} \to \varphi^{w^*} & \text{strongly in} & L^2([0,T]; H^1(\Omega)), \\ u_{\varepsilon_n}^{w_{\varepsilon_n}^*} \to u^{w^*} & \text{strongly in} & L^2([0,T]; H^1(\Omega)) \end{array}$$

where $(u_{\varepsilon_n}^{w_{\varepsilon_n}^*}, \varphi_{\varepsilon_n}^{w_{\varepsilon_n}^*}, w_{\varepsilon_n}^*)$ is the solution to (3.207)-(3.210) corresponding to $w = w_{\varepsilon_n}^*$ and $(u^{w^*}, \varphi^{w^*}, w^*)$ is the solution to (3.1)-(3.3) corresponding to $w = w^*$.

Since:

$$\varphi \to \frac{1}{2} \int_Q (\varphi(t,x) + 1 + \delta_1)^2 \cdot \chi_{Q_0} \, dtdx,$$

$$u \to \frac{1}{2} \int_Q \left((u(t,x) + \delta_2 - \delta_1)^+\right)^2 \cdot \chi_{Q_0} \, dtdx,$$

$$w \to \frac{1}{2} \int_\Sigma w^2(t,x) \, dtd\gamma$$

are convex continuous functions, it follows that these are weakly lower semicontinuous functions. Hence

$$j(w^*) \le \liminf_{n\to\infty} j^{\varepsilon_n}(w_{\varepsilon_n}^*). \tag{3.217}$$

Let $\tilde{w}^*$ be an optimal controller for problem (P). Since $w_{\varepsilon_n}^*$ is an optimal controller for problem (P^{ε_n}) it follows that

$$j^{\varepsilon_n}(w_{\varepsilon_n}^*) \le j^{\varepsilon_n}(\tilde{w}^*).$$

But

$$\varphi_{\varepsilon_n}^{\tilde{w}^*} \to \varphi^{\tilde{w}^*}, \quad u_{\varepsilon_n}^{\tilde{w}^*} \to u^{\tilde{w}^*} \qquad \text{strongly in} \quad L^2([0,T]; H^1(\Omega)),$$

and so the latter implies

$$\lim_{n\to\infty} j^{\varepsilon_n}(\tilde{w}^*) \le j(\tilde{w}^*). \tag{3.218}$$

From (3.217 and (3.218) we obtain

$$j(w^*) \le \liminf_{n\to\infty} j^{\varepsilon_n}(w^*_{\varepsilon_n}) \le \limsup_{n\to\infty} j^{\varepsilon_n}(w^*_{\varepsilon_n}) \le j(\tilde{w}^*).$$

Hence

$$\lim_{\varepsilon_n\to 0} \inf j^{\varepsilon_n}(w^*_{\varepsilon_n}) = j(\tilde{w}^*) = \inf \{j(w), \quad w \in \mathcal{U}\}$$

and then (3.215) holds.

To prove (3.216) we set: $\tilde{u}_\varepsilon = u^{w^*_\varepsilon}$, $\tilde{\varphi}_\varepsilon = \varphi^{w^*_\varepsilon}$ (w^*_ε optimal in (P^ε)). We have on a subsequence $\{\varepsilon_n\}$

$$\begin{aligned} w^*_{\varepsilon_n} &\to w^0 \quad \text{weakly in} \quad && L^\infty(\Sigma), \\ \tilde{u}_{\varepsilon_n} &\to u \quad \text{strongly in} \quad && L^2([0,T], H^1(\Omega)), \\ \tilde{\varphi}_{\varepsilon_n} &\to \varphi \quad \text{strongly in} \quad && L^2([0,T], H^1(\Omega)), \end{aligned}$$

where (u, φ, w^0) satisfy (3.1)-(3.3), i.e., $(u, \varphi) = (u^{w^0}, \varphi^{w^0})$. We have therefore

$$j(w^0) \le \inf P$$

and, since $\{\varepsilon_n\}$ was arbitrary choosen, (3.216) follows.

Now, since w^*_ε is an optimal controller for problem (P^ε) it follows that

$$j^\varepsilon(w^*_\varepsilon) \le j^\varepsilon(w) \quad \forall\, w \in \mathcal{U}.$$

But, as we have seen above (relation (3.217)),

$$j(w^*) \le \liminf_{\varepsilon\to 0} j^\varepsilon(w^*_\varepsilon)$$

and thus, along with above inequality, we may conclude that

$$j(w^*) \le \lim_{\varepsilon\to 0} j^\varepsilon(w) \quad \forall\, w \in \mathcal{U}.$$

Hence

$$j(w^*) \le j(w) \quad \forall\, w \in \mathcal{U}$$

i.e., the weak limit point w^* is a suboptimal controller for problem (P). This completes the proof of Theorem 3.6. **q.e.d.**

3.7c Necessary optimality conditions in (P^ε): Theorem 3.7

Let $(u^\varepsilon, \varphi^e, w)$ be the solution of (3.207)-(3.209) and let $\tilde{w} \in L^\infty(\Sigma)$ be arbitrary but fixed and $\lambda > 0$. Set $w^\lambda = w + \lambda\tilde{w}$ and let $(u^{\lambda,\varepsilon}, \varphi^{\lambda,\varepsilon})$ be the solution of (3.207)-(3.209) corresponding to w^λ. First of all we easily can derive from above that

$$w^\lambda \to w \qquad \text{strongly in } L^2(\Sigma) \text{ as } \lambda \to 0.$$

Corresponding to $(u^{\lambda,\varepsilon}, \varphi^{\lambda,\varepsilon}, w^\lambda)$, for $i = 0, 1, \cdots, M_\varepsilon - 1$, we have

$$\begin{cases} \rho V u_t^{\lambda,\varepsilon} + \dfrac{\ell}{2}\varphi_t^{\lambda,\varepsilon} = k\Delta u^{\lambda,\varepsilon} \\ \tau\varphi_t^{\lambda,\varepsilon} = \xi^2\Delta\varphi^{\lambda,\varepsilon} + \dfrac{1}{2a}\varphi^{\lambda,\varepsilon} + 2u^{\lambda,\varepsilon} \end{cases} \quad \text{in } Q_i^\varepsilon, \qquad (3.219)$$

$$\begin{cases} \dfrac{\partial}{\partial\nu}u^{\lambda,\varepsilon} + hu^{\lambda,\varepsilon} = w^\lambda \\ \dfrac{\partial}{\partial\nu}\varphi^{\lambda,\varepsilon} = 0 \end{cases} \quad \text{on } \Sigma_i^\varepsilon, \qquad (3.220)$$

$$\begin{cases} u_+^{\lambda,\varepsilon}(i\varepsilon, x) = u_-^{\lambda,\varepsilon}(i\varepsilon, x), \quad u_-^{\lambda,\varepsilon}(0,x) = u_0(x) \\ \varphi_+^{\lambda,\varepsilon}(i\varepsilon, x) = z^\lambda(\varepsilon, \varphi_-^{\lambda,\varepsilon}(i\varepsilon, x)) \end{cases} \quad \text{on } \Omega, \qquad (3.221)$$

where z^λ is the solution of the Cauchy problem:

$$\begin{cases} \left(z^\lambda(s)\right)' + \dfrac{1}{2a}\left(z^\lambda(s)\right)^3 = 0 \quad s \in (i\varepsilon, (i+1)\varepsilon), \\ z^\lambda(i\varepsilon) = \varphi_-^{\lambda,\varepsilon}(i\varepsilon, x) \quad \varphi_-^{\lambda,\varepsilon}(0,x) = \varphi_0(x). \end{cases} \qquad (3.222)$$

Subtracting (3.207)-(3.209) from (3.219)-(3.221) and dividing by $\lambda > 0$, we get

$$\begin{cases} \rho V\left(\frac{u^{\lambda,\varepsilon}-u^\varepsilon}{\lambda}\right)_t + \frac{\ell}{2}\left(\frac{\varphi^{\lambda,\varepsilon}-\varphi^e}{\lambda}\right)_t = k\Delta\left(\frac{u^{\lambda,\varepsilon}-u^\varepsilon}{\lambda}\right) \quad \text{in } Q_i^\varepsilon, \\ \tau\left(\frac{\varphi^{\lambda,\varepsilon}-\varphi^e}{\lambda}\right)_t = \xi^2\Delta\left(\frac{\varphi^{\lambda,\varepsilon}-\varphi^e}{\lambda}\right) \\ \qquad + \frac{1}{2a}\left(\frac{\varphi^{\lambda,\varepsilon}-\varphi^e}{\lambda}\right) + 2\left(\frac{u^{\lambda,\varepsilon}-u^e}{\lambda}\right) \quad \text{in } Q_i^\varepsilon, \end{cases} \qquad (3.223)$$

$$\begin{cases} \frac{\partial}{\partial\nu}\left(\frac{u^{\lambda,\varepsilon}-u^e}{\lambda}\right) + h\left(\frac{u^{\lambda,\varepsilon}-u^e}{\lambda}\right) = \frac{w^\lambda - w}{\lambda}, \quad \text{on } \Sigma_i^\varepsilon, \\ \frac{\partial}{\partial\nu}\left(\frac{\varphi^{\lambda,\varepsilon}-\varphi^e}{\lambda}\right) = 0 \quad \text{on } \Sigma_i^\varepsilon, \end{cases} \qquad (3.224)$$

$$\begin{cases} \frac{u_+^{\lambda,\varepsilon}(i\varepsilon,x)-u_+^{\varepsilon}(i\varepsilon,x)}{\lambda} = \frac{u_-^{\lambda,\varepsilon}(i\varepsilon,x)-u_-^{\varepsilon}(i\varepsilon,x)}{\lambda}, \\ \qquad u_-^{\lambda,\varepsilon}(0,x)-u^{\varepsilon}(0,x)=0 \quad \text{on } \Omega, \\ \frac{z^{\lambda}(\varepsilon,\varphi_-^{\lambda,\varepsilon}(i\varepsilon,x))-z(\varepsilon,\varphi_-^{\lambda,\varepsilon}(i\varepsilon,x))}{\lambda} = \frac{\varphi_+^{\lambda,\varepsilon}-\varphi_+^{e}}{\lambda} \quad \text{on } \Omega, \end{cases} \tag{3.225}$$

for $i = 0, 1, \cdots, M_\varepsilon - 1$.

Letting λ tend to zero in (3.223)-(3.225) we get the *system in variation* (3.226)-(3.228) below

$$\begin{cases} \rho V \tilde{u}_t^\varepsilon + \frac{\ell}{2}\tilde{\varphi}_t^\varepsilon = k\Delta\tilde{u}^\varepsilon & \text{in } Q_i^\varepsilon, \\ \tau\tilde{\varphi}_t^\varepsilon = \xi^2\Delta\tilde{\varphi}^\varepsilon + \frac{1}{2a}\tilde{\varphi}^\varepsilon + 2\tilde{u}^\varepsilon & \text{in } Q_i^\varepsilon, \end{cases} \tag{3.226}$$

$$\frac{\partial}{\partial\nu}\tilde{u}^\varepsilon + h\tilde{u}^\varepsilon = \tilde{w}, \qquad \frac{\partial}{\partial\nu}\tilde{\varphi}^\varepsilon = 0 \qquad \text{on } \Sigma_i^\varepsilon, \tag{3.227}$$

$$\begin{cases} \tilde{u}_+^\varepsilon(i\varepsilon,x) = \tilde{u}_-^\varepsilon(i\varepsilon,x) \quad \tilde{u}^\varepsilon(0,x) = 0 & \text{on } \Omega, \\ \tilde{\varphi}_+^\varepsilon(i\varepsilon,x) = \eta((i+1)\varepsilon,x) & \text{on } \Omega, \end{cases} \tag{3.228}$$

for $i = 0, 1, \cdots, M_\varepsilon - 1$, where

$$\tilde{u}^\varepsilon = \lim_{\lambda\to 0}\frac{u^{\lambda,\varepsilon}-u^\varepsilon}{\lambda}, \qquad \tilde{\varphi}^\varepsilon = \lim_{\lambda\to 0}\frac{\varphi^{\lambda,\varepsilon}-\varphi^\varepsilon}{\lambda},$$
$$\tilde{w} = \lim_{\lambda\to 0}\frac{w^\lambda - w}{\lambda}, \qquad \tilde{z} = \lim_{\lambda\to 0}\frac{z^\lambda - z}{\lambda}$$
$$\tilde{\varphi}_+^\varepsilon = \lim_{\lambda\to 0}\frac{\varphi_+^{\lambda,\varepsilon}-\varphi_+^\varepsilon}{\lambda}, \quad \tilde{u}_+^\varepsilon = \lim_{\lambda\to 0}\frac{u_+^{\lambda,\varepsilon}-u_+^\varepsilon}{\lambda},$$
$$\eta((i+1)\varepsilon,x) = \lim_{\lambda\to 0}\frac{z^\lambda(\varepsilon,\varphi_-^{\lambda,\varepsilon}(i\varepsilon,x)) - z(\varepsilon,\varphi_-^\varepsilon(i\varepsilon,x))}{\lambda}$$
$$= z(\varepsilon,\varphi_-^\varepsilon(i\varepsilon,x))\tilde{\varphi}_-^\varepsilon(i\varepsilon,x) + \tilde{z}(\varepsilon,\varphi_-^\varepsilon(i\varepsilon,x))$$

with $\eta(\cdot)$ the solution of Cauchy problem

$$\begin{cases} \eta'(s) + \dfrac{3}{2a} z^2 \eta(s) = 0 \quad s \in (i\varepsilon, (i+1)\varepsilon), \\ \\ \eta(i\varepsilon) = \tilde{\varphi}^{\varepsilon}_{-}(i\varepsilon, x) \quad \tilde{\varphi}^{\varepsilon}_{-}(0, x) = 0, \end{cases} \tag{3.229}$$

that is

$$\eta((i+1)\varepsilon) = \exp\Big(- \int\limits_{i\varepsilon}^{(i+1)\varepsilon} \frac{3}{2a} z(t, \cdot)^2 dt \Big) \tilde{\varphi}^{\varepsilon}_{-}(i\varepsilon, x). \tag{3.230}$$

We now introduce the *adjoint state system.* For this, the system (3.226) can be written in the following abstract form:

$$\frac{\partial}{\partial t}\begin{pmatrix} \tilde{u}^{\varepsilon} \\ \tilde{\varphi}^{\varepsilon} \end{pmatrix} = A \begin{pmatrix} \tilde{u}^{\varepsilon} \\ \tilde{\varphi}^{\varepsilon} \end{pmatrix} \qquad \text{in } Q_i^{\varepsilon}$$

where

$$A = \begin{pmatrix} \frac{1}{\rho V}\Big(k\Delta - \dfrac{\ell}{\tau}\Big) & -\dfrac{\ell}{2\tau\rho V}\Big(\xi^2\Delta + \dfrac{1}{2a}\Big) \\ \\ \dfrac{2}{\tau} & \dfrac{1}{\tau}\Big(\xi^2\Delta + \dfrac{1}{2a}\Big) \end{pmatrix}.$$

Then

$$A^* = \begin{pmatrix} \frac{1}{\rho V}\Big(k\Delta - \dfrac{\ell}{\tau}\Big) & \dfrac{2}{\tau} \\ \\ -\dfrac{\ell}{2\tau\rho V}\Big(\xi^2\Delta + \dfrac{1}{2a}\Big) & \dfrac{1}{\tau}\Big(\xi^2\Delta + \dfrac{1}{2a}\Big) \end{pmatrix}$$

and so, the *adjoint state system* corresponding to (3.226) is

$$\frac{\partial}{\partial t}\begin{pmatrix} p^{\varepsilon} \\ q^{\varepsilon} \end{pmatrix} = -A^* \begin{pmatrix} p^{\varepsilon} \\ q^{\varepsilon} \end{pmatrix} + \begin{pmatrix} \dfrac{\partial}{\partial u^{\varepsilon}} j^{\varepsilon}(w) \\ \\ \dfrac{\partial}{\partial \varphi^{\varepsilon}} j^{\varepsilon}(w) \end{pmatrix}$$

i.e.:

$$\begin{cases} p_t^{\varepsilon} + \dfrac{k}{\rho V}\Delta p^{\varepsilon} - \dfrac{\ell}{\tau\rho V} p^{\varepsilon} + \dfrac{2}{\tau} q^{\varepsilon} \\ \\ \quad = (u^{\varepsilon} + \delta_2 - \delta_1)^{+} \cdot \chi_{Q_0} & \text{in } Q_i^{\varepsilon}, \\ \\ \frac{\partial}{\partial \nu} p^{\varepsilon} + h p^{\varepsilon} = 0 & \text{on } \Sigma_i^{\varepsilon}, \\ \\ p^{\varepsilon}_{-}((i+1)\varepsilon, x) = 0, \quad p^{\varepsilon}_{-}(T, x) = 0 & x \in \Omega, \end{cases} \tag{3.231}$$

$$
\begin{cases}
q_t^\varepsilon - \frac{\ell\xi^2}{2\tau\rho V}\Delta p^\varepsilon - \frac{\ell}{4a\tau\rho V}p^\varepsilon + \frac{\xi^2}{\tau}\Delta q^\varepsilon + \frac{1}{2a\tau}q^\varepsilon \\
\quad = (\varphi^\varepsilon + 1 + \delta_1)\cdot \chi_{Q_0} & \text{in } Q_i^\varepsilon, \\
\frac{\partial}{\partial\nu}q^\varepsilon = \frac{2\rho V}{\ell}\frac{\partial p^\varepsilon}{\partial \nu} & \text{on } \Sigma_i^\varepsilon, \\
q_-^\varepsilon((i+1)\varepsilon, x) \\
\quad = \exp\Big(\int\limits_{i\varepsilon}^{(i+1)\varepsilon}\frac{3}{2a}z^2(t,\cdot)dt\Big)q_+^\varepsilon((i+1)\varepsilon, x), \\
q_-^\varepsilon(T,x) = 0 & x\in\Omega,
\end{cases}
\tag{3.232}
$$

for $i = M_\varepsilon - 2, M_\varepsilon - 3, ..., 1, 0$, where $z(t,\cdot)$ is the solution of (3.210).
Let us introduce the cost functional

$$
j_1^\varepsilon(w) = j^\varepsilon(w) + \frac{1}{2}I_{\mathcal{U}}(w)
$$

where $I_{\mathcal{U}}(w)$ is the indicator function of the set $\mathcal{U}$. If w^* is an optimal controller of problem (P^ε), then

$$
\frac{j_1^\varepsilon(w^* + \lambda\tilde{w}) - j_1^\varepsilon(w^*)}{\lambda} \geq 0 \qquad \forall\,\lambda > 0.
$$

Letting λ tend to zero in above inequality; we get

$$
\int\limits_Q (\varphi^\varepsilon + 1 + \delta_1)\cdot\chi_{Q_0}\,\tilde{\varphi}^\varepsilon\; dtdx + \int\limits_Q (u^\varepsilon + \delta_2 - \delta_1)^+\cdot\chi_{Q_0}\,\tilde{u}^\varepsilon\; dtdx
$$

$$
+ \int\limits_\Sigma w^*\;\tilde{w}\; dtd\gamma + I'_{\mathcal{U}}(w^*, \tilde{w}) \geq 0 \qquad \forall\,\tilde{w}\in T_{\mathcal{U}}(w^*). \tag{3.233}
$$

Multiplying $(3.231)_1$ by $\tilde{u}^\varepsilon$ and $(3.232)_1$ by $\tilde{\varphi}^\varepsilon$, using integration by parts and Green's formula, we derive

$$
\int\limits_{Q_i^\varepsilon} p_t^\varepsilon\tilde{u}^\varepsilon dt\; dx + \frac{k}{\rho V}\int\limits_{Q_i^\varepsilon} p^\varepsilon\Delta\tilde{u}^\varepsilon dt\; dx - \frac{\ell}{\tau\rho V}\int\limits_{Q_i^\varepsilon} p^\varepsilon\tilde{u}^\varepsilon dt\; dx \tag{3.234}
$$

$$+\frac{2}{\tau}\int\limits_{Q_i^\varepsilon} q^\varepsilon \tilde{u}^\varepsilon dt\ dx + \frac{k}{\rho V}\int\limits_{\Sigma_i^\varepsilon}\Big(p^\varepsilon\frac{\partial\tilde{u}^\varepsilon}{\partial\nu} - \frac{\partial p^\varepsilon}{\partial\nu}\tilde{u}^\varepsilon\Big)dt\ d\gamma$$

$$= \int\limits_{Q_i^\varepsilon}(u^\varepsilon + \delta_2 - \delta_1)^+ \cdot \chi_{Q_0}\ \tilde{u}^\varepsilon\ dtdx,$$

$$\int\limits_{Q_i^\varepsilon} q_t^\varepsilon \tilde{\varphi}^\varepsilon dt\ dx - \frac{\ell\xi^2}{2\tau\rho V}\int\limits_{Q_i^\varepsilon} p^\varepsilon \Delta\tilde{\varphi}^\varepsilon dt\ dx - \frac{\ell}{4a\tau\rho V}\int\limits_{Q_i^\varepsilon} p^\varepsilon\tilde{\varphi}^\varepsilon dt\ dx \tag{3.235}$$

$$+\frac{\ell\xi^2}{2\tau\rho V}\int\limits_{\Sigma_i^\varepsilon}\Big(\frac{\partial p^\varepsilon}{\partial\nu}\tilde{\varphi}^\varepsilon - p^\varepsilon\frac{\partial\tilde{\varphi}^\varepsilon}{\partial\nu}\Big)dt\ d\gamma + \frac{\xi^2}{\tau}\int\limits_{\Sigma_i^\varepsilon}\Big(q^\varepsilon\frac{\partial\tilde{\varphi}^\varepsilon}{\partial\nu} - \tilde{\varphi}^\varepsilon\frac{\partial q^\varepsilon}{\partial\nu}\Big)dt\ d\gamma$$

$$+\frac{\xi^2}{\tau}\int\limits_{Q_i^\varepsilon} q^\varepsilon\Delta\tilde{\varphi}^\varepsilon dt\ dx + \frac{1}{2a\tau}\int\limits_{Q_i^\varepsilon} q^\varepsilon\tilde{\varphi}^\varepsilon dt\ dx$$

$$= \int\limits_{Q_i^\varepsilon}(\varphi^\varepsilon + 1 + \delta_1)\cdot\chi_{Q_0}\ \tilde{\varphi}^\varepsilon\ dtdx.$$

Now we multiply $(3.227)_1$ by p^ε, $(3.231)_2$ by $\tilde{u}^\varepsilon$ and, by subtraction we get

$$\frac{\partial}{\partial\nu}p^\varepsilon\tilde{u}^\varepsilon - \frac{\partial}{\partial\nu}\tilde{u}^\varepsilon\ p^\varepsilon = -p^\varepsilon\tilde{w}. \tag{3.236}$$

Adding (3.234)-(3.235) and taking into account $(3.227)_2$, $(3.232)_2$, (3.236), we obtain

$$\int\limits_{Q_i^\varepsilon} p_t^\varepsilon\tilde{u}^\varepsilon\ dtdx + \int\limits_{Q_i^\varepsilon} q_t^\varepsilon\tilde{\varphi}^\varepsilon\ dtdx + \frac{k}{\rho V}\int\limits_{\Sigma_i^\varepsilon} p^\varepsilon\tilde{w}\ dtd\gamma$$

$$+\int\limits_{Q_i^\varepsilon} p^\varepsilon\Big[\frac{k}{\rho V}\Delta\tilde{u}^\varepsilon - \frac{\ell\xi^2}{2\tau\rho V}\Delta\tilde{\varphi}^\varepsilon - \frac{\ell}{4a\tau\rho V}\tilde{\varphi}^\varepsilon - \frac{\ell}{\tau\rho V}\tilde{u}^\varepsilon\Big]dtdx$$

$$+\int\limits_{Q_i^\varepsilon} q^\varepsilon\Big[\frac{\xi^2}{\tau}\Delta\tilde{\varphi}^\varepsilon + \frac{1}{2a\tau}\tilde{\varphi}^\varepsilon + \frac{2}{\tau}\tilde{u}^\varepsilon\Big]dtdx,$$

$$= \int_{Q_i^\varepsilon} (u^\varepsilon + \delta_2 - \delta_1)^+ \cdot \chi_{Q_0}\ \tilde{u}^\varepsilon\ dtdx + \int_{Q_i^\varepsilon} (\varphi^\varepsilon + 1 + \delta_1) \cdot \chi_{Q_0}\ \tilde{\varphi}^\varepsilon\ dtdx,$$

i.e., on the basis of equations in system (3.226), from the last equality we derive that

$$\int_{Q_i^\varepsilon} \left(p_t^\varepsilon \tilde{u}^\varepsilon + p^\varepsilon \tilde{u}_t^\varepsilon + q_t^\varepsilon \tilde{\varphi}^\varepsilon + q^\varepsilon \tilde{\varphi}_t^\varepsilon\right) dtdx + \frac{k}{\rho V} \int_{\Sigma_i^\varepsilon} p^\varepsilon \tilde{w}\ dtd\gamma$$

$$= \int_{Q_i^\varepsilon} (u^\varepsilon + \delta_2 - \delta_1)^+ \cdot \chi_{Q_0}\ \tilde{u}^\varepsilon\ dtdx + \int_{Q_i^\varepsilon} (\varphi^\varepsilon + 1 + \delta_1) \cdot \chi_{Q_0}\ \tilde{\varphi}^\varepsilon\ dtdx.$$

By Fubini's theorem and definition of distributional derivative, the latter relation leads to

$$\frac{k}{\rho V} \int_{\Sigma_i^\varepsilon} p^\varepsilon \tilde{w}\ dtd\gamma = \int_{Q_i^\varepsilon} \left[(u^\varepsilon + \delta_2 - \delta_1)^+ \cdot \chi_{Q_0}\ \tilde{u}^\varepsilon + (\varphi^\varepsilon + 1 + \delta_1) \cdot \chi_{Q_0}\ \tilde{\varphi}^\varepsilon\right] dtdx$$

and then (3.233) becomes

$$\frac{k}{\rho V} \int_{\Sigma} p^\varepsilon \tilde{w}\ dtd\gamma + \int_{\Sigma} w^* \tilde{w}\ dtd\gamma + I'_{\mathcal{U}}(w^*, \tilde{w}) \geq 0 \qquad \forall\, \tilde{w} \in T_{\mathcal{U}}(w^*)$$

or

$$\left(\frac{k}{\rho V} p^\varepsilon + w^*, \tilde{w}\right)_{L^2(\Sigma)\times L^2(\Sigma)} + I'_{\mathcal{U}}(w^*, \tilde{w}) \geq 0 \qquad \forall\, \tilde{w} \in T_{\mathcal{U}}(w^*).$$

The last inequality is equivalent with

$$-r(t,x) \in \partial I_{\mathcal{U}}(w^*) \qquad \text{a.p.t. } (t,x) \in \Sigma,$$

where $r(t,x) = \frac{k}{\rho V} p^\varepsilon(t,x) + w(t,x)$, and thus we can conclude that

$$w^*(t,x) = \begin{cases} R, & \text{if } r(t,x) > 0 \\ 0, & \text{if } r(t,x) < 0. \end{cases} \tag{3.237}$$

Summing, we have proved the following maximum principle for problem (P^ε) introduced at beginning of Section 3.7:[5]

Theorem 3.7. *Let $(u^{*,\varepsilon}, \varphi^{*,\varepsilon}, w^*)$ be optimal in problem (P^ε). Then the optimal control is given by (3.237) where $(p^\varepsilon, q^\varepsilon)$ satisfy along with $u^{*,\varepsilon}, \varphi^{*,\varepsilon}$ the dual system (3.231)-(3.232).*

---

[5] For a well-informed readers we add in the end the fact that some of the results presented in this chapter have been published recently in "J. Optim. Theory and Appl., vol. 148, no. 1, pp. 14-45, 2011"

Analysis and Optimal Control of Phase-Field Transition System: Fractional Steps Methods, 2012, 185-246 185

CHAPTER 4

METHODS OF APPROXIMATION AND ALGORITHMS

Costică MOROŞANU

Faculty of Mathematics, "Al.I. Cuza" University of Iasi, Romania

Abstract: Two topics will be covered in this Chapter:

- *numerical approximation of the solution of nonlinear phase-field transition system;*
- *numerical approximation of boundary optimal control, governed by phase-field transition system.*

For both subjects we consider the phase-field system with non-homogeneous Cauchy-Neumann boundary conditions:

$$\begin{cases} \rho c \frac{\partial}{\partial t} u + \frac{\ell}{2} \frac{\partial}{\partial t} \varphi = k \Delta u \\ \tau \frac{\partial}{\partial t} \varphi = \xi^2 \Delta \varphi + \frac{1}{2a}(\varphi - \varphi^3) + 2u \end{cases} \quad \text{in } Q, \quad (4.1)$$

$$\begin{cases} \frac{\partial}{\partial \nu} u + hu = w(t,x) \\ \frac{\partial}{\partial \nu} \varphi = 0 \end{cases} \quad \text{on } \Sigma, \quad (4.2)$$

and initial conditions:

$$u(0,x) = u_0(x), \quad \varphi(0,x) = \varphi_0(x) \qquad \text{on } \Omega, \quad (4.3)$$

where Q, Σ, Ω, u, φ, u_0, φ_0, w, ρ, c, τ, ξ, ℓ, k, h, a have the same meaning as in Chapter 2.

The first Section is dedicated to the approximation of solution to the phase-field system in 1D. The discrete state equations were constructed using a *First-order Implicit Backward Difference Formula* (**1-IMBDF**). Three numerical methods to compute the approximate solution (*Newton method*, the *fractional steps method* and *cubic spline method*) are developed and, corresponding, conceptual algorithms are presented (`algnewton1D`, `algfrac1D`, and `algspline1D`). Numerical investigations of the phase-field system can be found in works: V. Arnăutu, C. Moroşanu [8], G. Caginalp, X. Chen, C. Eck [64], G. Caginalp, J.T. Lin [65], P.C. Fife, G.S. Gill [82] and G. Fix [83]. The use of the *fractional steps method* simplifies the numerical computation due to its decoupling feature (compares the algorithms `algnewton1D` and `algfrac1D`). Stability conditions (**Proposition 4.1**, **Proposition 4.2**) for fractional and cubic spline approximation are established too.

Section 4.2 deals with the approximation of solution to the phase-field system in 2D via fractional steps method. An implicit (backward) finite difference scheme in time and a **f**inite **e**lement **m**ethod (**fem**) in space are used to construct the discrete equations. A conceptual algorithm (`algfracfem2D`) has been introduced in order to compute the approximate solution.

In the next Section we are concerning with the approximation of the boundary optimal control in (P^{ε}_{inv}) stated by Theorem 3.4 in Section 3.5. The main novelty brought by the conceptual algorithm `InvPHT1D` is that the computation of the approximate solution corresponding to the nonlinear system (4.1) is replaced with calculation of the approximate solution for an ordinary equation and a linear system (compare step **P1** in C. Moroşanu [14] with the steps **P1-P2** in present algorithm). Numerical experiments accompany the abstract treatment.

Two conceptual algorithms of gradient type (`scpth-2D`, `cpth-2D`) to compute the optimal control (see Theorem 3.5) and the suboptimal control (see Theorem 3.7) of problems (P_S) and (P^{ε}), introduced in Chapter 3, are presented at the end of this Chapter. Each algorithm is implemented and accompanied by numerical experiments.

Keywords: Nonlinear parabolic systems, free boundary problems for PDE, optimal control, inverse problems, methods of approximations based on necessary conditions, numerical analysis, finite difference methods, finite element method, fractional steps method (Lie-Trotter product formula), stability of numerical methods, computer science, analysis of algorithms, methods of gradient type, phase changes, mathematical modeling, applications in engineering and industry.

4.1 Approximation of phase-field transition system in 1D

Consider the phase-field transition system (4.1)-(4.3) in 1-D ($\Delta u = u_{xx}$, $\Delta\varphi = \varphi_{xx}$):

$$\begin{cases} \rho c \frac{\partial}{\partial t} u + \frac{\ell}{2}\frac{\partial}{\partial t}\varphi = k u_{xx} \\ \tau \frac{\partial}{\partial t}\varphi = \xi^2 \varphi_{xx} + \frac{1}{2a}(\varphi - \varphi^3) + 2u \end{cases} \quad \text{in } Q, \tag{4.4}$$

$$\begin{cases} \frac{\partial}{\partial \nu} u + hu = w(t,x) \\ \frac{\partial}{\partial \nu}\varphi = 0 \end{cases} \quad \text{on } \Sigma, \tag{4.5}$$

$$u(0,x) = u_0(x), \qquad \varphi(0,x) = \varphi_0(x) \qquad \text{on } \Omega. \tag{4.6}$$

For every $\varepsilon > 0$, we denote

$Q^\varepsilon = \{(t,x) \in Q,\ \varepsilon \leq t \leq T\}$ and $\Sigma^\varepsilon = \{(t,x) \in \Sigma,\ \varepsilon \leq t \leq T\}$, and we associate to the nonlinear problem (4.4)-(4.6) the following approximating scheme of fractional steps type (also known in literature as *the Lie-Trotter product formula*):

$$\begin{cases} \rho c \ \frac{\partial}{\partial t} u^\varepsilon + \frac{\ell}{2}\frac{\partial}{\partial t}\varphi^\varepsilon = k\Delta u^\varepsilon \\ \tau \ \frac{\partial}{\partial t}\varphi^\varepsilon = \xi^2 \Delta\varphi^\varepsilon + \frac{1}{2a}\varphi^\varepsilon + 2u^\varepsilon \end{cases} \quad \text{in } Q^\varepsilon, \tag{4.7}$$

$$\begin{cases} \frac{\partial}{\partial \nu} u^\varepsilon + hu^\varepsilon = w(t,x) \\ \frac{\partial}{\partial \nu}\varphi^\varepsilon = 0 \end{cases} \quad \text{on } \Sigma^\varepsilon, \tag{4.8}$$

$$\begin{cases} u^\varepsilon(\varepsilon, x) = u_0(x) \\ \varphi^\varepsilon(\varepsilon, x) = z(\varepsilon, \varphi_0(x)) \end{cases} \quad \text{on } \Omega, \tag{4.9}$$

where $z(\varepsilon, \varphi_0(x))$ is the solution of Cauchy problem:

$$\begin{cases} z'(s) + \frac{1}{2a} z^3(s) = 0 \qquad s \in (0, \varepsilon), \\ z(0) = \varphi_0(x) \end{cases} \tag{4.10}$$

(see and (2.153)-(2.156) in Section 2.6 with $M_\varepsilon = 2$).

The aim of this Section is to develop numerical algorithms for approximating the solution to the nonlinear parabolic system (4.4)-(4.6). To achieve this goal, we use two numerical methods: *Newton Method* and *the fractional steps method* (Lie-Trotter product formula) (4.7)-(4.10). The second method is based on our convergence result given by Theorem 2.6 in Chapter 2.

4.1a Newton method. Algorithm `algnewton1D`

To fix the ideas of this Subsection, let the space interval $\Omega = [0, b] \subset I\!R_+$. We introduce over it the grid with N equidistant nodes

$$x_j = (j-1)dx \qquad j = 1, 2, \ldots, N, \qquad dx = b/(N-1),$$

and we consider the time interval $[0, T]$ is divided into M equal parts

$$t_i = (i-1)dt \qquad i = 1, 2, \ldots, M, \qquad dt = T/(M-1).$$

Let's denote by u_j^i, φ_j^i the approximate values in the point (t_i, x_j) of the unknown functions u, φ in (4.4)-(4.6); more precisely:

$$\begin{array}{l} u_j^i = u(t_i, x_j) \\ \\ \varphi_j^i = \varphi(t_i, x_j) \end{array} \qquad i = 1, 2, \ldots, M, \quad j = 1, 2, \ldots, N,$$

or, for later use

$$u^i \overset{\text{not}}{=} \left(u_1^i, u_2^i, \ldots, u_N^i\right)^T \quad \varphi^i \overset{\text{not}}{=} \left(\varphi_1^i, \varphi_2^i, \ldots, \varphi_N^i\right)^T \qquad i = \overline{1, M}. \quad (4.11)$$

The derivatives u_{xx}, φ_{xx} in (4.4) (the *diffusion term*) will be approximated by a second order centered finite differences:

$$\begin{array}{l} u_{xx}(t_i, x_j) = \Delta_{dx} u_j^i \approx \frac{u_{j-1}^i - 2u_j^i + u_{j+1}^i}{dx^2} \\ \\ \varphi_{xx}(t_i, x_j) = \Delta_{dx} \varphi_j^i \approx \frac{\varphi_{j-1}^i - 2\varphi_j^i + \varphi_{j+1}^i}{dx^2} \end{array} \qquad i = \overline{1, M}, \quad j = \overline{1, N}, \quad (4.12)$$

(Δ_{dx} is the discrete Laplacian depending on the step-size dx).

Corresponding to the initial conditions (4.6) we have

$$\begin{cases} u_j^1 = u(t_1, x_j) = u_0(x_j) \\ \\ \varphi_j^1 = \varphi(t_1, x_j) = \varphi_0(x_j) \end{cases} \qquad j = 1, 2, \ldots, N. \tag{4.13}$$

The boundary $\partial\Omega$ corresponding to Ω already chosen in 1D is reduced to the set $\{0, b\}$. So the boundary condition $(4.5)_1$ becomes

$$\begin{cases} -u_x(0) + h\ u(0) = w(t, 0) \\ \\ \quad u_x(b) + h\ u(b) = w(t, b), \end{cases} \tag{4.14}$$

where the sign in front of $\dfrac{\partial}{\partial \nu} u = u_x$ is $\mp$ because the normal to $[0, b]$ at 0 (b) points in the negative (positive) direction.

Using in (4.14) a forward (backward) finite difference to approximate $u_x(0)$ ($u_x(b)$), we get

$$\begin{cases} -\dfrac{u_2^i - u_1^i}{dx} + h\ u_1^i = w^i(0) \\ \dfrac{u_N^i - u_{N-1}^i}{dx} + h\ u_N^i = w^i(b) \end{cases} \qquad i = 1, 2, \ldots, M,$$

i.e.

$$\begin{cases} (1 + dx\ h)u_1^i - u_2^i = dx\ w^i(0) \\ \\ -u_{N-1}^i + (1 + dx\ h)u_N^i = dx\ w^i(b) \end{cases} \qquad i = 1, 2, \ldots, M, \tag{4.15}$$

where $w^i(0) = w(t_i, 0)$, $w^i(b) = w(t_i, b)$, $i = 1, 2, ..., M$.

To approximate $\varphi_x(0)$ ($\varphi_x(b)$) we will use a backward (forward) finite difference; this leads to

$$\varphi_0^i = \varphi_1^i, \quad \varphi_{N+1}^i = \varphi_N^i \quad i = 1, 2, \ldots, M, \tag{4.16}$$

where φ_0^i and φ_{N+1}^i are dummy variables.

For approximating the partial derivative with respect to time we employed a *first-order scheme*, namely:

$$\frac{\partial}{\partial t} u(t_i, x_j) \approx \frac{u_j^i - u_j^{i-1}}{dt}, \quad \frac{\partial}{\partial t} \varphi(t_i, x_j) \approx \frac{\varphi_j^i - \varphi_j^{i-1}}{dt} \tag{4.17}$$

$i = 1, 2, \ldots, M, \quad j = 1, 2, \ldots, N.$

To approximate the *reaction term* in (4.4), we will involve an implicit formula, i.e.:

$$\frac{1}{2a}\left[\varphi(t_i, x_j) - \varphi^3(t_i, x_j)\right] + 2u(t_i, x_j) \approx \frac{1}{2a}\left[\varphi_j^i - (\varphi_j^i)^3\right] + 2u_j^i, \quad (4.18)$$

$i = 1, 2, \ldots, M,\ j = 1, 2, \ldots, N.$

We are now ready to build the discrete state equations for (4.4)-(4.6) using a *First-order Implicit Backward Difference Formula* (**1-IMBDF**). To easily track this construction, we will do this in two steps (see, for example, S.J. Ruuth [48]). We begin by replaying in (4.4) the approximations stated in (4.12), (4.17) and (4.18); we get:

$$\begin{cases} \rho c\ \frac{u_j^i - u_j^{i-1}}{dt} + \frac{\ell}{2}\frac{\varphi_j^i - \varphi_j^{i-1}}{dt} = k\Delta_{dx} u_j^i \\ \tau\ \frac{\varphi_j^i - \varphi_j^{i-1}}{dt} = \xi^2 \Delta_{dx}\varphi_j^i + \frac{1}{2a}\left[\varphi_j^i - (\varphi_j^i)^3\right] + 2u_j^i, \end{cases}$$

for $i = 1, 2, \ldots, M,\ j = 1, 2, \ldots, N.$

Secondly, we will substitute in above the assigned equalities in (4.12) for discrete Laplacian and, arranging suitable, we conclude that the system (4.4) is discretized as

$$\begin{cases} -k\frac{dt}{dx^2}u_{j-1}^i + \left[\rho c + 2k\frac{dt}{dx^2}\right]u_j^i - k\frac{dt}{dx^2}u_{j+1}^i \\ \qquad\qquad + \frac{\ell}{2}\varphi_j^i = \rho c u_j^{i-1} + \frac{\ell}{2}\varphi_j^{i-1}, \\ -\xi^2\frac{dt}{dx^2}\varphi_{j-1}^i + \left[\tau + 2\xi^2\frac{dt}{dx^2} - \frac{dt}{2a}\right]\varphi_j^i - \xi^2\frac{dt}{dx^2}\varphi_{j+1}^i \\ \qquad\qquad + \frac{dt}{2a}(\varphi_j^i)^3 - 2dt\ u_j^i = \tau\varphi_j^{i-1}, \end{cases} \quad (4.19)$$

for $i = 1, 2, \ldots, M,\ j = 1, 2, \ldots, N.$

Taking in $(4.19)_1$ $u_0^i = u_1^i$ and $u_{N+1}^i = u_N^i$ (corresponding to $j = 1$ and $j = N$, respectively) and setting

$$c_1 = -k\frac{dt}{dx^2}, \quad c_2 = \rho c - 2c_1, \quad c_3 = \frac{\ell}{2},$$
$$c_5 = -\xi^2\frac{dt}{dx^2}, \quad c_6 = \tau - 2c_5 - \frac{dt}{2a},$$

than the discrete system (4.19), coupled with (4.15)-(4.16), can be written in matrix form as follows

$$Av^i + \frac{dt}{2a}\begin{pmatrix} 0 & 0 \\ 0 & diag\left((\varphi_j^i)^3\right)_{j=\overline{1,N}} \end{pmatrix} = Bv^{i-1} + d^i, \qquad (4.20)$$

for $i = 2, 3, ..., M$, where $v^i = \begin{pmatrix} u^i \\ \varphi^i \end{pmatrix}_{2N\times 1}$, $d^i = \begin{pmatrix} d_1^i \\ d_2^i \end{pmatrix}_{2N\times 1}$,

$$A = \begin{pmatrix} A_{11} & A_{12} \\ -2A_{21} & A_{22} \end{pmatrix}_{2N\times 2N} \qquad B = \begin{pmatrix} B_{11} & A_{12} \\ 0 & B_{22} \end{pmatrix}_{2N\times 2N}$$

with A_{11}, A_{12}, A_{21}, A_{22}, B_{11}, B_{22} having the same size $N \times N$, and

$$A_{11} = \begin{pmatrix} a_1 & c_1 - 1 & 0 & \cdots & 0 & 0 & 0 \\ c_1 & c_2 & c_1 & \cdots & 0 & 0 & 0 \\ \vdots & \vdots & \vdots & \ddots & \vdots & \vdots & \vdots \\ 0 & 0 & 0 & \cdots & c_1 & c_2 & c_1 \\ 0 & 0 & 0 & \cdots & 0 & c_1 - 1 & a_1 \end{pmatrix}$$

$a_1 = c_1 + c_2 + 1 + dx \cdot h$,

$$A_{22} = \begin{pmatrix} a_2 & c_5 & 0 & \cdots & 0 & 0 & 0 \\ c_5 & c_6 & c_5 & \cdots & 0 & 0 & 0 \\ \vdots & \vdots & \vdots & \ddots & \vdots & \vdots & \vdots \\ 0 & 0 & 0 & \cdots & c_5 & c_6 & c_5 \\ 0 & 0 & 0 & \cdots & 0 & c_5 & a_2 \end{pmatrix}$$

$a_2 = c_5 + c_6$,

$$A_{12} = \begin{pmatrix} c_3 & 0 & \cdots & 0 & 0 \\ 0 & c_3 & \cdots & 0 & 0 \\ \vdots & \vdots & \ddots & \vdots & \vdots \\ 0 & 0 & \cdots & c_3 & 0 \\ 0 & 0 & \cdots & 0 & c_3 \end{pmatrix} \qquad A_{21} = \begin{pmatrix} dt & 0 & \cdots & 0 & 0 \\ 0 & dt & \cdots & 0 & 0 \\ \vdots & \vdots & \ddots & \vdots & \vdots \\ 0 & 0 & \cdots & dt & 0 \\ 0 & 0 & \cdots & 0 & dt \end{pmatrix}$$

$$B_{11} = \begin{pmatrix} \rho c & 0 & \cdots & 0 & 0 \\ 0 & \rho c & \cdots & 0 & 0 \\ \vdots & \vdots & \ddots & \vdots & \vdots \\ 0 & 0 & \cdots & \rho c & 0 \\ 0 & 0 & \cdots & 0 & \rho c \end{pmatrix} \qquad B_{22} = \begin{pmatrix} \tau & 0 & \cdots & 0 & 0 \\ 0 & \tau & \cdots & 0 & 0 \\ \vdots & \vdots & \ddots & \vdots & \vdots \\ 0 & 0 & \cdots & \tau & 0 \\ 0 & 0 & \cdots & 0 & \tau \end{pmatrix}$$

$$d_1^i = \begin{pmatrix} dx \cdot w^i(0) \\ 0 \\ \vdots \\ 0 \\ dx \cdot w^i(b) \end{pmatrix}_{N\times 1} \qquad d_2^i = \begin{pmatrix} 0 \\ \vdots \\ 0 \end{pmatrix}_{N\times 1}.$$

We shall now present the usual iterative **Newton method** in order to approximate the vector-solution v^i in (4.20) for time level i (we omit the upper index i), i.e.

$$v = (v_1, v_2, \dots, v_{2N})^T = (u_1, \dots, u_N, \varphi_1, \dots, \varphi_N)^T.$$

From (4.20) it follows that we have to solve a nonlinear equation of the form $Y(v) = 0$. We recall that the iterative Newton method was

$$v^{(k+1)} = v^{(k)} - [Y'(v^{(k)})]^{-1}_{2N\times 2N} Y(v^{(k)}), \tag{4.21}$$

where k is the iteration index. In our situation (4.20), (4.21) we have

$$Y(v) = \begin{pmatrix} A_{11}u^i + A_{12}\varphi^i - B_{11}u^{i-1} - A_{12}\varphi^{i-1} - d_1^i \\ -2A_{21}u^i + A_{22}\varphi^i + \frac{dt}{2a} diag\,((\varphi_j^i)^3)_{j=\overline{1,N}} - B_{22}\varphi^{i-1} \end{pmatrix},$$

$$Y'(v) = \begin{pmatrix} A_{11} & A_{12} \\ -2A_{21} & A_{22} + 3\frac{dt}{2a} diag\,((\varphi_j^i)^2)_{j=\overline{1,N}} \end{pmatrix},$$

with

$$Y = (Y_1, Y_2, \dots, Y_{N-1}, Y_N, Y_{N+1}, Y_{N+2}, \dots, Y_{2N-1}, Y_{2N})^T$$

given by

$$Y_1(v) = a_1 u_1^i + (c_1 - 1)u_2^i + c_3\varphi_1^i - \rho c u_1^{i-1} - c_3\varphi_1^{i-1} - dx\; w^i(0),$$

$$Y_j(v) = c_1 u_{j-1}^i + c_2 u_j^i + c_1 u_{j+1}^i + c_3\varphi_j^i - \rho c u_j^{i-1} - c_3 \varphi_j^{i-1}, \quad j = \overline{2, N-1},$$

$$Y_N(v) = (c_1-1)u^i_{N-1} + a_1u^i_N + c_3\varphi^i_N - \rho c u^{i-1}_N - c_3\varphi^{i-1}_N - dx\ w^i(b),$$

$$Y_{N+1}(v) = -2dtu^i_1 + a_2\varphi^i_1 + c_5\varphi^i_2 + \tfrac{dt}{2a}(\varphi^i_1)^3 - \tau\varphi^{i-1}_1,$$

$$Y_{N+j}(v) = -2dtu^i_j + c_5\varphi^i_{j-1} + c_6\varphi^i_j + c_5\varphi^i_{j+1} + \tfrac{dt}{2a}(\varphi^i_j)^3 - \tau\varphi^{i-1}_j, \quad j = \overline{2, N-1},$$

$$Y_{2N}(v) = -2dtu^i_N + c_5\varphi^i_{N-1} + a_2\varphi^i_N + \tfrac{dt}{2a}(\varphi^i_N)^3 - \tau\varphi^{i-1}_N.$$

Thus, $v^{(k+1)}$ from (4.21) is the solution v of the linear system

$$[Y'(v^{(k)})]v = [Y'(v^{(k)})]v^{(k)} - Y(v^{(k)}). \tag{4.22}$$

The numerical algorithm to calculate the approximate solution by Newton method is the following one (i denotes the time level)

```
Begin algnewton1D
  i := 1  →  u^1, φ^1 from the initial conditions (4.13);
  For i := 2 to M do
    Establish the system (4.20) to pass from i − 1 to i;
    Repeat (k)
      Newton iteration (4.22)
    Until convergence;
  End-for;
End.
```

4.1b Fractional steps method. Algorithm `algfrac1D`

In this Subsection we are concerned with the numerical approximation of the nonlinear parabolic system (4.4)-(4.6) by *fractional steps method* (4.7)-(4.10). To construct the discrete state equations we will use the space-interval $[0, b]$ the same grid as in previous Subsection and, considering M as the number of equidistant nodes in which is divided the time-interval $[\varepsilon, T]$, we set

$$t_i = \varepsilon + (i-1)dt \qquad i = 1, 2, \ldots, M, \quad dt = (T-\varepsilon)/(M-1).$$

Also, we denote by u^i_j, φ^i_j the estimated values in the point (t_i, x_j) of the unknown functions u^ε, φ^ε in (4.7)-(4.10):

$$\begin{aligned} u^i_j &= u^\varepsilon(t_i, x_j) \\ \varphi^i_j &= \varphi^\varepsilon(t_i, x_j) \end{aligned} \qquad i = 1, 2, \ldots, M, \quad j = 1, 2, \ldots, N,$$

or (see (4.11))

$$u^i \overset{\text{not}}{=} \left(u_1^i, u_2^i, \ldots, u_N^i\right)^T \qquad \varphi^i \overset{\text{not}}{=} \left(\varphi_1^i, \varphi_2^i, \ldots, \varphi_N^i\right)^T \qquad i = \overline{1, M}.$$

For the derivatives u^ε_{xx}, φ^ε_{xx}, $\frac{\partial}{\partial t}u^\varepsilon$, and $\frac{\partial}{\partial t}\varphi^\varepsilon$ in (4.7) and for the boundary conditions (4.8) we will keep the same approximations as in (4.12), (4.17), (4.15) and (4.16), respectively, while, for the *reaction term* in (4.7) we will involve the implicit formula:

$$\frac{1}{2a}\varphi^\varepsilon(t_i, x_j) + 2u^\varepsilon(t_i, x_j) \approx \frac{1}{2a}\varphi_j^i + 2u_j^i. \tag{4.23}$$

Corresponding to the initial conditions (4.9) we have

$$u_j^1 = u^\varepsilon(t_1, x_j) = u_0(x_j) \qquad j = 1, 2, \ldots, N, \tag{4.24}$$

and, involving the separation of variables method to solve the Cauchy problem (4.10) (see Morosanu [15]), we get

$$\varphi_j^1 = \varphi^\varepsilon(t_1, x_j) = z(\varepsilon, \varphi_0(x_j)) = \varphi_0(x_j)\sqrt{\frac{a}{a + \varepsilon \cdot \varphi_0^2(x_j)}}, \tag{4.25}$$

for $j = 1, 2, \ldots, N$.

Using again a *First-order Implicit Backward Difference Formula* (**1-IMBDF**) we conclude that the system (4.7) is discretized as

$$\begin{cases} -k\frac{dt}{dx^2}u_{j-1}^i + \left[\rho c + 2k\frac{dt}{dx^2}\right]u_j^i - k\frac{dt}{dx^2}u_{j+1}^i \\ \qquad\qquad + \frac{\ell}{2}\varphi_j^i = \rho c u_j^{i-1} + \frac{\ell}{2}\varphi_j^{i-1}, \\ -\xi^2\frac{dt}{dx^2}\varphi_{j-1}^i + \left[\tau + 2\xi^2\frac{dt}{dx^2} - \frac{dt}{2a}\right]\varphi_j^i - \xi^2\frac{dt}{dx^2}\varphi_{j+1}^i \\ \qquad\qquad -2dt\, u_j^i = \tau\varphi_j^{i-1}, \end{cases} \tag{4.26}$$

for $i = 1, 2, \ldots, M$, $j = 1, 2, \ldots, N$.

Keeping the same settings: $u_0^i = u_1^i$, $u_{N+1}^i = u_N^i$ (for $j = 1$ and $j = N$, respectively), $c_1 = -k\frac{dt}{dx^2}$, $c_2 = \rho c - 2c_1$, $c_3 = \frac{\ell}{2}$, $c_5 = -\xi^2\frac{dt}{dx^2}$,

$c_6 = \tau - 2c_5 - \frac{dt}{2a}$, the discrete system (4.26), coupled with (4.15)-(4.16), can be written in matrix form as:

$$A\begin{pmatrix} u^i \\ \\ \varphi^i \end{pmatrix} = B\begin{pmatrix} u^{i-1} \\ \\ \varphi^{i-1} \end{pmatrix} + \begin{pmatrix} d_1^i \\ \\ d_2^i \end{pmatrix} \qquad i = 2, 3, ..., M, \tag{4.27}$$

where A, B, d_1^i, d_2^i was defined in (4.20).

The general algorithm to compute the approximate solution by means of a *fractional steps method* is obtained by the following sequence (again, i denotes the time level)

```
Begin algfrac1D
  Choose ε > 0, T > ε, b > 0;
  Choose M > 0, N > 0 and compute dt, dx;
  Choose u0, φ0, w;
  Compute z(ε,·) from (4.25);
  i := 1 → u^1, φ^1 from initial conditions (4.24)-(4.25);
  For i = 2 to M do
    Compute u^i, φ^i solving the linear system (4.27);
  End-for;
End.
```

Stability conditions. To establish conditions of stability for the linear difference equations (4.27), we will use in our analysis the Lax-Richtmyer definition of stability, expressed in terms of norm $\|\cdot\|_\infty$ (see Smith [84], pp. 48). We start by rewriting the linear system to be analyzed in a more convenient form, namely

$$\begin{pmatrix} u^i \\ \\ \varphi^i \end{pmatrix} = A^{-1}B\begin{pmatrix} u^{i-1} \\ \\ \varphi^{i-1} \end{pmatrix} + A^{-1}\begin{pmatrix} d_1^i \\ \\ d_2^i \end{pmatrix} \qquad i = 2, 3, ..., M \tag{4.28}$$

(the existence of A^{-1} will be proved in the proof of Proposition 4.1 below). In addition, the matrix A can be written in the form

$$A = D_1(I + D_1^{-1}G_1) \tag{4.29}$$

where $D_1 = diag(a_1, c_2, \cdots, c_2, a_1, a_2, c_6, \cdots, c_6, a_2)$ and $G_1 = A - D_1$.

Thus we conclude that the matrix $D_1^{-1}G_1$ is expressed as

$$
\begin{pmatrix}
0 & \frac{c_1-1}{a_1} & 0 & \cdots & 0 & 0 & \frac{c_3}{a_1} & 0 & 0 & \cdots & 0 & 0 & 0 \\
\frac{c_1}{c_2} & 0 & \frac{c_1}{c_2} & \cdots & 0 & 0 & 0 & \frac{c_3}{c_2} & 0 & \cdots & 0 & 0 & 0 \\
\vdots & \vdots & \vdots & \ddots & \vdots & \vdots & \vdots & \vdots & \vdots & \ddots & \vdots & \vdots & \vdots \\
0 & 0 & 0 & \cdots & 0 & \frac{c_1}{c_2} & 0 & 0 & 0 & \cdots & 0 & \frac{c_3}{c_2} & 0 \\
0 & 0 & 0 & \cdots & \frac{c_1-1}{a_1} & 0 & 0 & 0 & 0 & \cdots & 0 & 0 & \frac{c_3}{a_1} \\
-\frac{2dt}{a_2} & 0 & 0 & \cdots & 0 & 0 & 0 & \frac{c_5}{a_2} & 0 & \cdots & 0 & 0 & 0 \\
0 & -\frac{2dt}{c_6} & 0 & \cdots & 0 & 0 & \frac{c_5}{c_6} & 0 & \frac{c_5}{c_6} & \cdots & 0 & 0 & 0 \\
\vdots & \vdots & \vdots & \ddots & \vdots & \vdots & \vdots & \vdots & \vdots & \ddots & \vdots & \vdots & \vdots \\
0 & 0 & 0 & \cdots & -\frac{2dt}{c_6} & 0 & 0 & 0 & 0 & \cdots & \frac{c_5}{c_6} & 0 & \frac{c_5}{c_6} \\
0 & 0 & 0 & \cdots & 0 & -\frac{2dt}{a_2} & 0 & 0 & 0 & \cdots & 0 & \frac{c_5}{a_2} & 0
\end{pmatrix}
$$

and, a simple analysis of all lines in array above, allows us to deduce that we only have four distinct lines. The sum of each such line is written in vector v^1 below (recall that $a_1 = c_1 + c_2 + 1 + dx \cdot h$ and $a_2 = c_5 + c_6$)

$$
v^1 = \left[\frac{c_1 + c_3 - 1}{a_1}, \frac{2c_1 + c_3}{c_2}, \frac{-2dt + c_5}{a_2}, \frac{-2dt + 2c_5}{c_6} \right]. \tag{4.30}
$$

Let's denote by

$$0 < v^1_{max} = \max\{|c_1 + c_3 - 1|, |2c_1 + c_3|, |-2dt + c_5|, |-2dt + 2c_5|\},$$

and

$$0 < v^1_{min} = \min\{|c_1 + c_2 + 1 + dx \cdot h|, |c_2|, |c_5 + c_6|, |c_6|\}.$$

Now we are able to prove the following result with respect to the stability in matrix equation (4.28).

Proposition 4.1. *Suppose that* $v^1_{min} - v^1_{max} > 0$. *If one of the following conditions is true:*

i) $\rho c + \frac{\ell}{2} > \tau \quad \& \quad \dfrac{\rho c + \frac{\ell}{2}}{v^1_{min} - v^1_{max}} < 1$

or

$$ii)\ \rho c + \tfrac{\ell}{2} \leq \tau \quad \& \quad \frac{\tau}{v^1_{min} - v^1_{max}} < 1,$$

then the equation (4.28) *is stable. Otherwise, it is unstable.*

Proof. The proof is reduced to checking the condition of stability which, based on the Lax-Richtmyer definition mentioned above and taking into account the relation (4.28), it reduces to check the inequality

$$\|A^{-1}B\|_\infty < 1.$$

We begin our analysis by determining an estimate for $\|D_1^{-1}G_1\|_\infty$. As we have already noted (see relation (4.30)), this is equal with the following equality: $\|D_1^{-1}G_1\|_\infty = \max|v^1|$, wherefrom we easily derive the estimate

$$\|D_1^{-1}G_1\|_\infty < \frac{v^1_{max}}{v^1_{min}}. \tag{4.31}$$

The estimate (4.31) allows us to prove the existence of A^{-1}. Indeed, since by hypothesis we have assumed that $v^1_{max} < v^1_{min}$ than $\|D_1^{-1}G_1\|_\infty < 1$ which guarantees that there exists $(I + D_1^{-1}G_1)^{-1}$. Moreover, there exist A^{-1} and $A^{-1} = (I + D_1^{-1}G_1)^{-1}D_1^{-1}$. Using the well known inequality: $\|(I + D_1^{-1}G_1)^{-1}\|_\infty \leq \frac{1}{1-\|D_1^{-1}G_1\|_\infty}$ and making use of (4.29), it follows that

$$\begin{aligned} \|A^{-1}\|_\infty &\leq \|(I + D_1^{-1}G_1)^{-1}\|_\infty \|D_1^{-1}\|_\infty \\ &\leq \frac{1}{1 - \|D_1^{-1}G_1\|_\infty}\|D_1^{-1}\|_\infty. \end{aligned} \tag{4.32}$$

How $\|D_1^{-1}G_1\|_\infty \leq 1$ imply that $1 - \|D_1^{-1}G_1\|_\infty \geq 1 - \frac{v^1_{max}}{v^1_{min}} > 0$, we easily deduce from this that

$$0 < \frac{1}{1 - \|D_1^{-1}G_1\|_\infty} \leq \frac{v^1_{min}}{v^1_{min} - v^1_{max}}.$$

Since $\|D_1^{-1}\|_\infty \leq \frac{1}{v^1_{min}}$ and involving the above estimate, from (4.32) we finally obtain

$$\|A^{-1}\|_\infty < \frac{1}{v^1_{min} - v^1_{max}}. \tag{4.33}$$

Now we turn our attention to matrix B. Analyzing matrix B lines it follows that

$$\|B\|_\infty = \max\left\{\rho c + \frac{\ell}{2}, \tau\right\}. \tag{4.34}$$

Summing up and making use of (4.33), (4.34) we derive the following estimate

$$\|A^{-1}B\|_\infty \le \|A^{-1}\|_\infty \|B\|_\infty < \frac{1}{v^1_{min} - v^1_{max}}\|B\|_\infty,$$

which, in either cases i) or ii) leads us to the estimate $\|A^{-1}B\|_\infty < 1$ as we claimed at beginning of proof.

4.1c Cubic spline method. Algorithm `algspline1D`

Here we will be concerning with the cubic spline approximation of the unknowns u^ε, φ^ε in the linear system (4.7)-(4.10), obtained by a *fractional steps method* in order to approximate the nonlinear phase-field transition system. In the following, for $[0, b]$ and $[\varepsilon, T]$, we will use the same grids as in previous Subsections and, for the unknowns $u^\varepsilon(t_i, x_j)$, $\varphi^\varepsilon(t_i, x_j)$, $i = \overline{1, M}$, $j = \overline{1, N}$, we will keep the notations introduced in IV.1b.

Let us denote by $S_i(x)$ and $V_i(x)$ the *cubic splines* that interpolate $u^\varepsilon(t, x)$ and $\varphi^\varepsilon(t, x)$, respectively, in nodes x_j, $j = \overline{1, N}$ at the time level i, $i = \overline{1, M}$. Then, making use of (4.17) - a *first-order scheme* to approximate the partial derivatives with respect to t, and (4.23) - an implicit formula to approximate the *reaction term*, we associate to (4.7) at the point (t_i, x_j), $i = \overline{1, M}$, $j = \overline{1, N}$, the following general approximating scheme:

$$\begin{cases} \rho c \frac{u_j^{i+1} - u_j^i}{dt} + \frac{\ell}{2}\frac{\varphi_j^{i+1} - \varphi_j^i}{dt} = k\left[\theta M^{i+1,j} + (1-\theta)M^{i,j}\right], \\ \tau \frac{\varphi_j^{i+1} - \varphi_j^i}{dt} = \xi^2\left[\theta N^{i+1,j} + (1-\theta)N^{i,j}\right] + \frac{1}{2a}\varphi_j^i + 2u_j^i, \end{cases} \tag{4.35}$$

where $0 \le \theta \le 1$ and

$$M^{i,j} = S_i''(x_j) \quad N^{i,j} = V_i''(x_j), \quad i = \overline{1, M},\ j = \overline{1, N}.$$

It is known that at the time level $t = t_i$ the cubic splines $S_i(x)$ and $V_i(x)$ have the following expression (see Chapter I.9 with $\Delta : 0 =$

$x_1 < \cdots < x_N = b)$

$$S_i(x) = \left(u^i_{j-1} - \frac{dx^2}{6}M^{i,j-1}\right)\frac{x_j - x}{dx} + \left(u^i_j - \frac{dx^2}{6}M^{i,j}\right)\frac{x - x_{j-1}}{dx}$$
$$+ M^{i,j-1}\frac{(x_j - x)^3}{6dx} + M^{i,j}\frac{(x - x_{j-1})^3}{6dx}, \quad x \in [x_{j-1}, x_j],$$

$$V_i(x) = \left(\varphi^i_{j-1} - \frac{dx^2}{6}N^{i,j-1}\right)\frac{x_j - x}{dx} + \left(\varphi^i_j - \frac{dx^2}{6}N^{i,j}\right)\frac{x - x_{j-1}}{dx}$$
$$+ N^{i,j-1}\frac{(x_j - x)^3}{6dx} + N^{i,j}\frac{(x - x_{j-1})^3}{6dx}, \quad x \in [x_{j-1}, x_j],$$

for $i = \overline{1, M}$, $j = \overline{2, N}$.

We now turn our attention only on the relation $(4.35)_1$. In other words, we will deal with $S_i(x)$. So, let's consider the intervals $[x_{j-1}, x_j]$, $[x_j, x_{j+1}]$. Correspondingly, we have

$$S_i'(x) = \left(u^i_{j-1} - \frac{dx^2}{6}M^{i,j-1}\right)\cdot\frac{-1}{dx} + \left(u^i_j - \frac{dx^2}{6}M^{i,j}\right)\cdot\frac{1}{dx}$$
$$-M^{i,j-1}\frac{(x_j - x)^2}{2dx} + M^{i,j}\frac{(x - x_{j-1})^2}{2dx},$$
$$x \in [x_{j-1}, x_j], \quad j = \overline{2, N},$$

$$S_i'(x) = \left(u^i_j - \frac{dx^2}{6}M^{i,j}\right)\cdot\frac{-1}{dx} + \left(u^i_{j+1} - \frac{dx^2}{6}M^{i,j+1}\right)\cdot\frac{1}{dx}$$
$$-M^{i,j}\frac{(x_{j+1} - x)^2}{2dx} + M^{i,j+1}\frac{(x - x_j)^2}{2dx},$$
$$x \in [x_j, x_{j+1}], \quad j = \overline{1, N-1}.$$

From the continuity of $S_i'(x)$ at the interior nodes x_j, $j = \overline{2, N-1}$, expressed below:

$$S_i'(x_j - 0) = \frac{1}{dx}\left[-u^i_{j-1} + \frac{dx^2}{6}M^{i,j-1} + u^i_j - \frac{dx^2}{6}M^{i,j} + \frac{dx^2}{2}M^{i,j}\right]$$

$$= \frac{1}{dx}\left[-u_j^i + \frac{dx^2}{6}M^{i,j} + u_{j+1}^i - \frac{dx^2}{6}M^{i,j+1} - \frac{dx^2}{2}M^{i,j}\right] = S_i'(x_j+0),$$

we get $(j = \overline{2, N-1})$

$$M^{i,j-1} + 4M^{i,j} + M^{i,j+1} = \frac{6}{dx^2}[u_{j-1}^i - 2u_j^i + u_{j+1}^i]. \qquad (4.36)$$

Doing the same computations relative to the time level $t = t_{i+1}$, we derive $(j = \overline{2, N-1})$

$$M^{i+1,j-1} + 4M^{i+1,j} + M^{i+1,j+1} = \frac{6}{dx^2}[u_{j-1}^{i+1} - 2u_j^{i+1} + u_{j+1}^{i+1}]. \quad (4.37)$$

Now, multiplying (4.36) by $k(1-\theta)$, (4.37) by $k\theta$ and summing the results, we obtain $(j = \overline{2, N-1})$

$$k\left[\theta M^{i+1,j-1} + (1-\theta)M^{i,j-1}\right] + 4k[\theta M^{i+1,j} + (1-\theta)M^{i,j}]$$
$$+k\left[\theta M^{i+1,j+1} + (1-\theta)M^{i,j+1}\right]$$
$$= 6\frac{k}{dx^2}\left[\theta\left(u_{j-1}^{i+1} - 2u_j^{i+1} + u_{j+1}^{i+1}\right) + (1-\theta)\left(u_{j-1}^i - 2u_j^i + u_{j+1}^i\right)\right].$$

Let us observe that the left term in the expression above represents the right side of the equality $(4.35)_1$, expressed in nodes x_{j-1}, x_j, x_{j+1}, at the level t_i of time. Thus, the latter can be written equivalent as $(j = \overline{2, N-1})$

$$\rho c\left[(u_{j-1}^{i+1} - u_{j-1}^i) + 4(u_j^{i+1} - u_j^i) + (u_{j+1}^{i+1} - u_{j+1}^i)\right] \qquad (4.38)$$
$$+\frac{\ell}{2}\left[(\varphi_{j-1}^{i+1} - \varphi_{j-1}^i) + 4(\varphi_j^{i+1} - \varphi_j^i) + (\varphi_{j+1}^{i+1} - \varphi_{j+1}^i)\right]$$
$$= 6\frac{dt}{dx^2}k\left[\theta(u_{j-1}^{i+1} - 2u_j^{i+1} + u_{j+1}^{i+1}) + (1-\theta)(u_{j-1}^i - 2u_j^i + u_{j+1}^i)\right].$$

Setting now $r = \frac{dt}{dx^2}$ and performing some straightforward computation, the relation (4.38) leads to $(j = \overline{2, N-1})$

$$(\rho c - 6rk\theta)u_{j-1}^{i+1} + (4\rho c + 12rk\theta)u_j^{i+1} + (\rho c - 6rk\theta)u_{j+1}^{i+1} \qquad (4.39)$$

$$+\frac{\ell}{2}\varphi_{j-1}^{i+1}+2\ell\varphi_j^{i+1}+\frac{\ell}{2}\varphi_{j+1}^{i+1}$$

$$=[\rho c+6rk(1-\theta)]\,u_{j-1}^i+[4\rho c-12rk(1-\theta)]\,u_j^i+[\rho c+6rk(1-\theta)]\,u_{j+1}^i$$

$$+\frac{\ell}{2}\varphi_{j-1}^{i}+2\ell\varphi_j^{i}+\frac{\ell}{2}\varphi_{j+1}^{i}.$$

Following the same sequence of computations as in $(4.35)_1$, from the second equation in relation (4.35) one obtains ($j=\overline{2,N-1}$)

$$N^{i,j-1}+4N^{i,j}+N^{i,j+1}=\frac{6}{dx^2}[\varphi_{j-1}^i-2\varphi_j^i+\varphi_{j+1}^i], \qquad (4.40)$$

$$\left(\tau-6r\xi^2\theta\right)\varphi_{j-1}^{i+1}+\left(4\tau+12r\xi^2\theta\right)\varphi_j^{i+1}+\left(\tau-6r\xi^2\theta\right)\varphi_{j+1}^{i+1} \qquad (4.40')$$

$$=\left(\tau+6r\xi^2(1-\theta)+\frac{dt}{2a}\right)\varphi_{j-1}^i+\left(4\tau-12r\xi^2(1-\theta)+\frac{dt}{a}\right)\varphi_j^i$$

$$+\left(\tau+6r\xi^2(1-\theta)+\frac{dt}{2a}\right)\varphi_{j+1}^i+2dt(u_{j-1}^i+u_j^i+u_{j+1}^i).$$

To treat the boundary conditions (4.8), we have to consider the nodes x_1 ($j=1$) and x_N ($j=N$). In the sequel, we will present a detailed computation only for $(4.8)_1$ and $j=1$; for the remaining cases ($(4.8)_1$ and $j=N$, $(4.8)_2$ and $j=1$, $(4.8)_2$ and $j=N$) we will give only the final result.

Now, let's deal with the boundary conditions $(4.8)_1$. Invoking relation (4.14), we have (we remember that in the grid associated to $[0,b]$ we have noted $0=x_1$ and $b=x_N$)

$$\begin{cases} -S_i'(0)+hu_1^i=w^i(0) \\ S_i'(b)+hu_N^i=w^i(b), \end{cases} \qquad (4.41)$$

where $w^i(0)=w(t_i,0)$, $w^i(b)=w(t_i,b)$, $i=1,2,...,M$.

If we limit our study in relation (4.41) to the case $j=1$ (so, on $[x_1,x_2]$) at the level t_i of time, we know that

$$S_i'(0)=\tfrac{1}{dx}\left[-u_1^i+\tfrac{dx^2}{6}M^{i,1}+u_2^i-\tfrac{dx^2}{6}M^{i,2}-\tfrac{dx^2}{2}M^{i,1}\right]$$
$$=-\tfrac{1}{dx}\left[\tfrac{dx^2}{6}\left(2M^{i,1}+M^{i,2}\right)+u_1^i-u_2^i\right],$$

while, at the level t_{i+1} of time

$$S'_{i+1}(0) = \frac{1}{dx}\left[-u_1^{i+1} + \frac{dx^2}{6}M^{i+1,1} + u_2^{i+1} - \frac{dx^2}{6}M^{i+1,2} - \frac{dx^2}{2}M^{i+1,1}\right]$$
$$= -\frac{1}{dx}\left[\frac{dx^2}{6}\left(2M^{i+1,1} + M^{i+1,2}\right) + u_1^{i+1} - u_2^{i+1}\right].$$

Substituting the expressions $S'_i(0)$ and $S'_{i+1}(0)$ in $(4.41)_1$, one obtains

$$\begin{cases} 2M^{i,1} + M^{i,2} = \left(-\frac{6}{dx^2} - \frac{6}{dx}h\right)u_1^i + \frac{6}{dx^2}u_2^i + \frac{6}{dx}w^i(0), \\ 2M^{i+1,1} + M^{i+1,2} = \left(-\frac{6}{dx^2} - \frac{6}{dx}h\right)u_1^{i+1} + \frac{6}{dx^2}u_2^{i+1} + \frac{6}{dx}w^{i+1}(0). \end{cases} \tag{4.42}$$

Multiplying $(4.42)_1$ by $k(1-\theta)$, $(4.42)_2$ by $k\theta$ and summing, we deduce the equality

$$2k\left[\theta M^{i+1,1} + (1-\theta)M^{i,1}\right] + k[\theta M^{i+1,2} + (1-\theta)M^{i,2}]$$
$$+\left(\frac{6}{dx^2} + \frac{6}{dx}h\right)k\theta u_1^{i+1} - \frac{6}{dx^2}k\theta u_2^{i+1}$$
$$= -\left(\frac{6}{dx^2} + \frac{6}{dx}h\right)k(1-\theta)u_1^i + \frac{6}{dx^2}k(1-\theta)u_2^i$$
$$+\frac{6}{dx}\left[k(1-\theta)w^i(0) + k\theta w^{i+1}(0)\right].$$

Combining the last relation with $(4.35)_1$, we conclude that $(4.41)_1$ takes the discretized form $(i = \overline{1, M-1})$

$$\left(2\rho c + 6rk\theta + 6\frac{dt}{dx}kh\theta\right)u_1^{i+1} + (\rho c - 6rk\theta)\,u_2^{i+1} + \ell\varphi_1^{i+1} + \frac{\ell}{2}\varphi_2^{i+1}$$
$$= \left[2\rho c - 6rk(1-\theta) - 6\frac{dt}{dx}kh(1-\theta)\right]u_1^i + [\rho c + 6rk(1-\theta)]\,u_2^i$$
$$+\ell\varphi_1^i + \frac{\ell}{2}\varphi_2^i + 6\frac{dt}{dx}\left[k(1-\theta)w^i(0) + k\theta w^{i+1}(0)\right], \tag{4.43}$$

for $i = 1, 2, ..., M-1$, where $r = \frac{dt}{dx^2}$.

In the case $j = N$ (i.e. on $[x_{N-1}, x_N]$), we obtain from $(4.41)_2$ that

$$M^{i,N-1} + 2M^{i,N} = \frac{6}{dx^2}u_{N-1}^i - \left(\frac{6}{dx^2} + \frac{6}{dx}h\right)u_N^i + \frac{6}{dx}w^i(b), \tag{4.44}$$

$$(\rho c - 6rk\theta)\, u_{N-1}^{i+1} + \left(2\rho c + 6rk\theta + 6\frac{dt}{dx}kh\theta\right) u_N^{i+1} + \frac{\ell}{2}\varphi_{N-1}^{i+1} + \ell\varphi_N^{i+1}$$

$$= [\rho c + 6rk(1-\theta)]\, u_{N-1}^{i} + \left[2\rho c - 6rk(1-\theta) - 6\frac{dt}{dx}kh(1-\theta)\right] u_N^{i}$$

$$+\frac{\ell}{2}\varphi_{N-1}^{i} + \ell\varphi_N^{i} + 6\frac{dt}{dx}\left[k(1-\theta)w^i(b) + k\theta w^{i+1}(b)\right], \qquad (4.44')$$

for $i = 1, 2, ..., M-1$, where $r = \frac{dt}{dx^2}$.

Treating the boundary conditions $(4.8)_2$ in the same manner, we arrive at

$$\begin{cases} 2N^{i,1} + N^{i,2} = -\frac{6}{dx^2}\left(\varphi_1^i - \varphi_2^i\right), \\ N^{i,N-1} + 2N^{i,N} = \frac{6}{dx^2}\left(\varphi_{N-1}^i - \varphi_N^i\right), \end{cases} \qquad (4.45)$$

$$\left(2\tau + 6r\xi^2\theta\right)\varphi_1^{i+1} + \left(\tau - 6r\xi^2\theta\right)\varphi_2^{i+1} = \left[2\tau - 6r\xi^2(1-\theta) + \frac{dt}{2a}\right]\varphi_1^i$$

$$+\left[\tau + 6r\xi^2(1-\theta) + \frac{dt}{2a}\right]\varphi_2^i + 2dt(u_1^i + u_2^i), \qquad (4.46)$$

for $j = 1$, $i = 1, 2, ..., M-1$, and

$$\left(\tau - 6r\xi^2\theta\right)\varphi_{N-1}^{i+1} + \left(2\tau + 6r\xi^2\theta\right)\varphi_N^{i+1} = \left[\tau + 6r\xi^2(1-\theta) + \frac{dt}{2a}\right]\varphi_{N-1}^i$$

$$+\left[2\tau - 6r\xi^2(1-\theta) + \frac{dt}{2a}\right]\varphi_N^i + 2dt(u_{N-1}^i + u_N^i), \qquad (4.47)$$

for $j - N$, $i = 1, 2, ..., M \quad 1$.

Setting

$$e_1 = 6rk\theta \qquad e_2 = 6r\xi^2\theta \qquad e_3 = 6rk(1-\theta) \quad e_4 = 6r\xi^2(1-\theta),$$

than the discrete system (4.39)-(4.40), coupled with (4.43)-(4.46), can be written in matrix form as:

$$E\begin{pmatrix} u^{i+1} \\ \varphi^{i+1} \end{pmatrix} = F\begin{pmatrix} u^{i} \\ \varphi^{i} \end{pmatrix} + \begin{pmatrix} d_1^{i} \\ d_2^{i} \end{pmatrix} \qquad i = 1, 2, ..., M-1, \qquad (4.48)$$

where

$$E = \begin{pmatrix} E_{11} & E_{12} \\ 0 & E_{22} \end{pmatrix}_{2N\times 2N} \qquad F = \begin{pmatrix} F_{11} & E_{12} \\ F_{21} & F_{22} \end{pmatrix}_{2N\times 2N}$$

with E_{11}, E_{12}, E_{22}, F_{11}, F_{21}, F_{22} having the same size $N \times N$, and

$$E_{11}=\begin{pmatrix} ea & \rho c - e_1 & 0 & \cdots & 0 & 0 & 0 \\ \rho c - e_1 & eb & \rho c - e_1 & \cdots & 0 & 0 & 0 \\ \vdots & \vdots & \vdots & \ddots & \vdots & \vdots & \vdots \\ 0 & 0 & 0 & \cdots & \rho c - e_1 & eb & \rho c - e_1 \\ 0 & 0 & 0 & \cdots & 0 & \rho c - e_1 & ea \end{pmatrix}$$

$ea = 2\rho c + e_1 + 6\frac{dt}{dx} kh\theta > 0,$
$eb = 4\rho c + 2e_1 > 0,$

$$E_{22}=\begin{pmatrix} 2\tau + e_2 & \tau - e_2 & 0 & \cdots & 0 & 0 & 0 \\ \tau - e_2 & ec & \tau - e_2 & \cdots & 0 & 0 & 0 \\ \vdots & \vdots & \vdots & \ddots & \vdots & \vdots & \vdots \\ 0 & 0 & 0 & \cdots & \tau - e_2 & ec & \tau - e_2 \\ 0 & 0 & 0 & \cdots & 0 & \tau - e_2 & 2\tau + e_2 \end{pmatrix}$$

$ec = 4\tau + 2e_2 > 0,$

$$E_{12}=\begin{pmatrix} 2c_3 & c_3 & 0 & \cdots & 0 & 0 & 0 \\ c_3 & 4c_3 & c_3 & \cdots & 0 & 0 & 0 \\ \vdots & \vdots & \vdots & \ddots & \vdots & \vdots & \vdots \\ 0 & 0 & 0 & \cdots & c_3 & 4c_3 & c_3 \\ 0 & 0 & 0 & \cdots & 0 & c_3 & 2c_3 \end{pmatrix}$$

$c_3 = \frac{\ell}{2},$

$$F_{11}=\begin{pmatrix} fa & \rho c + e_3 & 0 & \cdots & 0 & 0 & 0 \\ \rho c + e_3 & fb & \rho c + e_3 & \cdots & 0 & 0 & 0 \\ \vdots & \vdots & \vdots & \ddots & \vdots & \vdots & \vdots \\ 0 & 0 & 0 & \cdots & \rho c + e_3 & fb & \rho c + e_3 \\ 0 & 0 & 0 & \cdots & 0 & \rho c + e_3 & fa \end{pmatrix}$$

$fa = 2\rho c - e_3 - 6\frac{dt}{dx}kh(1-\theta),$
$fb = 4\rho c - 2e_3,$

$$F_{22} = \begin{pmatrix} fc & fd & 0 & \cdots & 0 & 0 & 0 \\ fd & fe & fd & \cdots & 0 & 0 & 0 \\ \vdots & \vdots & \vdots & \ddots & \vdots & \vdots & \vdots \\ 0 & 0 & 0 & \cdots & fd & fe & fd \\ 0 & 0 & 0 & \cdots & 0 & fd & fc \end{pmatrix}$$

$fc = 2\tau - e_4 + \frac{dt}{2a},$
$fd = \tau + e_4 + \frac{dt}{2a},$
$fe = 4\tau - 2e_4 + \frac{dt}{a},$

$$F_{21} = \begin{pmatrix} dt & dt & 0 & \cdots & 0 & 0 & 0 \\ dt & dt & dt & \cdots & 0 & 0 & 0 \\ \vdots & \vdots & \vdots & \ddots & \vdots & \vdots & \vdots \\ 0 & 0 & 0 & \cdots & dt & dt & dt \\ 0 & 0 & 0 & \cdots & 0 & dt & dt \end{pmatrix}$$

$$d_1^i = \begin{pmatrix} 6\frac{dt}{dx}\left[k(1-\theta)w^i(0) + k\theta w^{i+1}(0)\right] \\ 0 \\ \vdots \\ 0 \\ 6\frac{dt}{dx}\left[k(1-\theta)w^i(b) + k\theta w^{i+1}(b)\right] \end{pmatrix}_{N\times 1} \qquad d_2^i = \begin{pmatrix} 0 \\ \vdots \\ 0 \end{pmatrix}_{N\times 1}.$$

The following sequence (again, i denotes the time level) can be used to calculate the approximate solution (u^{i+1}, φ^{i+1}):

Begin **algspline1D**
  Choose $\epsilon > 0$, $T > \varepsilon$, $b > 0$;
  Choose $M > 0$, $N > 0$ and compute dt, dx;
  Choose u_0, φ_0, w;
  Compute $z(\epsilon, \cdot)$ from (4.25);
  $i := 1 \rightarrow u^1, \varphi^1$ from initial conditions (4.24)-(4.25);
  For $j = \overline{1, N}$:
    Compute $M^{1,j}$, solving (4.36)+(4.42)$_1$+(4.44);
    Compute $N^{1,j}$, solving (4.40)+(4.45) $(i = 1)$;

```
    End-for;
    For i := 1 to M - 1 do
      Compute u^{i+1}, φ^{i+1} solving the linear system (4.48);
      For j = 1,N, compute (see (4.35)):
```

$$M^{i+1,j} = \frac{\theta-1}{\theta} M^{i,j} + \frac{\rho c}{k\theta dt}(u_j^{i+1} - u_j^i) + \frac{\ell}{2}\frac{1}{k\theta dt}(\varphi_j^{i+1} - \varphi_j^i);$$

$$N^{i+1,j} = \frac{\theta-1}{\theta} N^{i,j} + \frac{\tau}{\xi^2\theta dt}(\varphi_j^{i+1} - \varphi_j^i) - \frac{1}{2a\xi^2\theta}\varphi_j^i - \frac{2}{\xi^2\theta}u_j^i;$$

```
      End-for;
    End-for;
  End.
```

Stability conditions. To establish conditions of stability for the discrete equation (4.48) we will use again the Lax-Richtmyer definition of stability, expressed in terms of norm $\|\cdot\|_\infty$ (see G.D. Smith [84], pp. 48). Let us rewrite the linear system (4.48) in a more suitable form, namely

$$\begin{pmatrix} u^{i+1} \\ \\ \varphi^{i+1} \end{pmatrix} = E^{-1}F \begin{pmatrix} u^{i} \\ \\ \varphi^{i} \end{pmatrix} + E^{-1} \begin{pmatrix} d_1^{i} \\ \\ d_2^{i} \end{pmatrix} \qquad i = 1, 2, ..., M-1 \quad (4.49)$$

(the existence of E^{-1} will be proved in the proof of Proposition 4.2 below). Also, the matrix E can be written in the form

$$E = D_2(I + D_2^{-1}G_2) \tag{4.50}$$

where $D_2 = diag(ea, eb, \cdots, eb, ea, 2\tau + e_2, ec, \cdots, ec, 2\tau + e_2)$ and

$G_2 = A - D_2$. Thus we conclude that the matrix $D_2^{-1}G_2$ is expressed as

$$\begin{pmatrix}
0 & \frac{g_1}{ea} & 0 & .. & 0 & 0 & 0 & \frac{2c_3}{ea} & \frac{c_3}{ea} & 0 & .. & 0 & 0 & 0 \\
\frac{g_1}{eb} & 0 & \frac{g_1}{eb} & .. & 0 & 0 & 0 & \frac{c_3}{eb} & \frac{4c_3}{eb} & \frac{c_3}{eb} & .. & 0 & 0 & 0 \\
\vdots & \vdots & \vdots & \vdots & \vdots & \vdots & \vdots & \vdots & \vdots & \vdots & \ddots & \vdots & \vdots & \vdots \\
0 & 0 & 0 & .. & \frac{g_1}{eb} & 0 & \frac{g_1}{eb} & 0 & 0 & 0 & .. & \frac{c_3}{eb} & \frac{4c_3}{eb} & \frac{c_3}{eb} \\
0 & 0 & 0 & .. & 0 & \frac{g_1}{ea} & 0 & 0 & 0 & 0 & .. & 0 & \frac{c_3}{ea} & \frac{2c_3}{ea} \\
0 & 0 & 0 & .. & 0 & 0 & 0 & 0 & \frac{g_2}{g_3} & 0 & .. & 0 & 0 & 0 \\
0 & 0 & 0 & .. & 0 & 0 & 0 & \frac{g_2}{ec} & 0 & \frac{g_2}{ec} & .. & 0 & 0 & 0 \\
\vdots & \vdots & \vdots & \vdots & \vdots & \vdots & \vdots & \vdots & \vdots & \vdots & \ddots & \vdots & \vdots & \vdots \\
0 & 0 & 0 & .. & 0 & 0 & 0 & 0 & 0 & 0 & .. & \frac{g_2}{ec} & 0 & \frac{g_2}{ec} \\
0 & 0 & 0 & .. & 0 & 0 & 0 & 0 & 0 & 0 & .. & 0 & \frac{g_2}{g_3} & 0
\end{pmatrix}$$

with $g_1 = \rho c - e_1$, $g_2 = \tau - e_2$, $g_3 = 2\tau + e_2$.

The sum of each distinct line in array above is written in vector v^2 below

$$v^2 = \left[\frac{\rho c - e_1 + 3c_3}{ea},\ 2\frac{\rho c - e_1 + 3c_3}{eb},\ \frac{\tau - e_2}{2\tau + e_2},\ 2\frac{\tau - e_2}{ec} \right]. \tag{4.51}$$

Let us denote by

$$0 < v_{max}^2 = \max\{|\rho c - e_1 + 3c_3|, |\tau - e_2|\},$$

and

$$0 < v_{min}^2 = \min\{ea, eb, 2\tau + e_2, ec\}.$$

Now we are able to prove the following result with respect to the stability in matrix equation (4.49).

Proposition 4.2. *Suppose that* $v_{min}^2 - v_{max}^2 > 0$.

I. *If* $4\rho c + 3\frac{\ell}{2} \leq 6\frac{dt}{dx}kh(\theta - 1)$ *and one of the following conditions is true:*

I_1. $2\rho c + 3\frac{\ell}{2} - 6\frac{dt}{dx}kh(1-\theta) > 6\tau + 3dt + 2\frac{dt}{a}$ &

$$\frac{2\rho c + 3\frac{\ell}{2} - 6\frac{dt}{dx}kh(1-\theta)}{v_{min}^2 - v_{max}^2} < 1,$$

or

I_2. $2\rho c + 3\frac{\ell}{2} - 6\frac{dt}{dx}kh(1-\theta) \leq 6\tau + 3dt + 2\frac{dt}{a}$ &

$$\frac{6\tau + 3dt + 2\frac{dt}{a}}{v_{min}^2 - v_{max}^2} < 1,$$

then the equation (4.49) *is stable.*

II. *If* $4\rho c + 3\frac{\ell}{2} > 6\frac{dt}{dx}kh(\theta - 1)$ *and one of the following conditions is true:*

II_1. $6\rho c + 3\ell > 6\tau + 3dt + 2\frac{dt}{a}$ & $\dfrac{6\rho c + 3\ell}{v_{min}^2 - v_{max}^2} < 1,$

or

II_2. $6\rho c + 3\ell \leq 6\tau + 3dt + 2\frac{dt}{a}$ & $\dfrac{6\tau + 3dt + 2\frac{dt}{a}}{v_{min}^2 - v_{max}^2} < 1,$

then the equation (4.49) *is stable, otherwise unstable.*

Proof. As usual, the proof is reduced to checking the condition of stability which, based on the Lax-Richtmyer definition mentioned above and taking into account the relation (4.49), it reduces to check the inequality

$$\|E^{-1}F\|_\infty < 1.$$

First of all let us determine an estimate for $\|D_2^{-1}G_2\|_\infty$. Based on relation (4.51), this is equivalent with: $\|D_2^{-1}G_2\|_\infty = \max|v^2|$, wherefrom we can easily derive the estimate

$$\|D_2^{-1}G_2\|_\infty < \frac{v_{max}^2}{v_{min}^2}. \tag{4.52}$$

The estimate (4.52) allows us to prove the existence of E^{-1}. Indeed, since by hypothesis we have assumed that $v_{max}^2 < v_{min}^2$, then making use of (4.52) we get $\|D_2^{-1}G_2\|_\infty < 1$ which guarantees that there exists $(I + D_2^{-1}G_2)^{-1}$. Moreover, there also exist E^{-1} and $E^{-1} = (I + D_2^{-1}G_2)^{-1}D_2^{-1}$. Using the well known inequality: $\|(I + D_2^{-1}G_2)^{-1}\|_\infty \leq \dfrac{1}{1 - \|D_2^{-1}G_2\|_\infty}$ and making use of (4.50), it follows that

$$\|E^{-1}\|_\infty \leq \|(I + D_2^{-1}G_2)^{-1}\|_\infty \|D_2^{-1}\|_\infty \tag{4.53}$$

$$\leq \frac{1}{1 - \|D_2^{-1} G_2\|_\infty} \|D_2^{-1}\|_\infty.$$

How $\|D_2^{-1} G_2\|_\infty \leq 1$ implies that $1 - \|D_2^{-1} G_2\|_\infty \geq 1 - \dfrac{v_{max}^2}{v_{min}^2} > 0$, we easily deduce from this that

$$0 < \frac{1}{1 - \|D_2^{-1} G_2\|_\infty} \leq \frac{v_{min}^2}{v_{min}^2 - v_{max}^2}.$$

Since $\|D_2^{-1}\|_\infty \leq \frac{1}{v_{min}^2}$ and involving the above estimate, from (4.53) we obtain

$$\|E^{-1}\|_\infty < \frac{1}{v_{min}^2 - v_{max}^2}$$

which allows us to deduce

$$\|E^{-1}F\|_\infty \leq \|E^{-1}\|_\infty \|F\|_\infty < \frac{1}{v_{min}^2 - v_{max}^2} \|F\|_\infty, \tag{4.54}$$

where

$$\begin{aligned}
\|F\|_\infty &= \max\{fa + \rho c + e_3 + 3c_3, fb + 2\rho c + 2e_3 + 6c + 3, \\
&\qquad 2dt + fd + fc, 3dt + 2fd + fe\} \\
&= \max\left\{2\rho c + 3\frac{\ell}{2} - 6\frac{dt}{dx}kh(1-\theta),\ 6\rho c + 3\ell,\ 6\tau + 3dt + 2\frac{dt}{a}\right\} \qquad (4.55) \\
&= \begin{cases}
2\rho c + 3\frac{\ell}{2} & \text{if } 4\rho c + 3\frac{\ell}{2} \leq 6\frac{dt}{dx}kh(\theta - 1)\ \& \\
\quad -6\frac{dt}{dx}kh(1-\theta) & \quad 2\rho c + 3\frac{\ell}{2} - 6\frac{dt}{dx}kh(1-\theta) > 6\tau + 3dt + 2\frac{dt}{a}, \\
6\tau + 3dt + 2\frac{dt}{a} & \text{if } 4\rho c + 3\frac{\ell}{2} \leq 6\frac{dt}{dx}kh(\theta - 1)\ \& \\
 & \quad 2\rho c + 3\frac{\ell}{2} - 6\frac{dt}{dx}kh(1-\theta) \leq 6\tau + 3dt + 2\frac{dt}{a}, \\
6\rho c + 3\ell & \text{if } 4\rho c + 3\frac{\ell}{2} > 6\frac{dt}{dx}kh(\theta - 1)\ \& \\
 & \quad 6\rho c + 3\ell > 6\tau + 3dt + 2\frac{dt}{a}, \\
6\tau + 3dt + 2\frac{dt}{a} & \text{if } 4\rho c + 3\frac{\ell}{2} > 6\frac{dt}{dx}kh(\theta - 1)\ \& \\
 & \quad 6\rho c + 3\ell \leq 6\tau + 3dt + 2\frac{dt}{a}.
\end{cases}
\end{aligned}$$

The last equality, which gives us $\|F\|_\infty$, was established after comparing the first two elements in the set of (4.55). Substituting now in (4.54) the corresponding value of $\|F\|_\infty$, we find that in either cases I or II we get the estimate $\|E^{-1}F\|_\infty < 1$ as we claimed at beginning of proof.

4.2 Approximation of phase-field transition system in 2D by finite element method. Algorithm `algfracfem2D`

In this Section we are concerned with the numerical approximation of the weak solution corresponding with (4.7)-(4.10) (see Definition 2.6 in Section 2.6) by finite element method (**fem**) i.e. with the numerical approximation of the weak solution of the following equations (see ANNEXE B - **P1.**):

$$\left(\rho c\, u_t^\varepsilon + \frac{\ell}{2}\varphi_t^\varepsilon, \psi\right) + k(\nabla u^\varepsilon, \nabla\psi) + kh \int_{\partial\Omega} u^\varepsilon \psi \, dxdy \tag{4.56}$$

$$= k \int_{\partial\Omega} w(\cdot, x, y)\psi \, dxdy \quad \forall\, \psi \in H^1(\Omega), \;\; a.e.\ in\ [\varepsilon, T],$$

$$\tau(\varphi_t^\varepsilon, \zeta) + \xi^2(\nabla\varphi^\varepsilon, \nabla\zeta) = \frac{1}{2a}(\varphi^\varepsilon, \zeta) + 2(u^\varepsilon, \zeta), \tag{4.57}$$

$$\forall\, \zeta \in H^1(\Omega), \;\; a.e.\ in\ [\varepsilon, T],$$

together with the initial conditions

$$u(0, x) = u_0(x), \quad \varphi(0, x) = \varphi_0(x), \quad x \in \Omega. \tag{4.58}$$

In (4.56)-(4.57), $(\cdot, \cdot)$ denotes the scalar product in $L^2(\Omega)$.

Considering M as the number of equidistant nodes in which is divided the time-interval $[\varepsilon, T]$, we set

$$t_i = \varepsilon + (i-1)dt \qquad i = 1, 2, \ldots, M, \;\; dt = (T - \varepsilon)/(M-1).$$

We assume that $\Omega \subset I\!R^2$ is a polygonal domain. Let T_ρ be the triangulation (mesh) over Ω and $\bar{\Omega} = \cup\{K,\ K \in T_\rho\}$ and let $N_j = (x_k, y_l)$, $j = \overline{1, nn}$, be the nodes associated to T_ρ. If we denote by V_{nn}

the corresponding finite element space to T_ρ then, the basis functions $\{b_j\}_{j=1}^{nn}$ of V_{nn} are defined by

$$b_j(N_i) = \delta_{ji}, \quad i, j = \overline{1, nn},$$

and

$$V_{nn} = \text{SPAN } \{b_1, b_2, ..., b_{nn}\}.$$

We say that function z belongs to V_{nn} only if it can be expressed as

$$z(x, y) = \sum_{l=1}^{nn} c_l b_l(x, y), \quad (x, y) \in \bar{\Omega}.$$

For $i = \overline{1, M}$, we denote by u^i and φ^i the V_{nn} interpolant of u^ε and φ^ε, respectively. Then u^i, $\varphi^i \in V_{nn}$ and

$$u^i(x, y) = \sum_{l=1}^{nn} u_l^i b_l(x, y), \quad i = \overline{1, M}, \tag{4.59}$$

$$\varphi^i(x, y) = \sum_{l=1}^{nn} \varphi_l^i b_l(x, y), \quad i = \overline{1, M}, \tag{4.60}$$

where the unknowns $u_l^i = u^\varepsilon(t_i, N_l)$, $\varphi_l^i = \varphi^\varepsilon(t_i, N_l)$, $i = \overline{1, M}$, $l = \overline{1, nn}$, represents the discrete solution of (4.65)-(4.66) below.

Let now U, $\Phi \in V_{nn}$ be two arbitrary functions, i.e.

$$U(x, y) = \sum_{l=1}^{nn} U_l b_l(x, y), \tag{4.61}$$

$$\Phi(x, y) = \sum_{l=1}^{nn} \Phi_l b_l(x, y). \tag{4.62}$$

Using an implicit (backward) finite difference scheme in time and taking into account the above notations, we introduce the discrete equations corresponding to (4.56)-(4.57) as follows ($i = \overline{1, M}$)

$$\rho c(u^i, U) + \frac{\ell}{2}(\varphi^i, U) + dt\ k(\nabla u^i, \nabla U) + dt\ kh \int_{\partial\Omega} u^i U\ dxdy \tag{4.63}$$

$$= (\rho c\ u^{i-1} + \frac{\ell}{2}\varphi^{i-1}, U) + dt\ k \int_{\partial\Omega} w^i U\ dxdy,$$

$$\tau(\varphi^i, \Phi) + dt\xi^2(\nabla\varphi^i, \nabla\Phi) - \frac{dt}{2a}(\varphi^i, \Phi) - 2dt(u^i, \Phi) = \tau(\varphi^{i-1}, \Phi), \quad (4.64)$$

$\forall U,\ \Phi \in V_{nn}$, with $w^i = w(t_i, \cdot)$.

Replaying (4.59)-(4.62) in (4.63)-(4.64), we get

$$\rho c \sum_{k=1}^{nn} U_k \Big(\sum_{l=1}^{nn} u_l^i \int_\Omega b_k b_l\ dxdy\Big) + \frac{\ell}{2} \sum_{k=1}^{nn} U_k \Big(\sum_{l=1}^{nn} \varphi_l^i \int_\Omega b_k b_l\ dxdy\Big) \quad (4.65)$$

$$+dt\ k \sum_{k=1}^{nn} U_k \Big(\sum_{l=1}^{nn} u_l^i \int_\Omega \nabla b_k \nabla b_l\ dxdy\Big)$$

$$+dt\ kh \sum_{k=1}^{nn} U_k \Big(\sum_{l=1}^{nn} u_l^i \int_{\partial\Omega} b_k b_l\ dxdy\Big)$$

$$= \sum_{k=1}^{nn} U_k \Big(\sum_{l=1}^{nn} (\rho c\ u_l^{i-1} + \frac{\ell}{2}\varphi_l^{i-1}) \int_\Omega b_k b_l\ dxdy\Big)$$

$$+dt\ k \sum_{k=1}^{nn} U_k \Big(\sum_{l=1}^{nn} w_l^i \int_{\partial\Omega} b_k b_l\ dxdy\Big),$$

$$\tau \sum_{k=1}^{nn} \Phi_k \Big(\sum_{l=1}^{nn} \varphi_l^i \int_\Omega b_k b_l\ dxdy\Big) \quad (4.66)$$

$$+dt\ \xi^2 \sum_{k=1}^{nn} \Phi_k \Big(\sum_{l=1}^{nn} \varphi_l^i \int_\Omega \nabla b_k \nabla b_l\ dxdy\Big)$$

$$-\frac{dt}{2a} \sum_{k=1}^{nn} \Phi_k \Big(\sum_{l=1}^{nn} \varphi_l^i \int_\Omega b_k b_l\ dxdy\Big) - 2dt \sum_{k=1}^{nn} \Phi_k \Big(\sum_{l=1}^{nn} u_l^i \int_\Omega b_k b_l\ dxdy\Big)$$

$$= \tau \sum_{k=1}^{nn} \Phi_k \Big(\sum_{l=1}^{nn} \varphi_l^{i-1} \int_{\Omega} b_k b_l \; dxdy \Big),$$

for $i = \overline{1, M}$, where $w_l^i = w(t_i, N_l)$, $l = \overline{1, nn}$.

Setting

$$b_{kl} = \int_{\Omega} b_k(x,y) b_l(x,y) dxdy,$$

$$g_{kl} = \int_{\Omega} \nabla b_k(x,y) \nabla b_l(x,y) dxdy,$$

$$B = (b_{kl})_{k,l=\overline{1,nn}},$$

$$R = (r_{kl})_{k,l=\overline{1,nn}}, \qquad r_{kl} = \rho c b_{kl} + dt \; k g_{kl},$$

$$S = (s_{kl})_{k,l=\overline{1,nn}}, \qquad s_{kl} = \tau b_{kl} + dt \xi^2 g_{kl} - \tfrac{dt}{2a} b_{kl},$$

$$FR = (fr_{kl})_{k,l=\overline{1,nn}}, \qquad fr_{kl} = \int_{\partial\Omega} b_k(x,y) b_l(x,y) dxdy,$$

the system (4.65)-(4.66) can be rewritten in the matrix form as

$$\begin{cases} R \; u_l^i + \frac{\ell}{2} B \; \varphi_l^i + dt \; kh FR \; u_l^i \\ \qquad\qquad = B \left(\rho c \; u_l^{i-1} + \frac{\ell}{2} \varphi_l^{i-1} + dt \; k w^i \right), \\ S \; \varphi_l^i - 2dt \; B \; u_l^i = \tau B \; \varphi_l^{i-1}, \end{cases} \tag{4.67}$$

where u_l^i, φ_l^i, and $l = \overline{1, nn}$ are unknown vectors corresponding to time level i.

From the initial conditions (4.58), we have

$$u^0(x,y) \overset{\text{not}}{=} u_0(x,y) = \sum_{l=1}^{nn} u_0(N_l) b_l(x,y),$$

$$\varphi^0(x,y) \overset{\text{not}}{=} \varphi_0(x,y) = \sum_{l=1}^{nn} \varphi_0(N_l) b_l(x,y)$$

and then (see (4.59)-(4.60))

$$\begin{cases} u_l^0 = u_0(N_l), \\ \varphi_l^0 = \varphi_0(N_l), \end{cases} \qquad l = \overline{1, nn}. \tag{4.68}$$

The numerical algorithm to calculate the approximate solution by fractional steps method can be obtained by the following sequence (i denotes the time level)

Begin **algfracfem2D**
  Choose $\epsilon > 0$, $T > \varepsilon$, Ω;
  Choose $M > 0$, $nn > 0$ and compute dt, dx;
  Choose u_0, φ_0, w;
  $i := 0 \ \rightarrow u_l^0, \ \varphi_l^0, \quad l = \overline{1, nn}$ from (4.68)
  Compute $z_l := z(\varepsilon, N_l), \quad l = \overline{1, nn}$ from (4.25);
  $\varphi^{i-1} := z_l, \quad l = \overline{1, nn}$;
  For $i := 1$ to M do
    Compute $u_l^i, \ \varphi_l^i, \quad l = \overline{1, nn}$, solving
    the linear system (4.68);
  End-for;
End.

4.3 Approximation of the solid region in 1D. Algorithm InvPHT1D

In Section 3.5 we have considered the following optimal control problem

$$(\mathrm{P}^{\varepsilon}_{inv}) \ \mathit{Minimize} \ L_0^{\varepsilon}(w) = \frac{\beta}{2} \int\limits_Q \left[(u^{\varepsilon}(t,x) - \delta_2)^+ \right]^2 \cdot \chi_{Q_0} \, dt dx$$

$$+ \frac{1}{2} \int\limits_Q (\varphi^{\varepsilon}(t,x) - 1 - \delta_1)^2 \cdot \chi_{Q_0} \, dt dx + \frac{1}{2} \int\limits_0^T w^2(t) \, dt,$$

for all $(u^{\varepsilon}, \varphi^{\varepsilon})$ solutions of the system (4.69)-(4.72) below ($\beta > 0$ is a given constant)

$$\begin{cases} \rho c \ u_t^{\varepsilon} + \frac{\ell}{2} \varphi_t^{\varepsilon} = k \ u_{xx}^{\varepsilon} \\ \tau \ \varphi_t^{\varepsilon} = \xi^2 \varphi_{xx}^{\varepsilon} + \frac{1}{2a} \varphi^{\varepsilon} + 2 u^{\varepsilon} \end{cases} \quad \text{in } Q_0^{\varepsilon} = \{(t,x) \in Q, \ \varepsilon \le t \le T\}, \quad (4.69)$$

$$u_x^{\varepsilon} + h u^{\varepsilon} = w(t), \quad \varphi_x^{\varepsilon} = 0 \quad \text{on } \ \Sigma^{\varepsilon} = [\varepsilon, T] \times \{b_1\}, \qquad (4.70)$$

$$-u_x^{\varepsilon} = 0, \quad -\varphi_x^{\varepsilon} = 0 \quad \text{on } \ \Sigma_0^{\varepsilon} = \{(t,x) \in Q, \ \varepsilon \le t \le T\}, \qquad (4.71)$$

$$u^\varepsilon(\varepsilon, x) = u_0(x), \quad \varphi_+^\varepsilon(\varepsilon, x) = z(\varepsilon, \varphi_-^\varepsilon(\varepsilon, x)) \quad \text{on } \Omega_\varepsilon, \tag{4.72}$$

where $z(\varepsilon, \varphi_-^\varepsilon(\varepsilon, x))$ is the solution of the Cauchy problem:

$$\begin{cases} z'(s) + \dfrac{1}{2a} z^3(s) = 0 & s \in (0, \varepsilon), \\ z(0) = \varphi_-^\varepsilon(\varepsilon, x), \quad \varphi_-^\varepsilon(0, x) = \varphi_0(x), \end{cases} \tag{4.73}$$

$\varphi_+^\varepsilon(\varepsilon, x) = \lim_{t\downarrow\varepsilon} \varphi^\varepsilon(t, x)$, $\varphi_-^\varepsilon(\varepsilon, x) = \lim_{t\uparrow\varepsilon} \varphi^\varepsilon(t, x)$, and for all $w \in \mathcal{U}$, with

$$\mathcal{U} = \{w \in L^\infty([0, T]), \ -R \leq w(t) \leq 0 \ \ a.e. \ \ t \in [0, T]\}. \tag{4.74}$$

On the basis of Theorem 3.4 in Section 3.5 we know that the optimal control in $(\mathrm{P}_{inv}^\varepsilon)$ is given by

$$w(t) = \begin{cases} 0 & \text{if } r(t) > 0 \\ -R & \text{if } r(t) \leq 0, \end{cases} \tag{4.75}$$

where

$$r(t) = \frac{k}{\rho c} p^\varepsilon(t, b_1) + w(t), \quad t \in [0, T],$$

and $(p^\varepsilon, q^\varepsilon)$ satisfy, along with $(u^\varepsilon, \varphi^\varepsilon)$ the *adjoint state system*

$$\begin{cases} p_t^\varepsilon + \dfrac{k}{\rho c} p_{xx}^\varepsilon - \dfrac{\ell}{\tau \rho c} p^\varepsilon + \dfrac{2}{\tau} q^\varepsilon \\ \quad = \beta(u^\varepsilon - \delta_2)^+ \cdot \chi_{Q_0} & \text{in } Q_0^\varepsilon, \\ p_x^\varepsilon + h p^\varepsilon = 0 & \text{on } \Sigma^\varepsilon, \\ p_x^\varepsilon = 0 & \text{on } \Sigma_0^\varepsilon, \\ p_-^\varepsilon(\varepsilon, x) = 0, \quad p_-^\varepsilon(T, x) = 0 & x \in \Omega_T, \end{cases} \tag{4.76}$$

$$\begin{cases} q_t^\varepsilon - \frac{\ell\xi^2}{2\tau\rho c} p_{xx}^\varepsilon - \frac{\ell}{4a\tau\rho c} p^\varepsilon + \frac{\xi^2}{\tau} q_{xx}^\varepsilon + \frac{1}{2a\tau} q^\varepsilon \\ \quad = (\varphi^\varepsilon - 1 - \delta_1) \cdot \chi_{Q_0} & \text{in } Q_0^\varepsilon, \\ q_x^\varepsilon = \dfrac{\ell}{2\rho c} p_x^\varepsilon & \text{on } \Sigma^\varepsilon, \\ q_x^\varepsilon = 0 & \text{on } \Sigma_0^\varepsilon, \\ q_-^\varepsilon(\varepsilon, x) = \exp\Big(\int\limits_0^\varepsilon \frac{3}{2a} z^2(t, \cdot) dt \Big) q_+^\varepsilon(\varepsilon, x), \\ q_-^\varepsilon(T, x) = 0 & \Omega_T. \end{cases} \tag{4.77}$$

In the sequel we will construct the discrete form of the problem $(\mathrm{P}_{inv}^\varepsilon)$. Before we start, let us remark that in the present case, the space-interval on which we work is changeable, denoted by Ω_i, $i = \overline{2, M}$ in Fig. (**1**) ($\Omega_1 = \Omega_\varepsilon$, $\Omega_M = \Omega_T$). The constant case was treated in Section 4.1b.

We begin by establishing the discrete form of the state system (4.69)-(4.73). Given the values ε, T ($\varepsilon << T$) and considering M as the number of equidistant nodes in which $[\varepsilon, T]$ is divided, we set

$$t_i = \varepsilon + (i-1) \cdot dt \quad i = 1, 2, \dots, M, \quad dt = (T - \varepsilon)/(M-1).$$

In order to define Σ_0, let $b_0 = \sqrt{T} = \sqrt{t_M}$ and choose $jb_0 = M+1$. We consider now $\Omega_0 = [b_0, b_1] \subset I\!R_+$ ($b_1 >> T$) and we introduce over it the grid with N equidistant nodes

$$x_j = b_0 + (j-1) \cdot dx \quad j = 1, 2, \dots, N, \quad dx = (b_1 - b_0)/(N-1).$$

On the subinterval $[0, b_0)$ we take the set of nodes obtained as follows

$$\bar{x}_{M-i+1} = b_0 - \sqrt{t_i} \qquad i = 1, 2, \dots, M,$$

that is, in other words (see Fig. (**1**) in Chapter 3)

$$t_i = f(\bar{x}_{M-i+1}) = \sqrt{b_0 - \bar{x}_{M-i+1}} \qquad i = 1, 2, \dots, M.$$

Thus, on the entire domain $\Omega = [0, b_1]$ we will consider in our numerical experiments the following set of points:

$$[\bar{x}_1, \bar{x}_2, \dots, \bar{x}_M, \bar{x}_{M+1} = x_{jb_0} = x_1, x_2, \dots, x_N],$$

i.e.

$$[0 = b_0 - \sqrt{t_M}, b_0 - \sqrt{t_{M-1}}, \ldots, b_0 - \sqrt{\varepsilon}, b_0, b_0 + dx, \ldots, b_1].$$

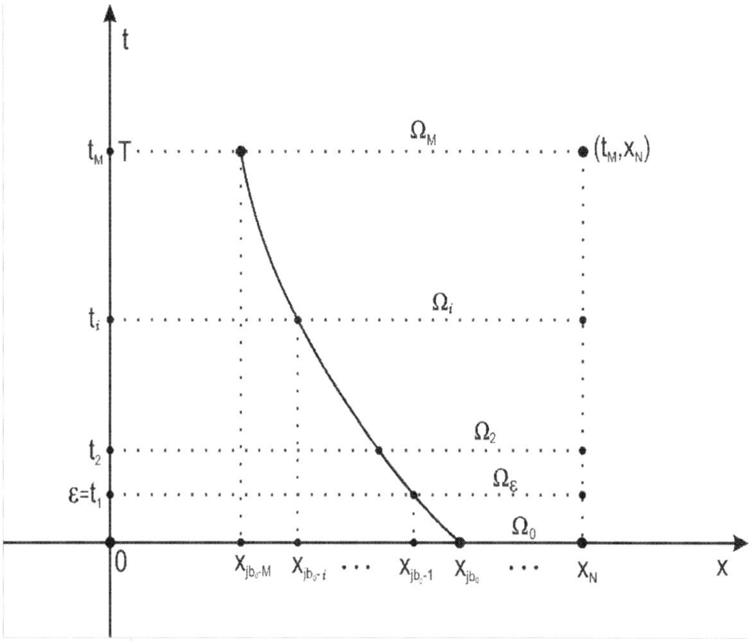

Figure 1 *The discrete form of Q_0, Σ_0, Σ in $(\mathbf{P}^{\varepsilon}_{inv})$*

For each time-level i, the mesh's of Ω_{t_i}, $i = \overline{2, M}$, consists in the points illustrated in Fig. (**2**) ($\Omega_{t_1} = \Omega_\varepsilon$):

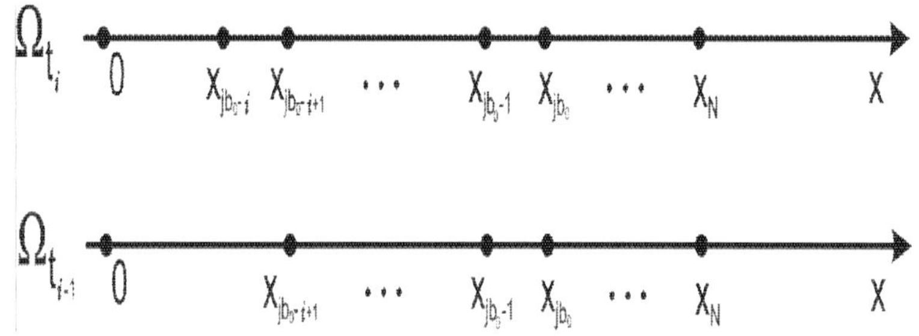

Figure 2 *Two adjacent space intervals*

and then the unknowns vectors, denoted in the sequel by $\bar{u}^i$, $\bar{\varphi}^i$, are

$$u^i = (u^i_{jb_0-i}, u^i_{jb_0-i+1}, \ldots, u^i_{jb_0}, u^i_{jb_0+1}, \ldots, u^i_N)^T,$$

$$\varphi^i = (\varphi^i_{jb_0-i}, \varphi^i_{jb_0-i+1}, \ldots, \varphi^i_{jb_0}, \varphi^i_{jb_0+1}, \ldots, \varphi^i_N)^T.$$

As usually, we denote by u^i_j, φ^i_j the approximate values in the point (t_i, x_j) of the state functions u^ε, φ^ε in (4.69)-(4.72):

$$\begin{matrix} u^i_j = u^\varepsilon(t_i, x_j) \\ \varphi^i_j = \varphi^\varepsilon(t_i, x_j) \end{matrix} \qquad i = 1, 2, ..., M, \quad j = jb_0 - i, ..., N.$$

Using a *First-order Implicit Backward Difference Formula* (**1-IMBDF**), the system (4.69) is discretized as

$$\begin{cases} -k\frac{dt}{dx^2}u^i_{j-1} + \left[\rho c + 2k\frac{dt}{dx^2}\right]u^i_j - k\frac{dt}{dx^2}u^i_{j+1} \\ \qquad\qquad + \frac{\ell}{2}\varphi^i_j = \rho c u^{i-1}_j + \frac{\ell}{2}\varphi^{i-1}_j, \\ -\xi^2\frac{dt}{dx^2}\varphi^i_{j-1} + \left[\tau + 2\xi^2\frac{dt}{dx^2} - \frac{dt}{2a}\right]\varphi^i_j - \xi^2\frac{dt}{dx^2}\varphi^i_{j+1} \\ \qquad\qquad -2dt\ u^i_j = \tau\varphi^{i-1}_j, \end{cases} \tag{4.78}$$

for $i = 1, 2, \ldots, M$, $j = jb_0 - i, \ldots, N$.

As we simply can constat, corresponding to Ω_{t_i}, the boundary is reduced to the set $\{x_{jb_0-i}, x_N\}$, $i = 1, 2, ..., M$. So, to approximate $u_x(x_{jb_0-i})$ $(u_x(b_1))$ we will use in (4.70)-(4.71) a forward (backward) finite difference formula, while, to approximate $\varphi_x(x_{jb_0-i})$ $(\varphi_x(b_1))$ we will use a backward (forward) finite difference formula; this leads us to:

$$\begin{cases} u^i_{jb_0-i} = u^i_{jb_0-i+1}, \\ -u^i_{N-1} + (1 + dx\ h)u^i_N = dx\ w^i, \end{cases} \tag{4.79}$$

$$\begin{cases} \varphi^i_{jb_0-i-1} = \varphi^i_{jb_0-i}, \\ \varphi^i_{N+1} = \varphi^i_N, \end{cases} \tag{4.80}$$

for $i = 1, 2, ..., M$, where $w^i = w(t_i)$, and $\varphi^i_{jb_0-i-1}$, φ^i_{N+1} are dummy variables.

From the initial conditions (4.72) we have

$$u^0_j = u_0(x_j) \qquad j = jb_0, ..., N, \tag{4.81}$$

and, involving the separation of variables method to solve the Cauchy problem (4.73), we derive

$$\varphi_j^0 = z(\varepsilon, \varphi_-^\varepsilon(0, x_j)) = \varphi_-^\varepsilon(0, x_j)\sqrt{\frac{a\tau}{a\tau + \varepsilon \cdot \varphi_-^\varepsilon(0, x_j)}} \tag{4.82}$$

for $j = jb_0, ..., N$.

Keeping the same settings as in previous Sections: $u_{N+1}^i = u_N^i$ (for $j = N$), $c_1 = -k\frac{dt}{dx^2}$, $c_2 = \rho c - 2c_1$, $c_3 = \frac{\ell}{2}$, $c_5 = -\xi^2\frac{dt}{dx^2}$, $c_6 = \tau - 2c_5 - \frac{dt}{2a}$, the discrete system (4.78), coupled with (4.79)-(4.80), can be written in matrix form as:

$$\begin{pmatrix} AI_{11} & A_{12} \\ -2A_{21} & A_{22} \end{pmatrix}\begin{pmatrix} u^i \\ \varphi^i \end{pmatrix} = \begin{pmatrix} B_{11} & A_{12} \\ 0 & B_{22} \end{pmatrix}\begin{pmatrix} u^{i-1} \\ \varphi^{i-1} \end{pmatrix} + \begin{pmatrix} d_1^i \\ d_2^i \end{pmatrix}, \tag{4.83}$$

with AI_{11}, A_{12}, A_{21}, A_{22}, B_{11}, B_{22} having the same dimension

$$Dim_i == N + i \quad i = 1, 2, ..., M, \tag{4.84}$$

A_{12}, A_{21}, A_{22}, B_{11}, B_{22}, d_2^i defined in (4.20), and

$$AI_{11} = \begin{pmatrix} c_1 + c_2 & c_1 & 0 & \cdots & 0 & 0 & 0 \\ c_1 & c_2 & c_1 & \cdots & 0 & 0 & 0 \\ \vdots & \vdots & \vdots & \ddots & \vdots & \vdots & \vdots \\ 0 & 0 & 0 & \cdots & c_1 & c_2 & c_1 \\ 0 & 0 & 0 & \cdots & 0 & c_1 - 1 & a_1 \end{pmatrix}_{Dim_i \times Dim_i}$$

$a_1 = c_1 + c_2 + 1 + dx \cdot h$,

$$d_1^i = \begin{pmatrix} 0 \\ 0 \\ \vdots \\ 0 \\ dx \cdot w^i \end{pmatrix}_{Dim_i \times 1}$$

for $i = 1, 2, ..., M$.

We continue now by treating the adjoint states p^ε, q^ε in $(4.76)_1$ and $(4.77)_1$, respectively. Denote by p_j^i, q_j^i the approximate matrix for p^ε and q^ε, respectively, where

$$\begin{matrix} p_j^i = p^\varepsilon(t_i, x_j) \\ q_j^i = q^\varepsilon(t_i, x_j) \end{matrix} \qquad i = 1, 2, ..., M, \qquad j = jb_0 - i, ..., N + i\,.$$

Applying the same **1-IMBDF** scheme to $(4.76)_1$-$(4.77)_1$, we get

$$\begin{cases} \frac{k}{\rho c}\frac{dt}{dx^2}p_{j-1}^i - \left[\frac{\ell}{\rho c\tau}dt + 2\frac{k}{\rho c}\frac{dt}{dx^2} + 1\right]p_j^i + \frac{k}{\rho c}\frac{dt}{dx^2}p_{j+1}^i \\ \quad +\frac{2}{\tau}dt\ q_j^i = -p_j^{i+1} + \beta dt(u_j^{i+1} - \delta_2)^+ \cdot \chi_{Q_0}, \\ -\frac{\ell\xi^2}{2\rho c\tau}\frac{dt}{dx^2}p_{j-1}^i + \left[2\frac{\ell\xi^2}{2\rho c\tau}\frac{dt}{dx^2} - \frac{\ell}{4\rho c\tau a}dt\right]p_j^i - \frac{\ell\xi^2}{2\rho c\tau}\frac{dt}{dx^2}p_{j+1}^i \\ \quad +\frac{\xi}{\tau}\frac{dt}{dx^2} \cdot q_{j-1}^i - \left[2\frac{\xi^2}{\tau}\frac{dt}{dx^2} - \frac{1}{2\tau a}dt + 1\right]q_j^i + \frac{\xi^2}{\tau}\frac{dt}{dx^2}q_{j+1}^i \\ \quad = -q_j^{i+1} + (\varphi_j^{i+1} - 1 - \delta_1) \cdot \chi_{Q_0}, \end{cases} \tag{4.85}$$

for $i = M-1, ..., 1, 0$, $j = jb_0 - i, ..., N + i$.

From $(4.76)_2$-$(4.77)_2$, using a backward finite difference formula, we obtain

$$\begin{cases} p_N^i - p_{N-1}^i + dx \cdot h \cdot p_N^i = 0 \\ q_N^i - q_{N-1}^i = \frac{\ell}{2\rho c}(p_N^i - p_{N-1}^i), \end{cases} \tag{4.86}$$

for $i = M-1, ..., 1, 0$, while from $(4.76)_3$-$(4.77)_3$, making use of a forward finite difference formula, we get

$$p_{jb_0-i+1}^i = p_{jb_0-i}^i, \qquad q_{jb_0-i+1}^i = q_{jb_0-i}^i \qquad i = M-1, \ldots, 1, 0. \tag{4.87}$$

From the final relations in (4.76)-(4.77) we derive

$$p_j^M = 0 \qquad q_j^M = 0, \qquad j = 1, ..., M+N, \tag{4.88}$$

and

$$q_{-,j}^i = \exp\Big(\int_0^\varepsilon \frac{3}{2a}z^2(t,\cdot)dt\Big)q_{+,j}^i \qquad q_{+,j}^M = q_j^M, \tag{4.89}$$

for $i = M-1, M-2, \ldots, 1, 0$.

Replaying (4.86)-(4.87) in (4.85) and setting

$$c_{11} = \frac{k}{\rho c}\frac{dt}{dx^2}, \qquad c_{12} = 2\frac{dt}{\tau},$$

$$c_{10} = \ell\frac{dt}{\rho c\tau} + 2c_{11} + 1,$$

$$c_{13} = \frac{\xi^2}{\tau}\frac{dt}{dx^2}, \qquad c_{14} = -\ell\frac{c_{13}}{2\rho c},$$

$$c_{15} = -2c_{14} - \ell\frac{dt}{4\rho c\tau a} \qquad c_{16} = -2c_{13} + \frac{dt}{2\tau a} + 1,$$

then (4.85) can be rewritten in matrix form as

$$\begin{pmatrix} P_{11} & Q_{12} \\ P_{21} & Q_{22} \end{pmatrix} \begin{pmatrix} p^i \\ q^i \end{pmatrix} = \begin{pmatrix} d_3^{i+1} \\ d_4^{i+1} \end{pmatrix} \tag{4.90}$$

where P_{11}, P_{21}, Q_{12}, and Q_{22} having the same dimension Dim_i stated in (4.84), are given by

$$P_{11} = \begin{pmatrix} c_{10}+c_{11} & c_{11} & 0 & \cdots & 0 & 0 & 0 \\ c_{11} & c_{10} & c_{11} & \cdots & 0 & 0 & 0 \\ \vdots & \vdots & \vdots & \ddots & \vdots & \vdots & \vdots \\ 0 & 0 & 0 & \cdots & c_{11} & c_{10} & c_{11} \\ 0 & 0 & 0 & \cdots & 0 & c_{11}-1 & p_1 \end{pmatrix}$$

$p_1 = c_{10} + c_{11} + dx \cdot h + 1,$

$$Q_{12} = \begin{pmatrix} c_{12} & 0 & \cdots & 0 & 0 \\ 0 & c_{12} & \cdots & 0 & 0 \\ \vdots & \vdots & \ddots & \vdots & \vdots \\ 0 & 0 & \cdots & c_{12} & 0 \\ 0 & 0 & \cdots & 0 & c_{12} \end{pmatrix}$$

$$P_{21} = \begin{pmatrix} c_{14}+c_{15} & c_{14} & 0 & \cdots & 0 & 0 & 0 \\ c_{14} & c_{15} & c_{14} & \cdots & 0 & 0 & 0 \\ \vdots & \vdots & \vdots & \ddots & \vdots & \vdots & \vdots \\ 0 & 0 & 0 & \cdots & c_{14} & c_{15} & c_{14} \\ 0 & 0 & 0 & \cdots & 0 & c_{14}-1 & p_2 \end{pmatrix}$$

$p_2 = c_{14} + c_{15} + dx \cdot h + 1,$

$$Q_{22} = \begin{pmatrix} c_{13}+c_{16} & c_{13} & 0 & \cdots & 0 & 0 & 0 \\ c_{13} & c_{16} & c_{13} & \cdots & 0 & 0 & 0 \\ \vdots & \vdots & \vdots & \ddots & \vdots & \vdots & \vdots \\ 0 & 0 & 0 & \cdots & c_{13} & c_{16} & c_{13} \\ 0 & 0 & 0 & \cdots & 0 & c_{13}-\frac{\ell}{2\rho c} & q_2 \end{pmatrix}$$

$q_2 = c_{13} + c_{16} + \frac{\ell}{2\rho c} + dx \cdot h$, and the column vectors elements in right side of (4.90) given by

$$d_3^{i+1} = -p_j^{i+1} + \beta dt(u_j^{i+1} - \delta_2)^+ \chi_{Q_0} \qquad j = \overline{jb_0 - i, N + i + 1},$$

$$bq^{i+1} = \begin{cases} -q_j^{i+1} + (\varphi_j^{i+1} - 1 - \delta_1)\chi_{Q_0}, & j = jb_0 - (i+1), N+i, \\ -q_{M+N}^{i+1} + (\varphi_{M+N}^{i+1} - 1 - \delta_1)\chi_{Q_0} - dx \cdot h \cdot p_{M+N}^{i+1}, \end{cases}$$

for $i = M-1, M-2, \ldots, 1, 0$.

Now we will present a numerical algorithm of gradient type in order to compute the approximating optimal control stated by Theorem 3.4.

Algorithm **InvPHT1D** (**Inv**ers **PH**ase **T**hransition case **1D**)

P0. Choose $w^0 \in \mathcal{U}$ and set $iter = 0$;
Choose $\varepsilon > 0$, $T > \varepsilon$, $M > 0$, and compute dt;
Choose $b_1 >> T$ and compute b_0;
Choose $N > 0$ and compute dx;
Set $jb_0 = M + 1$;
Choose u_0, φ_0 and compute $\varphi_{-}^{\varepsilon}(0, \cdot)$;

P1. Compute $z(\varepsilon, \cdot)$ from (4.82);

P2. Determine the matrices u^{iter}, φ^{iter} whose rows are line vectors u^i, φ^i, $i = 1, ..., M$ - the solution of linear system (4.83);

P3. Determine the matrices p^{iter}, q^{iter} whose rows are line vectors p^i, q^i, $i = M-1, ..., 0$ - the solution of linear system (4.85);

P4. For $t \in [0, T]$, compute

$$r^{iter}(t) = \frac{k}{\rho c} p^{iter}(t, b_1) + w^{iter}(t);$$

P5. Set (see (4.75))

$$\tilde{w}^{iter}(t) = \begin{cases} 0 & \text{if } r^{iter}(t) > 0, \\ -R & \text{if } r^{iter}(t) \le 0. \end{cases}$$

P6. Compute $\lambda_{iter} \in [0, 1]$ solution of the minimization process:

$$\min \{L_0^{\varepsilon}(\lambda w^{iter} + (1-\lambda)\tilde{w}^{iter}, \quad \lambda \in [0, 1];$$

Set $w^{iter+1} = \lambda_{iter} w^{iter} + (1 - \lambda_{iter})\tilde{w}^{iter}$;

P7. If $\| w^{iter+1} - w^{iter} \| \leq \eta$ /* the stopping criterion */

then STOP

else iter:= iter+1; Go to **P2.**

In the above, the variable *iter* represents the number of iterations after which the algorithm **InvPHT1D** found the optimal value of the cost functional $L_0^\varepsilon(w)$ in (P_{inv}^ε).

Remark 3.2. The *stopping criterion* in **P7** could be

$$\| L_0^\varepsilon(w^{iter+1}) - L_0^\varepsilon(w^{iter}) \| \leq \eta,$$

where η is a prescribed precision.

The main novelty brought by this work is that the computation of the approximate solution corresponding to the nonlinear system (1.1) is replaced with calculation of the approximate solution for an ordinary equation and a linear system (compare step **P1** in C. Moroşanu [14] with the steps **P1-P2** in the algorithm **InvPHT1D**).

Implementation of the conceptual algorithm **InvPHT1D** and numerical results are given in the sequel (see Fig. **3-8**).

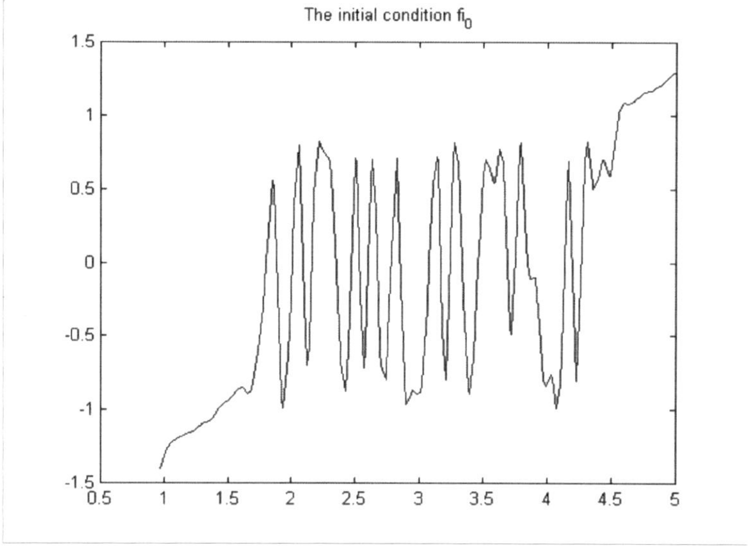

Figure 3 *The initial condition* φ_0

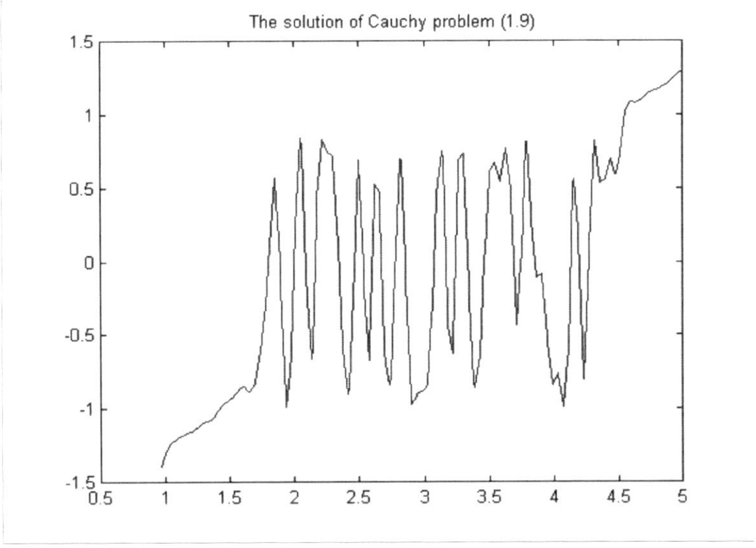

Figure 4 *The solution $z(\varepsilon, \cdot)$ of Cauchy problem (1.9)*

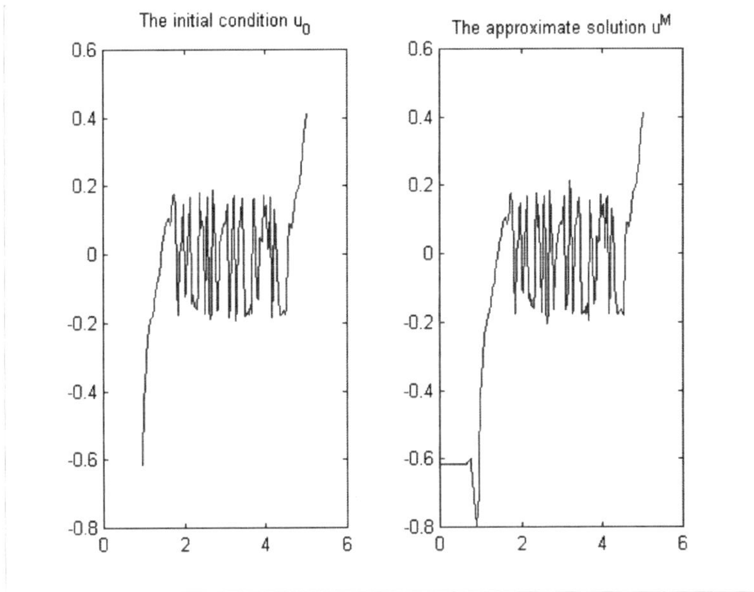

Figure 5 *The approximate solution u^0 and u^M*

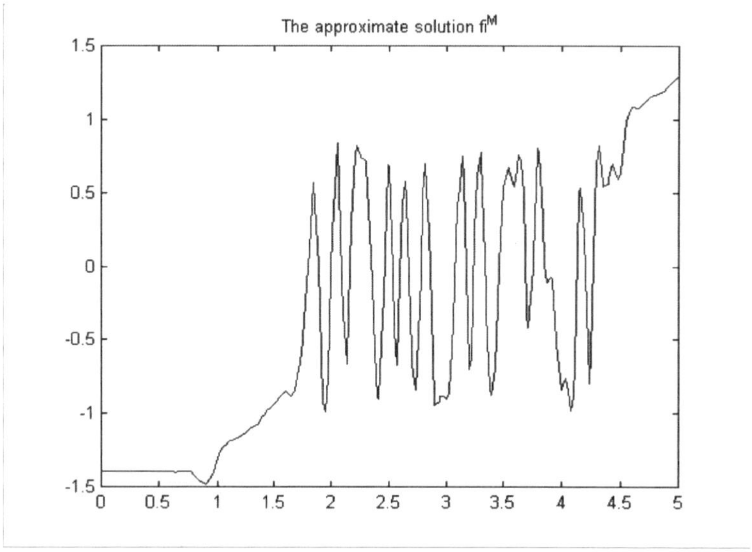

Figure 6 *The approximate solution* φ^M

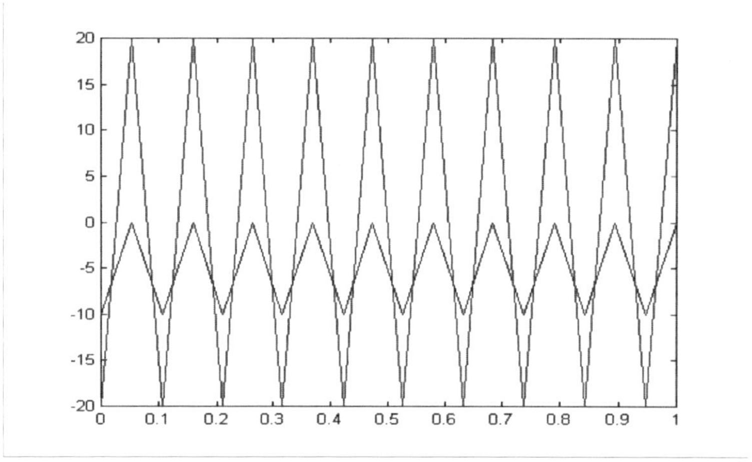

Figure 7 *The initial control* w^0 *and optimal control* w^{10}

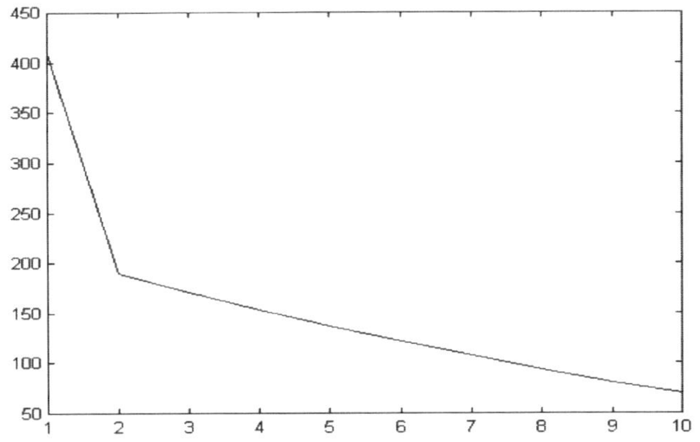

Figure 8 *The values of functional $j^{\varepsilon}(w)$*

We wish only to draw attention to the type of boundary condition considered here (see (3.98)) namely that they fully cover industrial application proposed by us for numerical simulations - a matter for further investigation.

4.4 Algorithm `scpht-2D`

In this section we will present a conceptual algorithm of gradient type, in order to calculate the approximating optimal control v^*, w^*. We work in the two-dimensional case, i.e., $N = 2$ and we assume that $\Omega \subset I\!R^2$ is a polygonal domain.

Let $h = T/M$ be the time step size. Let T_ρ be the triangulation (mesh) over Ω, $\bar{\Omega} = \cup_{K \in T_\rho} K$, and let $N_j = (x_k, y_l)$, and $j = \overline{1, nn}$ be the nodes associated to T_ρ (see Annexe B).

Firstly, we will construct the discrete problem corresponding to state and adjoint systems. In this sense we define the weak solution of (3.137)-(3.138) and (3.197)-(3.198) being the pairs (u, φ) and (p, q), respectively, that satisfies the following equations

$$\left(u_t + \frac{\ell}{2}\varphi_t, \psi_1\right) + k(\nabla u, \nabla \psi_1) = (mv, \psi_1) \quad \forall \psi_1 \in H^1(\Omega), \quad \text{a.e. in } Q, \tag{4.91}$$

$$\tau(\varphi_t, \psi_2) + \xi^2(\nabla\varphi, \nabla\psi_2) + \frac{1}{2a}(\varphi^3 - \varphi, \psi_2) - 2(u, \psi_2) = (mw, \psi_2) \tag{4.92}$$

$$\forall \psi_2 \in H_0^1(\Omega), \quad \text{a.e. in } Q,$$

together with the initial conditions

$$u(x,0) = u_0(x), \qquad \varphi(x,0) = \varphi_0(x) \quad \text{on } \Omega, \tag{4.93}$$

and

$$\Big(p_t, \zeta_1\Big) - k(\nabla p, \nabla \zeta_1) - \frac{\ell}{\tau}(p, \zeta_1) + \frac{2}{\tau}(q, \zeta_1) = 2\alpha(\mu_0 u, \zeta_1) \tag{4.94}$$

$$\forall \zeta_1 \in H^1(\Omega), \quad \text{a.e. in } Q,$$

$$(q_t, \zeta_2) + \frac{\ell\xi^2}{2\tau}(\nabla p, \nabla \zeta_2) + \frac{\ell}{4a\tau}((3\varphi^2 - 1)p, \zeta_2) - \frac{\xi^2}{\tau}(\nabla q, \nabla \zeta_2) \tag{4.95}$$

$$+\frac{1}{2a\tau}((1 - 3\varphi^2)q, \zeta_2) = 0 \quad \forall \zeta_2 \in H_0^1(\Omega), \quad \text{a.e. in } Q,$$

together with the final conditions

$$p(x,T) = 0, \qquad q(x,T) = -b_0 \quad \text{on } \Omega. \tag{4.96}$$

Here $(\cdot, \cdot)$ denotes the scalar product in $L^2(\Omega)$.

If we denote by V_{nn} the corresponding finite element space for the triangulation T_ρ, then the V_{nn} interpolant of u, φ, p, q, denoted by u^i, φ^i, p^i, q^i, can be expressed as follows

$$u^i(x,y) = \sum_{l=1}^{nn} u_l^i b_l(x,y), \quad i = \overline{1, M}, \tag{4.97}$$

$$\varphi^i(x,y) = \sum_{l=1}^{nn} \varphi_l^i b_l(x,y), \quad i = \overline{1, M}, \tag{4.98}$$

$$p^i(x,y) = \sum_{l=1}^{nn} p_l^i b_l(x,y), \quad i = \overline{0, M-1}, \tag{4.99}$$

$$q^i(x,y) = \sum_{l=1}^{nn} q_l^i b_l(x,y), \quad i = \overline{0, M-1}, \tag{4.100}$$

where the unknown $u_l^i = u(t_i, N_l)$, $\varphi_l^i = \varphi(t_i, N_l)$, $p_l^{i-1} = p(t_{i-1}, N_l)$, $q_l^{i-1} = q(t_{i-1}, N_l)$, $i = \overline{1, M}$, $l = \overline{1, nn}$, represent the discrete solution of the equations (4.105)-(4.108) below and $\{b_j\}_{j=1}^{nn}$ is the basis functions of V_{nn} defined by $b_j(N_i) = \delta_{ij}$, $i, j = 1, 2, ..., nn$.

Let $U \in V_{nn}$ and $\Phi \in V_{nn}^0 = \{v \in V_{nn},\ v|_{\partial\Omega} = 0\})$ be two arbitrary functions, i.e., $U(x,y) = \sum_{l=1}^{nn} U_l b_l(x,y)$, $\Phi(x,y) = \sum_{l=1}^{nn} \Phi_l b_l(x,y)$. Using an implicit (backward) finite difference scheme in time, the discrete equations corresponding to (4.91)-(4.92) and (4.94)-(4.95) are, respectively

$$(u^i, U) + \frac{\ell}{2}(\varphi^i, U) + hk(\nabla u^i, \nabla U) \tag{4.101}$$

$$= h(mv, U) + (u^{i-1} + \frac{\ell}{2}\varphi^{i-1}, U) \quad \forall U \in V_{nn},$$

$$\tau(\varphi^i, \Phi) + h\xi^2(\nabla\varphi^i, \nabla\Phi) + \frac{h}{2a}((\varphi^i)^3 - \varphi^i, \Phi) - 2h(u^i, \Phi) \tag{4.102}$$

$$= (hmw + \tau\varphi^{i-1}, \Phi) \quad \forall \Phi \in V_{nn}^0,$$

for $i = \overline{1, M}$, and

$$(p^i, U) - hk(\nabla p^i, \nabla U) - h\frac{\ell}{\tau}(p^i, U) + h\frac{2}{\tau}(q^i, U) \tag{4.103}$$

$$= (p^{i+1}, U) + 2\alpha h(\mu_0 u^i, U) \quad \forall U \in V_{nn},$$

$$(q^i, \Phi) + h\frac{\ell\xi^2}{2\tau}(\nabla p^i, \nabla\Phi) + h\frac{\ell}{4a\tau}((3(\varphi^i)^2 - 1)p^i, \Phi) - h\frac{\xi^2}{\tau}(\nabla q^i, \nabla\Phi) \tag{4.104}$$

$$+\frac{h}{2a\tau}((1 - 3(\varphi^i)^2)q^i, \Phi) = (q^{i+1}, \Phi) \quad \forall \Phi \in V_{nn}^0,$$

for $i = \overline{0, M-1}$. Replaying (4.97)-(4.100) in (4.101)-(4.104), we get

$$\sum_{k=1}^{nn} U_k \Big(\sum_{l=1}^{nn} u_l^i \int_\Omega b_k b_l dxdy \Big) + \frac{\ell}{2} \sum_{k=1}^{nn} U_k \Big(\sum_{l=1}^{nn} \varphi_l^i \int_\Omega b_k b_l dxdy \Big) \tag{4.105}$$

$$+h\kappa \sum_{k=1}^{nn} U_k \Big(\sum_{l=1}^{nn} u_l^i \int_\Omega \nabla b_k \nabla b_l dxdy \Big)$$

$$= \sum_{k=1}^{nn} U_k \Big(\sum_{l=1}^{nn} (u_l^{i-1} + \frac{\ell}{2}\varphi_l^{i-1} + hm_l v_l^i) \int\limits_{\Omega} b_k b_l dxdy \Big),$$

$$\tau \sum_{k=1}^{nn} \Phi_k \Big(\sum_{l=1}^{nn} \varphi_l^i \int\limits_{\Omega} b_k b_l dxdy \Big) + h\xi^2 \sum_{k=1}^{nn} \Phi_k \Big(\sum_{l=1}^{nn} \varphi_l^i \int\limits_{\Omega} \nabla b_k \nabla b_l dxdy \Big) \tag{4.106}$$

$$+ \frac{h}{2a} \sum_{k=1}^{nn} \Phi_k \Big(\sum_{l=1}^{nn} ((\varphi_l^i)^3 - \varphi_l^i) \int_{\Omega} b_k b_l dxdy \Big)$$

$$-2h \sum_{k=1}^{nn} \Phi_k \Big(\sum_{l=1}^{nn} u_l^i \int\limits_{\Omega} b_k b_l dxdy \Big)$$

$$= \sum_{k=1}^{nn} \Phi_k \Big(\sum_{l=1}^{nn} (\tau \varphi_l^{i-1} + hm_l w_l^i) \int\limits_{\Omega} b_k b_l dxdy \Big),$$

for $i = \overline{1, M}$, where $m_l = m(N_l)$, $v_l^i = v(t_i, N_l)$, $w_l^i = w(t_i, N_l)$, $l = \overline{1, nn}$, $i = \overline{1, M}$, and

$$\sum_{k=1}^{nn} U_k \Big(\sum_{l=1}^{nn} p_l^i \int\limits_{\Omega} b_k b_l dxdy \Big) - h\kappa \sum_{k=1}^{nn} U_k \Big(\sum_{l=1}^{nn} p_l^i \int\limits_{\Omega} \nabla b_k \nabla b_l dxdy \Big) \tag{4.107}$$

$$-h\frac{\ell}{\tau} \sum_{k=1}^{nn} U_k \Big(\sum_{l=1}^{nn} p_l^i \int\limits_{\Omega} b_k b_l dxdy \Big) + h\frac{2}{\tau} \sum_{k=1}^{nn} U_k \Big(\sum_{l=1}^{nn} q_l^i \int\limits_{\Omega} \nabla b_k \nabla b_l dxdy \Big)$$

$$= \sum_{k=1}^{nn} U_k \Big(\sum_{l=1}^{nn} (p_l^{i+1} + 2\alpha h \mu_{0,l} u_l^i) \int\limits_{\Omega} b_k b_l dxdy \Big),$$

$$\sum_{k=1}^{nn} \Phi_k \Big(\sum_{l=1}^{nn} q_l^i \int\limits_{\Omega} b_k b_l dxdy \Big) + h\frac{\ell\xi^2}{2\tau} \sum_{k=1}^{nn} \Phi_k \Big(\sum_{l=1}^{nn} p_l^i \int\limits_{\Omega} \nabla b_k \nabla b_l dxdy \Big) \tag{4.108}$$

$$+h\frac{\ell}{4a\tau}\sum_{k=1}^{nn}\Phi_k\Big(\sum_{l=1}^{nn}(3(\varphi_l^i)^2-1)p_l^i\int_\Omega b_k b_l dxdy\Big)$$

$$-h\frac{\xi^2}{\tau}\sum_{k=1}^{nn}\Phi_k\Big(\sum_{l=1}^{nn}q_l^i\int_\Omega \nabla b_k\nabla b_l dxdy\Big)$$

$$+\frac{h}{2a\tau}\sum_{k=1}^{nn}\Phi_k\Big(\sum_{l=1}^{nn}(1-3(\varphi_l^i)^2)q_l^i\int_\Omega b_k b_l dxdy\Big)$$

$$=\sum_{k=1}^{nn}\Phi_k\Big(\sum_{l=1}^{nn}q_l^{i+1}\int_\Omega b_k b_l dxdy\Big),$$

for $i=\overline{0,M-1}$.

Setting

$$b_{kl}=\int_\Omega b_k(x,y)b_l(x,y)dxdy,$$

$$g_{kl}=\int_\Omega \nabla b_k(x,y)\nabla b_l(x,y)dxdy,$$

$$B=(b_{kl})_{k,l=\overline{1,nn}},\quad G=(g_{kl})_{k,l=\overline{1,nn}},$$

$$C=(c_{kl})_{k,l=\overline{1,nn}},\quad c_{kl}=b_{kl}+h\kappa g_{kl},$$

$$D=(d_{kl})_{k,l=\overline{1,nn}},\quad d_{kl}=(\tau-\tfrac{h}{2a})b_{kl}+h\xi^2 g_{kl},$$

$$E=(e_{kl})_{k,l=\overline{1,nn}},\quad e_{kl}=(1-h\tfrac{\ell}{\tau})b_{kl}-hkg_{kl},$$

$$F=(f_{kl})_{k,l=\overline{1,nn}},\quad f_{kl}=(1+\tfrac{h}{2a\tau}(1-3(\varphi_l^i)^2))b_{kl}-h\tfrac{\xi^2}{\tau}g_{kl},$$

$$R=(r_{kl})_{k,l=\overline{1,nn}},\quad r_{kl}=h\tfrac{\ell\xi^2}{2\tau}g_{kl}+h\tfrac{\ell}{4a\tau}(3(\varphi_l^i)^2-1)b_{kl},$$

we can rewrite (4.105)-(4.108) in the matrix form

$$\begin{cases} C\ u^i+\frac{\ell}{2}B\ \varphi^i=B(u^{i-1}+\frac{\ell}{2}\varphi^{i-1}+hmv^i),\\ D\ \varphi^i+\frac{h}{2a}B(\varphi^i)^3-2hB\ u^i=B(\tau\varphi^{i-1}+hmw^i),\qquad i=\overline{1,M}\end{cases}\tag{4.109}$$

$$\begin{cases} E\ p^i+h\frac{2}{\tau}G\ q^i=B\ (p^{i+1}+2\alpha h\mu_0 u^i),\\ F\ q^i+R\ p^i=Bq^{i+1},\qquad i=\overline{0,M-1}.\end{cases}\tag{4.110}$$

From the initial and the final conditions (4.93) and (4.96), we have

$$u_l^0 = u_0(N_l) \quad \varphi_l^0 = \varphi_0(N_l); \qquad p_l^M = 0 \quad q_l^M = -b_0(N_l), \qquad l = \overline{1, nn}.$$

Now, we will present a conceptual algorithm of gradient type to compute the optimal control v^*, w^* in (3.202).

Algorithm `scpht-2D` (`state` constrained for the `phase` `transition-2D`)

P0. Set *iter:=0*;
Choose $v^{iter} = \left(v_l^{i,iter}\right)_{\substack{i=\overline{1,M}\\ l=\overline{1,nn}}}, \quad w^{iter} = \left(w_l^{i,iter}\right)_{\substack{i=\overline{1,M}\\ l=\overline{1,nn}}};$
Choose $\omega \subset \Omega$ and the set $S \subset L^2(\Omega)$;
Choose $\varepsilon_0 > 0$;

P1. Compute the approximate matrix

$$u^{iter} = (u^1, ..., u^M), \quad \varphi^{iter} = (\varphi^1, ..., \varphi^M);$$

The column vectors u^i, φ^i, $i = \overline{1, M}$, are obtained solving the nonlinear system (4.109);

P2. Compute $d_S(\varphi^M)$; /* $\varphi^M = \varphi^{iter}(x, T)$ */
Compute $\mu_0 = \frac{\varepsilon_0}{\varepsilon_0 + d_S(\varphi^M)}$;
Compute (see (3.187))

$$b_0 = \begin{cases} 1 & \text{if } \varphi^M \notin S, \\ 0 & \text{if } \varphi^M \in S; \end{cases}$$

P3. Compute the approximate matrix

$$p^{iter} = (p^0, p^1, ..., p^{M-1}), \quad q^{iter} = (q^0, q^1, ..., q^{M-1});$$

The column vectors p^i, q^i, $i = \overline{0, M-1}$, are obtained solving the linear system (4.110);

P4. Compute $\bar{v}^{iter} = \left(\bar{v}_l^{i,iter}\right)_{\substack{i=\overline{1,M}\\ l=\overline{1,nn}}}, \quad \bar{w}^{iter} = \left(\bar{w}_l^{i,iter}\right)_{\substack{i=\overline{1,M}\\ l=\overline{1,nn}}}$
from (3.202);

P5. Compute $\lambda_{iter} \in [0, 1]$ (the steplength of the gradient method) solution of the minimization process

$$\min\{L(u^{iter}, \varphi^{iter}, \lambda v^{iter}+(1-\lambda)\bar{v}^{iter}, \lambda w^{iter}+(1-\lambda)\bar{w}^{iter}),\ \lambda \in [0,1]\};$$

$$\text{Set}\quad v^{iter+1} := \lambda_{iter} v^{iter} + (1-\lambda_{iter})\bar{v}^{iter}, \qquad w^{iter+1} := \lambda_{iter} w^{iter} + (1-\lambda_{iter})\bar{w}^{iter};$$

P6. (the "*Stopping Criterion*")

if $\|v^{iter+1} - v^{iter}\| \le \eta_1$ and $\|w^{iter+1} - w^{iter}\| \le \eta_2$

then STOP (the algorithm is convergent)

else $iter := iter + 1$; Go to **P1**.

In the above algorithm, the variable $iter$ represents the number of iterations after which the algorithm `scpht-2D` found the optimal value of the cost functional $L(u, \varphi, v, w)$ in (P_S).

Remark. The "*Stopping Criterion*" in **P6** could be

$$\|L(u^{iter}, \varphi^{iter}, v^{iter+1}, w^{iter+1}) - L(u^{iter}, \varphi^{iter}, v^{iter}, w^{iter})\| \le \eta$$

where η is a prescribed precision.

Let us briefly discuss the main steps in algorithm `scpht-2D`. For approximating the solution of the nonlinear parabolic system (4.19)-(4.20) from **P1**, we have used the iterative Newton (see Subsection 4.1a). The values of λ_{iter} from **P5** are chosen from the sequence

$$1-d,\quad 1-2*d,\quad 1-3*d, ..., 0$$

where $d > 0$ is prescribed (in the numerical example $0 < d < 0.05$). Since we have a finite number of options for λ, we choose $\lambda_{iter} = \lambda$ the value from the above sequence which minimizes the process given in step **P5**

Next, we illustrate our algorithm by a concrete numerical computation. For this let the domain Ω be $[0,1] \times [0,1] \subset I\!R^2$ and the time interval be $[0,2]$. For the time as well as for the spatial discretization we use uniform discretization. The time step was selected $\varepsilon = 0.4$ (i.e. $M = 5$). The triangulation of the spatial domain is obtained by dividing x- and y- axis into equidistant subintervals of length $dx = dy = 0.05$ and then dividing each of the squares results into two triangles.

We apply our algorithm for the following choices of data $\xi = 0.7$, $a = \sqrt{\xi}$, $\tau = 1.0e + 5 * \xi^2$, $\ell = 0.6$, $k = 0.9$, $\alpha = 1$. Fig. (**10**) below represent the initial value $\varphi_0(x, y)$.

We take the values of function $u_0(x, y)$ (see Fig. **9**) to be equal to those satisfying the stationary equation $(2a\tau)^{-1}(\varphi_0 - \varphi_0^3) + 2\tau^{-1}u_0 = 0$. For completeness, we take the following values for v^0 and w^0 (see step **P0**):

$$\begin{cases} v_l^{i,0} = u_l^i - \frac{\ell}{2}\varphi_l^i, \\ w_l^{i,0} = 0, \end{cases} \quad i = \overline{0, M}, l = \overline{1, nn};$$

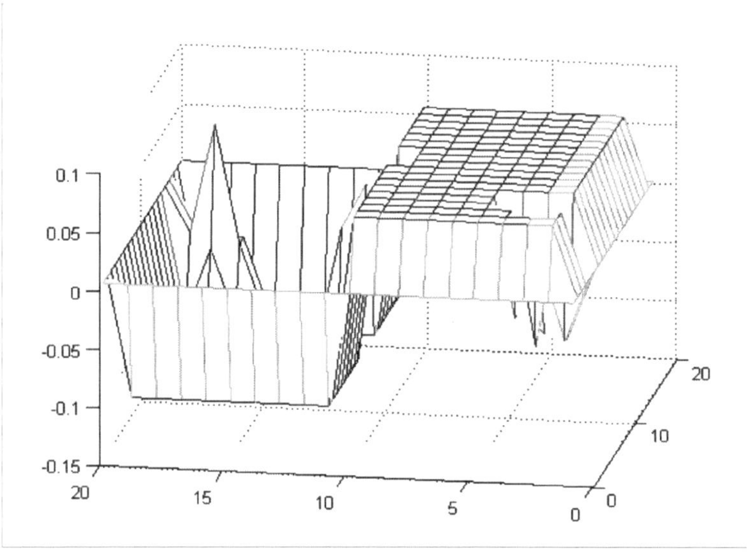

Figure 9 *The initial values $u_0(x, y)$*

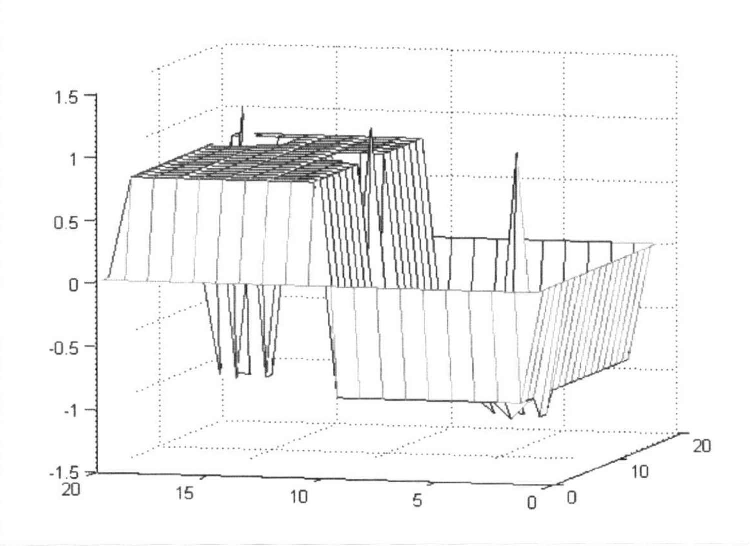

Figure 10 *The initial value* $\varphi_0(x, y)$

The subset $\omega \subset \Omega$ was chosen to be $\omega = supp\ N_{45} \cup supp\ N_{211} \cup supp\ N_{356}$, while $S = \{\varphi(x, y, T) \in L^2(\Omega); \|\varphi(x, y, T) - 1\| \leq 0.998\}$ and $\varepsilon_0 = 0.1$. Corresponding, $L(u^0, \varphi^0, v^0, w^0) = 3.066285e + 00$. The optimal value $L(u^*, \varphi^*, v^*, w^*) = 2.041611e + 00$ was obtained in 2 iterations.

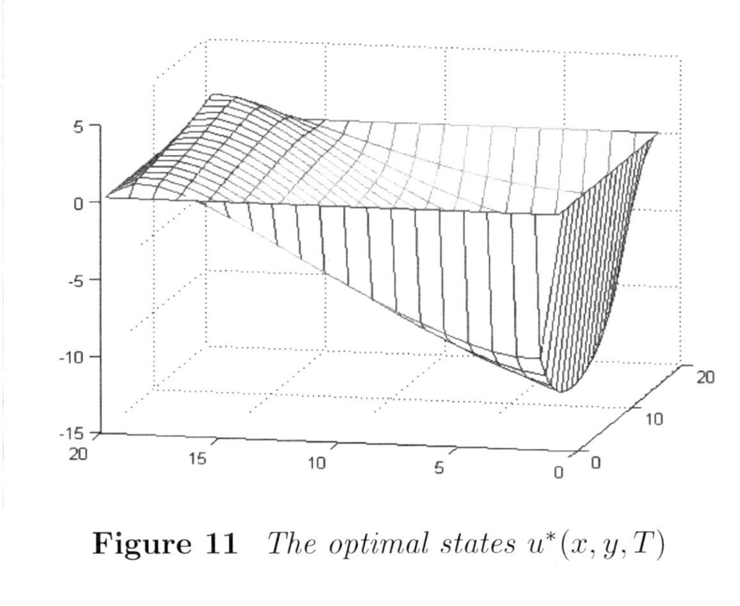

Figure 11 *The optimal states* $u^*(x, y, T)$

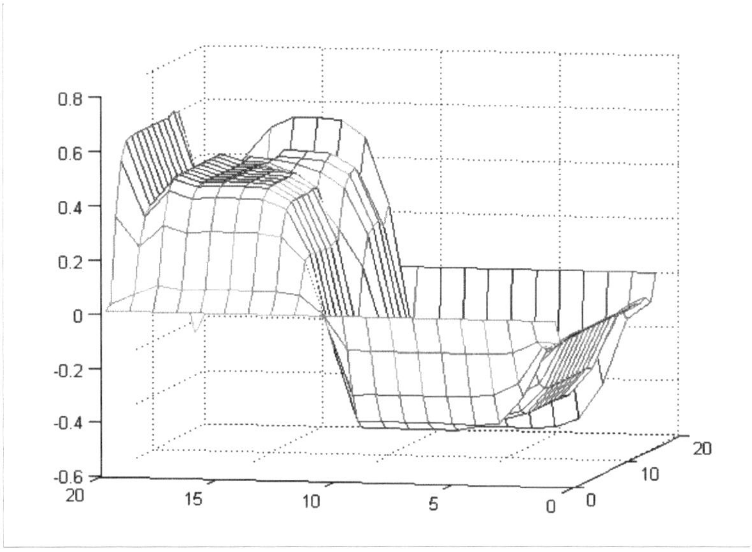

Figure 12 *The optimal states* $\varphi^*(x, y, T)$

Fig. (**11**) and (**12**) show the approximate solutions u^*, φ^* corresponding to the approximating optimal control v^*, w^* (see Theorem 3.5).

Remark. We underline that the numerical results obtained above do not have in view the physical aspects. They are given only to show that the algorithm `scpht-2D` may be implemented. Numerical results regarding physical nature of this problem are a matter for further investigation.

4.5 Approximation of the suboptimal control. Algorithm `cpht-2D`

The aim of this paragraph is to approximate the suboptimal controller corresponding to the approximating problem (P^ε) presented in Section 3.3, relation (3.37):

$$(P^\varepsilon) \quad \textit{Minimize } j^\varepsilon(v) = \int_Q \chi_0(\varphi^\varepsilon(t,x,y))^2 dxdydt + \int_0^T w^2(t)\, dt,$$

on all $(u^\varepsilon, \varphi^\varepsilon, w, v)$ subject to (4.111)-(4.113) below

$$\begin{cases} u_t^\varepsilon + \dfrac{\ell}{2}\varphi_t^\varepsilon - k\Delta u^\varepsilon = 0 & \text{in} \quad Q_i^\varepsilon = (i\varepsilon, (i+1)\varepsilon) \times \Omega, \\ \dfrac{\partial u^\varepsilon}{\partial \nu} + hu^\varepsilon = w(t)g(x,y) & \text{on} \quad \Sigma_i^\varepsilon = (i\varepsilon, (i+1)\varepsilon) \times \partial\Omega, \\ u^\varepsilon(0,x,y) = u_0(x,y) & (x,y) \in \Omega, \end{cases} \tag{4.111}$$

$$\begin{cases} w'(t) = \beta w(t) + v(t), \quad t \in [0,T], \\ w(0) = 0, \end{cases} \tag{4.112}$$

$$\begin{cases} \tau\varphi_t^\varepsilon - \xi^2\Delta\varphi^\varepsilon = \dfrac{1}{2a}\varphi^\varepsilon + 2u^\varepsilon + \dfrac{1}{2a} & \text{in} \quad Q_i^\varepsilon, \\ \varphi^\varepsilon = 0 & \text{on} \quad \Sigma_i^\varepsilon, \\ \varphi_+^\varepsilon(i\varepsilon,x,y) = z(\varepsilon, \varphi_-^\varepsilon(i\varepsilon,x,y)), \end{cases} \tag{4.113}$$

where $z(\cdot, \varphi_-^\varepsilon(i\varepsilon,x,y))$ is the solution of

$$\begin{cases} z'(s) + \frac{1}{2a}(z(s)+1)^3 = 0, \quad s \in [0,T], \\ z(0) = \varphi_-^\varepsilon(i\varepsilon,x,y)) \quad \varphi_-^\varepsilon(0,x,y)) = \varphi_0(x,y) - 1, \end{cases} \tag{4.114}$$

computed at $s = \varepsilon$, for $i = \overline{0, M_\varepsilon - 1}$, with $M_\varepsilon = \left[\dfrac{T}{\varepsilon}\right]$ and $Q_{M_\varepsilon-1}^\varepsilon = ((M_\varepsilon - 1)\varepsilon, T) \times \Omega$ and $v \in \mathcal{U}$,

$$\mathcal{U} = \{v \in L^\infty(0,T), \quad 0 \le v(t) \le R \quad a.e. \quad t \in [0,T]\}.$$

We have proved that the optimal control in problem (P^ε) is given by (see Theorem 3.2 in Section 3.4)

$$v_\varepsilon^*(t) = \begin{cases} R & \text{if } r(t) < 0 \\ 0 & \text{if } r(t) > 0, \end{cases} \tag{4.115}$$

where

$$r(t)=\int_t^T\Big(\int_{\partial\Omega}(w(s)-kp(s,x,y)g(x,y))dxdy\Big)e^{\beta(s-t)}ds \qquad (4.116)$$

and (p,q) satisfy along with u^ε, φ^ε the adjoint state system

$$\begin{cases} p_t^\varepsilon+k\Delta p^\varepsilon-\frac{\ell}{\tau}p^\varepsilon+\frac{2}{\tau}q^\varepsilon=0 & \text{in } Q_i^\varepsilon,\\ \frac{\partial p^\varepsilon}{\partial\nu}+hp^\varepsilon=0 & \text{on } \Sigma_i^\varepsilon,\\ p_-^\varepsilon((i+1)\varepsilon,x,y)=0,\quad p_-^\varepsilon(T,x,y)=0 & (x,y)\in\Omega,\end{cases} \qquad (4.117)$$

$$\begin{cases} \tau q_t^\varepsilon-\frac{\ell\xi^2}{2\tau}\Delta p^\varepsilon-\frac{\ell}{4a\tau}p^\varepsilon+\frac{\xi^2}{\tau}\Delta q^\varepsilon+\frac{1}{2a\tau}q^\varepsilon=\varphi^\varepsilon\chi_0 & \text{in } Q_i^\varepsilon,\\ q^\varepsilon=\frac{\ell}{2}p^\varepsilon & \text{on } \Sigma_i^\varepsilon,\\ q_-^\varepsilon((i+1)\varepsilon,\cdot) & \\ \quad=\exp\Big(\int_0^\varepsilon\frac{3}{2a}(z(t,\cdot)+1)^2dt\Big)q^+((i+1)\varepsilon,\cdot) & \text{on } \Omega,\\ q_-^\varepsilon(T,\cdot)=0, & \end{cases} \qquad (4.118)$$

for $i=M_\varepsilon-2,M_\varepsilon-3,...,1,0$, where $z(t,\cdot)$ is the solution of (3.36).

Next, we will construct the discrete form of the problem (P^ε). In this sense we define the weak solution of (4.111)-(4.113) and (4.117)-(4.118) being the pairs $(u^\varepsilon,\varphi^\varepsilon)$ and $(p^\varepsilon,q^\varepsilon)$, respectively, that satisfies the following equations

$$\Big(u_t^\varepsilon+\frac{\ell}{2}\varphi_t^\varepsilon,\psi_1\Big)+k(\nabla u^\varepsilon,\nabla\psi_1)+kh\int_{\partial\Omega}u^\varepsilon\psi_1dxdy=k\int_{\partial\Omega}wg\psi_1dxdy, \qquad (4.119)$$

$$\forall\,\psi_1\in H^1(\Omega),\quad a.e.\ in\ Q_i^\varepsilon,$$

$$\tau(\varphi_t^\varepsilon,\psi_2)+\xi^2(\nabla\varphi^\varepsilon,\nabla\psi_2)=\frac{1}{2a}(\varphi^\varepsilon,\psi_2)+2(u^\varepsilon,\psi_2)+(\frac{1}{2a},\psi_2), \qquad (4.120)$$

$$\forall\,\psi_2\in H_0^1(\Omega),\quad a.e.\ in\ Q_i^\varepsilon,$$

together with the initial conditions

$$u^\varepsilon(i\varepsilon,\cdot) = u^\varepsilon_+\varepsilon(i\varepsilon,\cdot), \quad u^\varepsilon(0,\cdot) = u_0(\cdot), \quad \text{on } \Omega, \tag{4.121}$$

$$\varphi^\varepsilon(i\varepsilon,\cdot) = \varphi^\varepsilon_+(i\varepsilon,\cdot), \quad \varphi^\varepsilon(0,\cdot) = \varphi_0(\cdot), \quad \text{on } \Omega, \tag{4.122}$$

for $i = \overline{0, M_\varepsilon - 1}$, with $M_\varepsilon = \left[\dfrac{T}{\varepsilon}\right]$, and

$$\left(p^\varepsilon_t,\zeta_1\right) - k(\nabla p^\varepsilon,\nabla\zeta_1) - kh\int_{\partial\Omega} p^\varepsilon\zeta_1 dxdy - \frac{\ell}{\tau}(p^\varepsilon,\zeta_1) + \frac{2}{\tau}(q^\varepsilon,\zeta_1) = 0,$$

$$\forall\zeta_1 \in H^1(\Omega), \quad a.e. \ in \ Q^\varepsilon_i, \tag{4.123}$$

$$\tau(q^\varepsilon_t,\zeta_2) + \frac{\ell\xi^2}{2\tau}(\nabla p^\varepsilon,\nabla\zeta_2) + h\int_{\partial\Omega} p^\varepsilon\zeta_2 dxdy \tag{4.124}$$

$$-\frac{\ell}{4a\tau}(p^\varepsilon,\zeta_2) - \frac{\xi^2}{\tau}(\nabla q^\varepsilon,\nabla\zeta_2) + \frac{\xi^2}{\tau}\frac{\ell}{2}\int_{\partial\Omega} p^\varepsilon\zeta_2 dxdy + \frac{1}{2a\tau}(q^\varepsilon,\zeta_2)$$

$$= (\varphi^\varepsilon\chi_0,\zeta_2), \qquad \forall\zeta_2 \in H^1(\Omega), \quad a.e. \ in \ Q^\varepsilon_i,$$

together with the final conditions

$$p^\varepsilon((i+1)\varepsilon,\cdot) = p^\varepsilon_-((i+1)\varepsilon,\cdot), \quad p^\varepsilon(T,\cdot) = p^\varepsilon_-(T,\cdot) = 0, \quad \text{on } \Omega, \tag{4.125}$$

$$q^\varepsilon((i+1)\varepsilon,\cdot) = q^\varepsilon_-((i+1)\varepsilon,\cdot), \quad q^\varepsilon(T,\cdot) = q^\varepsilon_-(T,\cdot) = 0, \quad \text{on } \Omega, \tag{4.126}$$

for $i = M_\varepsilon - 2, ..., 1, 0$.

Here $(\cdot,\cdot)$ denotes the scalar product in $L^2(\Omega)$.

In the sequel we will use the same notations as in the previous Section. So, V_{nn} interpolant of u^ε, φ^ε, p^ε, q^ε, denoted by $u^{\varepsilon,i}$, $\varphi^{\varepsilon,i}$, $p^{\varepsilon,i}$, $q^{\varepsilon,i}$, can be expressed as follows ($M \equiv M_\varepsilon$)

$$u^{\varepsilon,i}(x,y) = \sum_{l=1}^{nn} u_l^{\varepsilon,i} b_l(x,y), \quad i = \overline{1,M}, \tag{4.127}$$

$$\varphi^{\varepsilon,i}(x,y) = \sum_{l=1}^{nn} \varphi_l^{\varepsilon,i} b_l(x,y), \quad i = \overline{1,M}, \tag{4.128}$$

$$p^{\varepsilon,i}(x,y) = \sum_{l=1}^{nn} p_l^{\varepsilon,i} b_l(x,y), \quad i = \overline{0, M-1}, \tag{4.129}$$

$$q^{\varepsilon,i}(x,y) = \sum_{l=1}^{nn} q_l^{\varepsilon,i} b_l(x,y), \quad i = \overline{0, M-1}, \tag{4.130}$$

where the unknown $u_l^{\varepsilon,i} = u^{\varepsilon}(t_i, N_l)$, $\varphi_l^{\varepsilon,i} = \varphi^{\varepsilon}(t_i, N_l)$, $p_l^{\varepsilon,i} = p^{\varepsilon}(t_i, N_l)$, $q_l^{\varepsilon,i} = q^{\varepsilon}(t_i, N_l)$, $i = \overline{1, M}$, $l = \overline{1, nn}$, represent the discrete solution of the equations (4.135)-(4.138) below.

Let $U \in V_{nn}$ and $\Phi \in V_{nn}^0 = \{v \in V_{nn},\ v|_{\partial\Omega} = 0\})$ be two arbitrary functions (see (4.61), (4.62)). After using an implicit (backward) finite difference scheme in time, the discrete equations corresponding to (4.119)-(4.120) and (4.123)-(4.124) are, respectively

$$(u^{\varepsilon,i}, U) + \frac{\ell}{2}(\varphi^{\varepsilon,i}, U) + \varepsilon k(\nabla u^{\varepsilon,i}, \nabla U) + \varepsilon kh \int_{\partial\Omega} u^{\varepsilon,i} U dxdy \tag{4.131}$$

$$= \varepsilon k \int_{\partial\Omega} wgU dxdy + (u^{\varepsilon,i-1} + \frac{\ell}{2}\varphi^{\varepsilon,i-1}, U), \quad \forall U \in V_{nn},$$

$$\tau(\varphi^{\varepsilon,i}, \Phi) + \varepsilon\xi^2(\nabla\varphi^{\varepsilon,i}, \nabla\Phi) \tag{4.132}$$

$$= \frac{\varepsilon}{2a}(\varphi^{\varepsilon,i}, \Phi) + 2\varepsilon(u^{\varepsilon,i}, \Phi) + (\frac{\varepsilon}{2a}, \Phi), \quad \forall \Phi \in V_{nn}^0,$$

for $i = \overline{1, M}$, and

$$(p^{\varepsilon,i}, U) - \varepsilon k(\nabla p^{\varepsilon,i}, \nabla U) - \varepsilon kh \int_{\partial\Omega} p^{\varepsilon,i} U dxdy \tag{4.133}$$

$$= \varepsilon\frac{\ell}{\tau}(p^{\varepsilon,i}, U) - \varepsilon\frac{2}{\tau}(q^{\varepsilon,i}, U) + (p^{\varepsilon,i+1}, U), \quad \forall U \in V_{nn},$$

$$\tau(q^{\varepsilon,i}, U) + \varepsilon\frac{\ell\xi^2}{2\tau}(\nabla p^{\varepsilon,i}, \nabla U) \tag{4.134}$$

$$+\varepsilon h \int_{\partial\Omega} p^{\varepsilon,i} U dxdy - \varepsilon\frac{\ell}{4a\tau}(p^{\varepsilon,i}, U) - \varepsilon\frac{\xi^2}{\tau}(\nabla q^{\varepsilon,i}, \nabla U)$$

$$+\varepsilon\frac{\xi^2}{\tau}\frac{\ell}{2}\int_{\partial\Omega} p^{\varepsilon,i}U dxdy + \frac{\varepsilon}{2a\tau}(q^{\varepsilon,i}, U)$$

$$= (\varphi^{\varepsilon,i}\chi_0, U) + (p^{\varepsilon,i+1} + q^{\varepsilon,i+1}, U), \quad \forall U \in V_{nn},$$

for $i = \overline{0, M-1}$.

Replaying (4.127)-(4.130) in (4.131)-(4.134) and making use of (4.61)-(4.62), we get

$$\sum_{k=1}^{nn} U_k\Big(\sum_{l=1}^{nn} u_l^{\varepsilon,i}\int_\Omega b_k b_l dxdy\Big) \tag{4.135}$$

$$+\frac{\ell}{2}\sum_{k=1}^{nn} U_k\Big(\sum_{l=1}^{nn} \varphi_l^{\varepsilon,i}\int_\Omega b_k b_l dxdy\Big)$$

$$+\varepsilon\kappa\sum_{k=1}^{nn} U_k\Big(\sum_{l=1}^{nn} u_l^{\varepsilon,i}\int_\Omega \nabla b_k \nabla b_l dxdy\Big)$$

$$+\varepsilon\kappa h\sum_{k=1}^{nn} U_k\Big(\sum_{l=1}^{nn} u_l^{\varepsilon,i}\int_{\partial\Omega} b_k b_l dxdy\Big)$$

$$= \sum_{k=1}^{nn} U_k\Big(\sum_{l=1}^{nn}(u_l^{\varepsilon,i-1} + \frac{\ell}{2}\varphi_l^{\varepsilon,i-1} + \varepsilon k w^{i-1}\bar{g}_l)\int_\Omega b_k b_l dxdy\Big),$$

$$\tau\sum_{k=1}^{nn} \Phi_k\Big(\sum_{l=1}^{nn} \varphi_l^{\varepsilon,i}\int_\Omega b_k b_l dxdy\Big) \tag{4.136}$$

$$+\varepsilon\xi^2\sum_{k=1}^{nn} \Phi_k\Big(\sum_{l=1}^{nn} \varphi_l^{\varepsilon,i}\int_\Omega \nabla b_k \nabla b_l dxdy\Big)$$

$$-\frac{\varepsilon}{2a}\sum_{k=1}^{nn} \Phi_k\Big(\sum_{l=1}^{nn} \varphi_l^{\varepsilon,i}\int_\Omega b_k b_l dxdy\Big)$$

$$-2\varepsilon\sum_{k=1}^{nn} \Phi_k\Big(\sum_{l=1}^{nn} u_l^{\varepsilon,i}\int_\Omega b_k b_l dxdy\Big)$$

$$= \sum_{k=1}^{nn} \Phi_k \Big(\sum_{l=1}^{nn} (\tau \varphi_l^{\varepsilon,i-1} + \frac{\varepsilon}{2a}) \int_\Omega b_k b_l dxdy \Big),$$

for $i = \overline{1, M}$, where $\bar{g}_l = g(N_l)$, $l = \overline{1, nn}$, $w^i = w(t_i)$, $i = \overline{1, M}$, and

$$\sum_{k=1}^{nn} U_k \Big(\sum_{l=1}^{nn} p_l^{\varepsilon,i} \int_\Omega b_k b_l dxdy \Big) \tag{4.137}$$

$$-\varepsilon\kappa \sum_{k=1}^{nn} U_k \Big(\sum_{l=1}^{nn} p_l^{\varepsilon,i} \int_\Omega \nabla b_k \nabla b_l dxdy \Big)$$

$$-\varepsilon\kappa h \sum_{k=1}^{nn} U_k \Big(\sum_{l=1}^{nn} p_l^{\varepsilon,i} \int_{\partial\Omega} b_k b_l dxdy \Big)$$

$$-\varepsilon \frac{\ell}{\tau} \sum_{k=1}^{nn} U_k \Big(\sum_{l=1}^{nn} p_l^{\varepsilon,i} \int_\Omega b_k b_l dxdy \Big)$$

$$+\varepsilon \frac{2}{\tau} \sum_{k=1}^{nn} U_k \Big(\sum_{l=1}^{nn} q_l^{\varepsilon,i} \int_\Omega \nabla b_k \nabla b_l dxdy \Big)$$

$$= \sum_{k=1}^{nn} U_k \Big(\sum_{l=1}^{nn} p_l^{\varepsilon,i+1} \int_\Omega b_k b_l dxdy \Big),$$

$$\tau \sum_{k=1}^{nn} U_k \Big(\sum_{l=1}^{nn} q_l^{\varepsilon,i} \int_\Omega b_k b_l dxdy \Big) \tag{4.138}$$

$$+\varepsilon \frac{\ell \xi^2}{2\tau} \sum_{k=1}^{nn} U_k \Big(\sum_{l=1}^{nn} p_l^{\varepsilon,i} \int_\Omega \nabla b_k \nabla b_l dxdy \Big)$$

$$+(\varepsilon h + \varepsilon \frac{\xi}{\tau} \frac{\ell}{2} \sum_{k=1}^{nn} U_k \Big(\sum_{l=1}^{nn} p_l^{\varepsilon,i} \int_{\partial\Omega} b_k b_l dxdy \Big)$$

$$-\varepsilon \frac{\ell}{4a\tau} \sum_{k=1}^{nn} U_k \Big(\sum_{l=1}^{nn} p_l^{\varepsilon,i} \int_\Omega b_k b_l dxdy \Big)$$

$$-\varepsilon\frac{\xi^2}{\tau}\sum_{k=1}^{nn}U_k\Big(\sum_{l=1}^{nn}q_l^{\varepsilon,i}\int_\Omega\nabla b_k\nabla b_l dxdy\Big)$$

$$+\frac{\varepsilon}{2a\tau}\sum_{k=1}^{nn}U_k\Big(\sum_{l=1}^{nn}q_l^{\varepsilon,i}\int_\Omega b_k b_l dxdy\Big)$$

$$=\sum_{k=1}^{nn}U_k\Big(\sum_{l=1}^{nn}(\varphi_l^{\varepsilon,i}\chi_0+p_l^{\varepsilon,i+1}+q_l^{\varepsilon,i+1})\int_\Omega b_k b_l dxdy\Big),$$

for $i=\overline{0,M-1}$.

Setting

$$b_{kl}=\int_\Omega b_k(x,y)b_l(x,y)dxdy,$$

$$g_{kl}=\int_\Omega \nabla b_k(x,y)\nabla b_l(x,y)dxdy,$$

$$B=(b_{kl})_{k,l=\overline{1,nn}},$$

$$C=(c_{kl})_{k,l=\overline{1,nn}}, \qquad c_{kl}=b_{kl}+\varepsilon\kappa g_{kl},$$

$$D=(d_{kl})_{k,l=\overline{1,nn}}, \qquad d_{kl}=(\tau-\tfrac{\varepsilon}{2a})b_{kl}+\varepsilon\xi^2 g_{kl},$$

$$E=(e_{kl})_{k,l=\overline{1,nn}}, \qquad e_{kl}=(1-\varepsilon\tfrac{\ell}{\tau})b_{kl}-\varepsilon k g_{kl},$$

$$F=(f_{kl})_{k,l=\overline{1,nn}}, \qquad f_{kl}=(\tau+\tfrac{\varepsilon}{2a\tau})b_{kl}-\varepsilon\tfrac{\xi^2}{\tau}g_{kl},$$

$$R=(r_{kl})_{k,l=\overline{1,nn}}, \qquad r_{kl}=\varepsilon\tfrac{\ell\xi^2}{2\tau}g_{kl}-\varepsilon\tfrac{\ell}{4a\tau}b_{kl},$$

$$FR=(fr_{kl})_{k,l=\overline{1,nn}}, \qquad fr_{kl}=\int_{\partial\Omega} b_k(x,y)b_l(x,y)dxdy,$$

we can rewrite (4.135)-(4.138) in the matrix form

$$\begin{cases} C\ u_l^{\varepsilon,i}+\frac{\ell}{2}B\ \varphi_l^{\varepsilon,i}+\varepsilon kh\ FRu_l^{\varepsilon,i} \\ \quad =B(u^{\varepsilon,i-1}+\frac{\ell}{2}\varphi^{\varepsilon,i-1}+\varepsilon kw^{i-1}\bar{g}_l) \\ D\ \varphi_l^{\varepsilon,i}-2\varepsilon B\ u_l^{\varepsilon,i}=B(\tau\varphi_l^{\varepsilon,i-1}+\frac{\varepsilon}{2a}) \qquad i=\overline{1,M}, \end{cases} \tag{4.139}$$

$$\begin{cases} E\ p_l^{\varepsilon,i} + \varepsilon\frac{2}{\tau}B\ q_l^{\varepsilon,i} - \varepsilon kh F\!R\ p_l^{\varepsilon,i} = Bp_l^{\varepsilon,i+1} \\ F\ q_l^{\varepsilon,i} + R\ p_l^{\varepsilon,i} + (\varepsilon h + \varepsilon\frac{\xi^2}{\tau}\frac{\ell}{2})F\!R\ p_l^{\varepsilon,i} \\ \quad = B(\varphi_l^{\varepsilon,i}\chi_0 + p_l^{\varepsilon,i+1} + q_l^{\varepsilon,i+1}) \qquad i = \overline{0, M-1}. \end{cases} \tag{4.140}$$

From the initial conditions (4.121)-(4.122) we have

$$u_l^{\varepsilon,0} = u_0(N_l), \quad \varphi_l^{\varepsilon,0} = \varphi_0(N_l) - 1, \quad l = \overline{1, nn}. \tag{4.141}$$

We will now present a conceptual algorithm of gradient type (see Section 1.6) for the calculation of optimal control v_ε^* in (4.115) (also, see Barbu [85])

Algorithm `cpht-2D` (control phase transition-2D)

P0. Set *iter:=0*;
Choose $v^{\varepsilon,(iter)} = (v_0^{\varepsilon,(iter)}, v_1^{\varepsilon,(iter)}, ..., v_M^{\varepsilon,(iter)})$,
$v_i^{\varepsilon,(iter)} \in \mathcal{U}, i = \overline{0, M}$;

P1. Compute $w^{\varepsilon,(iter)} = (w_1^{\varepsilon,(iter)}, w_2^{\varepsilon,(iter)}, ..., w_M^{\varepsilon,(iter)})$ from (4.112), i.e.

$$\begin{cases} w_i^{\varepsilon,(iter)} = (1+\varepsilon\beta)w_{i-1}^{\varepsilon,(iter)} + \varepsilon v_i^{\varepsilon,(iter)} & i = \overline{1, M}, \\ w_0^{\varepsilon,(iter)} = 0 \end{cases};$$

P2. Compute the approximate matrix

$$u^{\varepsilon,(iter)} = (u_l^{\varepsilon,1}, ..., u_l^{\varepsilon,M}), \quad \varphi^{\varepsilon,(iter)} = (\varphi_l^{\varepsilon,1}, ..., \varphi_l^{\varepsilon,M});$$

i.e. (for $i = \overline{1, M}$):

- Compute $z(t_i, \cdot) = \pm\dfrac{|\varphi_-^\varepsilon(t_i, \cdot) + 1|}{\sqrt{1 + \frac{t}{a}(\varphi_-^\varepsilon(t_i, \cdot) + 1)^2}} - 1$, on Ω;
- Set $\varphi_+^\varepsilon(t_i, \cdot) = z(t_i, \cdot)$;
- Compute the column vectors $u_l^{\varepsilon,i}$, $\varphi_l^{\varepsilon,i}$, $l = \overline{1, nn}$, solving the linear system (4.139);

P3. Compute the approximate matrix

$$p^{\varepsilon,(iter)} = (p_l^{\varepsilon,0}, p_l^{\varepsilon,1}, ..., p_l^{\varepsilon,M-1}), \quad q^{\varepsilon,(iter)} = (q_l^{\varepsilon,0}, q_l^{\varepsilon,1}, ..., q_l^{\varepsilon,M-1});$$

The column vectors $p_l^{\varepsilon,i}$, $q_l^{\varepsilon,i}$, $i = \overline{0, M-1}$, and $l = \overline{1, nn}$, are obtained solving the linear system (4.140), respectively;

P4. For all $i \in \{0, 1, ..., M\}$ compute $r^{\varepsilon,(iter)}(t_i)$ and $\tilde{v}^{\varepsilon,(iter)}(t_i)$, from (4.116) and (4.115), respectively;

P5. Compute $\lambda_{iter} \in [0, 1]$ (the steplength of the gradient method) solution of the minimization process

$$\min\{j^\varepsilon(\lambda v^{\varepsilon,(iter)} + (1-\lambda)\tilde{v}^{\varepsilon,(iter)}), \quad \lambda \in [0, 1]\};$$

Set $v^{\varepsilon,(iter+1)} := \lambda_{iter} v^{\varepsilon,(iter)} + (1 - \lambda_{iter})\tilde{v}^{\varepsilon,(iter)}$;

P6. (the "*Stopping Criterion*")

if $\|v^{\varepsilon,(iter+1)} - v^{\varepsilon,(iter)}\| \leq \eta$

then STOP (the algorithm is convergent)

else $iter := iter + 1$; Go to **P1**.

As in `cpht-1D`, $iter$ represents the number of iterations after which the algorithm `cpht-2D` found the optimal value of the cost functional j^ε in (P^ε).

Remark 5.1 The "*Stopping Criterion*" in **P6** may be also

$$\|j^\varepsilon(v^{\varepsilon,(iter+1)}) - j^\varepsilon(v^{\varepsilon,(iter)})\| \leq \eta$$

where η is a prescribed precision.

Let us briefly discuss the main steps in algorithm `cpht-2D`. For approximating the solution of the nonlinear parabolic system (3.22)-(3.24) from Section 3.2, we have used a numerical method of fractional steps type (see (4.111)-(4.114)). This method, expressed in step **P2**, avoids the iterative process required by the classical approaches (e.g., Newton's type method) in passing from a time level to another (see also C. Moroşanu [19] for additional details).

The values of λ_{iter} from **P5** are chosen from the sequence

$$1 - d, \quad 1 - 2 * d, \quad 1 - 3 * d, ..., 0$$

where $d > 0$ is prescribed (in the numerical example $0 < d < 0.05$). Since we have a finite number of options for λ, we choose $\lambda_{iter} = \lambda$ the value from the above sequence which minimizes

$$j^{\varepsilon}(\lambda v^{\varepsilon,(iter)} + (1-\lambda)\tilde{v}^{\varepsilon,(iter)}), \quad \lambda \in [0,1].$$

Next, for the entirety, we illustrate our algorithm by a concrete numerical computation. For this let the domain $\Omega = \Omega_0$ be $[0,1] \times [0,1] \subset I\!R^2$ and the time interval be $[0,2]$. For the time as well as for the spatial discretization we use uniform discretization. The time step was chosen $\varepsilon = 0.4$ (i.e. $M_{\varepsilon} = 5$). The triangulation of the spatial domain is obtained by dividing x- and y- axis into equidistant subintervals of length $dx = dy = 0.048$ and then dividing each of the resulting squares into two triangles. The values of parameters are: $\xi = 0.5$, $a = \sqrt{\xi}$, $\tau = 1.0e + 5 * \xi^2$, $l = 0.6$, $k = 0.06$, $h = 1.2$, and $\beta = 1$.

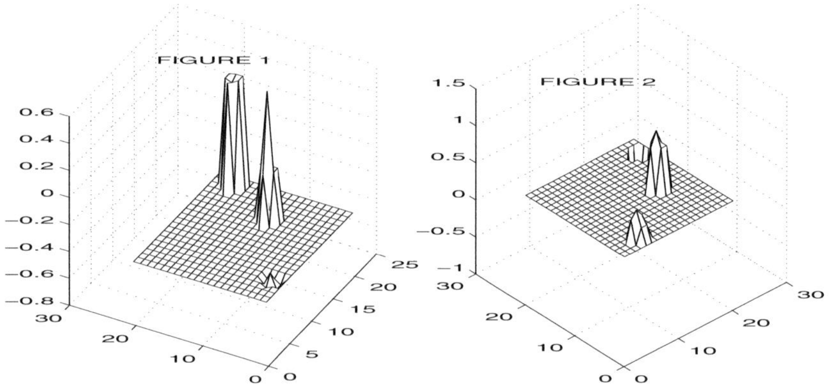

Figure 13 *The initial values* $u_0(x,y)$ *and* $\varphi_0(x,y)$

We shall present now a numerical example. Fig. (**13**) show the initial conditions. We choose $R = 200$ and $v_i^{\varepsilon,(0)} = R/2, \quad i = 1,2,...,M_{\varepsilon}$. In *Table* **1** we present the results obtained:

iter	$j^{\varepsilon}(v_{\varepsilon})$
1	4.162858e+03
3	2.091130e+03
5	8.291100e+02

Table 1

The optimal value $j^{\varepsilon}(v_{\varepsilon}^{*}) = j(v^{*}) = 5.127619e + 01$ was obtained in 8 iterations. Corresponding, the approximating optimal control is

$$v_{\varepsilon}^{*} = (9.739e - 01, 9.739e - 01, 9.739e - 01, 9.739e - 01, 9.739e - 01).$$

So the reader can easily appreciate that the algorithm `cpht-2D` is functional and, in consequence, may be implemented on a computer with higher performance that will permit to use a higher discretization.

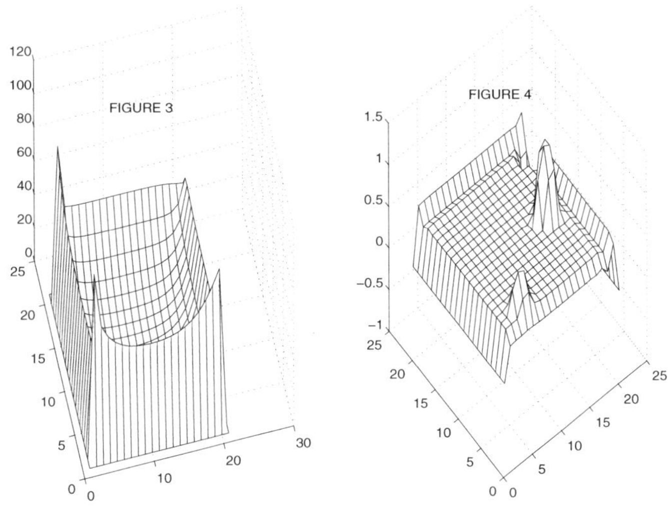

Figure 14 *The optimal states* $u_{\varepsilon}^{*}(x, y)$ *and* $\varphi_{\varepsilon}^{*}(x, y)$

Fig. (**14**) show the approximate solutions u_{ε}^{*} and $\varphi_{\varepsilon}^{*}$, respectively, corresponding to the approximating optimal control v_{ε}^{*} (see Theorem 3.2).

Again, we underline that the numerical results do not have in view the physical aspects. They are given only to implement the conceptual algorithm. The comparison between the algorithm `cpht-2D` and another numerical algorithms for approximation of optimal control in problem (P_0) (see Section 3.2), as well as numerical results regarding the physical nature of this problem, remains a matter for further investigation.

Analysis and Optimal Control of Phase-Field Transition System: Fractional Steps Methods, 2012, 247-282 247

CHAPTER 5

PROGRAM LISTINGS AND NUMERICAL RESULTS

Costică MOROŞANU

Faculty of Mathematics, "Al.I. Cuza" University of Iasi, Romania

Abstract: In this Chapter we give a description of the programs implementing the conceptual algorithms introduced in Chapter 4 to approximate the solution of nonlinear phase-field transition system. We also give some numerical results in order to illustrate the description of the programs and to convince readers of their correctness, in terms of syntactic and semantic.

All programs in this Chapter were written in MATLAB and have been endowed with a sufficient amount of comments.

Keywords: nonlinear parabolic systems, free boundary problems for PDE, computer science, analysis of algorithms, software, programming languages, phase changes, mathematical modeling, applications in engineering and industry, simulation.

5.1 The program implementing the algorithm `algnewton1D`

This Section describes the program implementing the algorithm `algnewton1D` stated in Section 4.1a. Further we give the program listing, calling *P_algnewton1D*, which implement the iterative *Newton method* (relation (4.22), Chapter 4) in order to compute the approximate solution $v^i = (u^i, \varphi^i)^T$, $i = 1, 2, ..., M$, of the nonlinear discrete phase-field transition system (4.20), Chapter 4.

Program *P_algnewton1D*

// Implementation of the conceptual algorithm "*algnewton1D*" to

// compute the approximate solution of nonlinear phase-field

// transition system by *Newton method.*

```
//
//    b,T  are input parameters, representing the length of the space
//              and the time intervals, respectively.
//    M     is an input parameter; it is the dimension of the uniform
//              grid {t_i}_{i=1}^M on [0,T] (t_1 = 0, t_M = T).
//    N     is an input parameter; it is the dimension of the uniform
//              grid {x_j}_{j=1}^N on [0,b] (x_1 = 0, x_N = b).
//    itnew is the number of loops performed (repeat (itnew)).
//    itmax is an input parameter; it is the maximum number
//              of iterations prescribed in the Newton method.
//    epsnw is an input parameter; it gives the accuracy of the
//              approximate solution.
//    R, prho, pc, pl, pcsi, ptau, ph, pk, pa are input
//              parameters; they are equal to the quantities R, ρ, c, ℓ,
//              ξ, τ, h, k, a, respectively.
//    dt, dx are input parameters; they are the time step size and
//              the space step size, respectively.
//    t     is the vector of length M containing the points t_i, i = 1,M.
//    x     is the vector of length N containing the points x_j, j = 1,N.
//    a1, a2, c1, c2, c4, c5, c6 are nonzero coefficients of
//              matrices A_11, A_22 (relation (4.20) in Chapter IV).
//    u0    is the vector of length N containing the initial values
//              u_0(x_j), j = 1,N.
//    fi0   is the vector of length 84 containing the initial values
//              φ_0(x_j), j = 1,84.
//    w is an input parameter; an array of length 2*M that contain
//              the boundary values in 0 and b, respectively.
//    u     is the approximate matrix for unknown u; an array
//              of length M*N.
//    fi    is the approximate matrix for unknown φ; an array
```

// of length M*N.
// d is the vector of length 2*N containing the right-hand side
// corresponding to the nonlinear system (4.20) in Ch. IV.
// F is a vector of length 2*N; it contains the quantities
// $F = (F_1, F_2, \cdots, F_{2N})^T$.
// Fpr is the matrix F' of dimension 2N*2N - see relation (4.21)
// in Chapter IV.
// v, yk, vmyk, tl are work vectors, each having the length
// equal with 2*N.
// dx1, x1, pa4, f0, t1, conv, tt, dist, t2, i, im1,
// j, jj are dummy variables.
// timp_CPU is an output parameter; CPU time spent.
//

```
format short e
clear
```

P0. Initial values

```
itmax = 20;
epsnew =.01;
```

P01. Physical parameters

```
R = 30;
prho= 7.85;
pc = 12.5;
pl = 65.28;
pcsi = .5;
ptau = 1.e+3*pcsi^2;
ph = 32.012;
pk = 7.8e-2;
```

P02. Construction of the time and space meshes

```
T = input('T=');
M = input('M=');
```

```
dt = T/(M-1);
t = 0:dt:T;
b = input('b=');
N = input('N=');
dx = b/(N-1);
x = 0:dx:b;
```

P03. Choose u_0, φ_0, w

```
dx1 = b/83;
x1 = 0:dx1:b;
fi0 = [-1.4 -1.4 -1.44 -1.42 -1.42 -1.44 -1.43 -1.43 -1.42 ...
    -1.42 -1.4 -1.4 -1.25 -1.2 -1.17 -1.15 -1.1 -1.08 ...
    -1.0 -.95 -.9 -.85 -.88 -.6 .0 .5 -.92 -.25 .8 -.7 .58 ...
    .75 .58 -.63 -.59 .69 -.72 .7 -.59 -.5 .7 -.79 -.87 -.88 ...
    .0 .72 -.8 .81 .0 -.89 .0 .7 .55 .68 -.49 .79 .0 -.1 -.8 ...
    -.78 -.83 .69 -.8 .68 .5 .7 .59 1. 1.08 1.1 1.15 1.17 1.2 ...
    1.25 1.3 1.3 1.25 1.24 1.3 1.31 1.3 1.32 1.3 1.3];
fi0cs = csapi(x1,fi0');
pa = sqrt(pcsi);
pa4 = 1./(4*pa);
for j=1:N
  f0(j) = fnval(fi0cs,x(j));
  u0(j) = (f0(j)*f0(j)*f0(j) - f0(j))*pa4;
end
u0cs = csapi(x,u0');
```

// choose w(t) from $\mathbb{R}$*,* $t\in [0,T]$

```
w = zeros(2,M);
w(1,:)  = -R/2;     % -2*R;
w(2,:)  = R/4;
```

// store the space for: u, fi, F, Fpr

```
u = zeros(M,N);
```

```
fi = zeros(M,N);
F = zeros(1,2*N);
Fpr = zeros(2*N,2*N);
  // u^1, fi^1 from initial conditions (4.13)
u(1,:)  = u0;
fi(1,:)  = f0;
  // Program Start
  t1 = cputime;
  c1 = -pk*dt/(dx*dx);
  c2 = prho*pc - 2*c1;
  a1 = c1 + c2 + 1 + dx*ph;
  c4 = dt/(2*pa);
  c5 = - pcsi^2*dt/(dx*dx);
  c6 = ptau - 2*c5 - c4;
  a2 = c5 + c6;

  P1. Solving the linear system (4.22)
  conv = 0;
    P11. Loop after the time level i
  for i = 2:M
    im1 = i - 1;
    // Start iteration
    yk = zeros(1,2*N);
    d = zeros(1,2*N);
    yk = [u(im1,:)  fi(im1,:)];        % yk ← v^(1), see (4.21)
    // the right side in  (4.20)
    for j = 2:N-1
      d(j) = prho*pc*u(im1,j) + (pl/2)*fi(im1,j);
    end
    d(1) = prho*pc*u(im1,1) ...
             + (pl/2)*fi(im1,1) + dx*w(1,im1);
```

```
d(N) = prho*pc*u(im1,N) ...
          + (pl/2)*fi(im1,N) + dx*w(2,im1);
for j = 1:N
  jj = N + j;
  d(jj) = ptau*fi(im1,j);
end
```

//

P12. Loop after `itnew` - for approximate computation of `v`

```
for itnew = 1:itmax                    % itnew ← k, see (4.21)
  Fpr(1,1) = a1;
  Fpr(1,2) = c1 - 1;
  Fpr(1,N+1) = pl/2;
  Fpr(N,N-1) = Fpr(1,2);
  Fpr(N,N) = Fpr(1,1);
  Fpr(N,2*N) = pl/2;
  Fpr(N+1,1) = -2*dt;
  Fpr(N+1,N+1) = a2 + 3.*c4*yk(N+1)*yk(N+1);
  Fpr(N+1,N+2) = c5;
  Fpr(2*N,N) = -2*dt;
  Fpr(2*N,2*N-1) = c5;
  Fpr(2*N,2*N) = a2 + 3.*c4*yk(2*N)*yk(N+N);
  for j=2:N-1
    Fpr(j,j-1) = c1;
    Fpr(j,j) = c2;
    Fpr(j,j+1) = c1;
    Fpr(j,N+j) = pl/2;
    Fpr(N+j,j) = dt;
    Fpr(N+j,N+j-1) = c5;
    Fpr(N+j,N+j) = c6 + 3.*c4*yk(N+j)*yk(N+j);
    Fpr(N+j,N+j+1) = c5;
```

```
        end
        // Calculates F(yk)
        F(1) = a1*yk(1) + (c1-1)*yk(2) ...
             + (pl/2)*yk(N+1) - d(1);
        for j = 2:N-1
          F(j) = c1*yk(j-1) + c1*yk(j+1) ...
               + c2*yk(j) + (pl/2)*yk(N+j) - d(j);
        end
        F(N) = a1*yk(N) + (c1-1)*yk(N-1) ...
             + (pl/2)*yk(2*N) - d(N);
        F(N+1) = -2*dt*yk(1) + a2*yk(N+1) ...
               + c4*yk(N+1)*yk(N+1)*yk(N+1) ...
               + c5*yk(N+2) - d(N+1);
        for j = 2:N-1
          F(N+j) = -2*dt*yk(j) + c5*yk(N+j-1) ...
            + c5*yk(N+j+1) + c6*yk(N+j) ...
            + c4*yk(N+j)*yk(N+j)*yk(N+j) - d(N+j);
        end
        F(2*N) = -2*dt*yk(N) + c5*yk(2*N-1) ...
               + c4*yk(2*N)*yk(2*N)*yk(2*N) ...
               + a2*yk(2*N) - d(2*N);
        // Calculates the right side of (4.22) in tl
        tl = zeros(1,2*N);
        for j = 1:2*N
          tt = 0.;
          for k = 1:2*N
            tt = tt + Fpr(j,k)*yk(k);
          end
          tl(j) = tt - F(j);
```

```
        end
        // New solution in v
        v = zeros(1,2*N);
        [L,U] = lu(Fpr);                        %see Remark 5.1
        v = U\(L\tl');
        // Computes the norm
        vmyk = zeros(1,2*N);
        vmyk(1,:) = v(1,:) - yk(1,:);
        dist = norm(vmyk,2);
        // Convergence test
        if dist < epsnew
        // Solution on level i
         for j = 1:N
          u(i,j) = v(j);
          fi(i,j) = v(N+j);
         end
          break;
          % found the solution on level i; leaving loop itnew
        end
        if itnew < itmax
          yk = v;
         else
          sprintf(
       '\n Newton Method not converged on level: %d\n',i);
          conv = -1;
        end
      end     % next itnew
      if conv == -1
        conv = 0;
        break;
```

```
                              % Program End; algnewton1D not converged
128         end
129      end     % next i
      // Program End
130      t2 = cputime;
131      timp_CPU = t2 - t1
```

Details about the functioning of the program *P_algnewton1D* are given in the sequel.

The lines 003-048 initialize the work arrays (parameters model, time and space grids, dimensions of arrays, initial conditions, etc).

The lines 049-129 implement the iterative *Newton method*:

- the lines 052-054 construct the iteration start. The starting iterate $v^{(itnew)}$ for the Newton method corresponding to the time level i is $v(t_{i-1})$ which was already computed;
- the lines 055-063 establish the system (4.20);
- the lines 064-124 repeat Newton iteration (4.22) until convergence;
- the line 107 computes, in vector v, the solution $v^{(itnew+1)}$, corresponding to the linear system (4.22);
- in the line 111 the convergence of the Newton method is checked;
- in the lines 112-117 the solution corresponding to the time-level i is constructed, founded by iterative Newton method in *itnew* iterations and with *epsnew* accuracy;
- in the lines 118-128 is treated the case when the program *P_algnewton1D* implementing the algorithm *algnewton1D* does not find the approximate solution on level i with accuracy *epsnew* in *itmax* iterations.

The program *P_algnewton1D* stops working by line 130.

The numerical results obtained through this program are presented in Section V.3.

Remark 5.1. *Noting that all elements of the main diagonal of block matrices in* (4.20) *are* $\neq 0$, *then the matrix coefficients* $[Y'(\cdot)]$ *in linear system* (4.22) *always have an LU factorization.*

5.2 The program implementing the algorithm `algfrac1D`

In this Section we will describe the program implementing the algorithm `algfrac1D` stated in Section 4.1b.

Below we give the program listing calling *P_algfrac1D* which implement the *fractional steps method* in order to compute the approximate solution u^i, φ^i, $i = 1, 2, ..., M$, of the linear system (4.27), Chapter 4.

Program *P_algfrac1D*

// Implementation of the conceptual algorithm "*algfrac1D*" to
// compute the approximate solution of nonlinear phase-field
// transition system by *fractional steps method.*
//
// `b,T` are input parameters representing the length of the space
// and the time intervals, respectively.
// `M` is an input parameter; it is the dimension of the uniform
// grid $\{t_i\}_{i=1}^{M}$ on $[0,T]$ $(t_1 = 0,\ t_M = T)$.
// `N` is an input parameter; it is the dimension of the uniform
// grid $\{x_j\}_{j=1}^{N}$ on $[0,b]$ $(x_1 = 0,\ x_N = b)$.
// `eps` is an input parameter, given by *fractional steps scheme.*
// `R, prho, pc, pl, pcsi, ptau, ph, pk, pa` are input
// parameters; they are equal with the quantities R, ρ, c,
// ℓ, ξ, τ, h, k, a, respectively.
// `dt, dx` are input parameters; they are the time step size and
// the space step size, respectively.
// `t` is the vector of length `M` containing the points t_i, $i = \overline{1,M}$.
// `x` is the vector of length `N` containing the points x_j, $j = \overline{1,N}$.
// `a1, a2, c1, c2, c3, c4, c5, c6` are nonzero coefficients of
// matrices A_{11}, A_{12}, A_{22} (see (4.20) in Chapter 4).
// `u0` is the vector of length `N` containing the initial values
// $u_0(x_j)$, $j = \overline{1,N}$.
// `fi0` is the vector of length `84` containing the initial values

// $\varphi_0(x_j)$, $j=\overline{1,84}$.
// z_eps is the vector of length N; the solution of Cauchy
// problem (4.25) on $[0,\varepsilon]$.
// w is an input parameter; an array of length 2*M that contain
// the boundary values in 0 and b, respectively.
// u is the approximate matrix for unknown u; an array
// of length M*N.
// fi is the approximate matrix for unknown φ; an array
// of length M*N.
// d is the vector of length 2*N containing the right-hand side
// corresponding to the linear system (4.27) in Ch. 4.
// aa is the matrix coefficients A of length 2N*2N
// - relation (4.27) in Chapter 4.
// z is a work array having the length equal to 2*N.
// dx1, x1, pa4, f0, t1, t2, i, im1, ip1, j, jj -
// are dummy variables.
// timp_CPU is an output parameter; CPU time spent.
//

```
format short e
clear
```

P0. Initial values

```
eps =.01;
```

P01. Physical parameters

```
R = 30;
prho= 7.85;
pc = 12.5;
pl = 65.28;
pcsi = .5;
ptau = 1.e+3*pcsi^2;
ph = 32.012;
```

```
pk = 7.8e-2;
```

P02. Construction of the time and space meshes

```
T = input('T=');
M = input('M=');
dt = (T-eps)/(M-1);
t = eps:dt:T;
b = input('b=');
N = input('N=');
dx = b/(N-1);
x = 0:dx:b;
```

P03. Choose u_0, φ_0, w

```
dx1 = b/83;
x1 = 0:dx1:b;
fi0 = [-1.4 -1.4 -1.44 -1.42 -1.42 -1.44 -1.43 -1.43 -1.42 ...
    -1.42 -1.4 -1.4 -1.25 -1.2 -1.17 -1.15 -1.1 -1.08 ...
    -1.0 -.95 -.9 -.85 -.88 -.6 .0 .5 -.92 -.25 .8 -.7 .58 ...
    .75 .58 -.63 -.59 .69 -.72 .7 -.59 -.5 .7 -.79 -.87 -.88 ...
    .0 .72 -.8 .81 .0 -.89 .0 .7 .55 .68 -.49 .79 .0 -.1 -.8 ...
    -.78 -.83 .69 -.8 .68 .5 .7 .59 1. 1.08 1.1 1.15 1.17 1.2 ...
    1.25 1.3 1.3 1.25 1.24 1.3 1.31 1.3 1.32 1.3 1.3];
fi0cs = csapi(x1,fi0');
pa = sqrt(pcsi);
pa4 = 1./(4*pa);
for j=1:N
  f0(j) = fnval(fi0cs,x(j));
  u0(j) = (f0(j)*f0(j)*f0(j) - f0(j))*pa4;
end
u0cs = csapi(x,u0');
```

// choose $w(t)$ *from* $\mathbb{R}$, $t \in [0,T]$

```
w = zeros(2,M);
```

```
w(1,:)  = -R/2;     % -2*R;
w(2,:)  = R/4;
  // store the space for: u, fi
u = zeros(M,N);
fi = zeros(M,N);

P1. Compute z(eps,.) from (4.25)
z_eps = f0.*sqrt(pa./(pa + eps.*f0.^2));
  // u^1, fi^1 from initial conditions (4.24)-(4.25)
u(1,:)  = u0;
fi(1,:)  = z_eps;
  // Program Start
  t1 = cputime;
  c1 = -pk*dt/(dx*dx);
  c2 = prho*pc - 2*c1;
  c3 = pl/2;
  a1 = c1 + c2 + 1 + dx*ph;
  c4 = dt/(2*pa);
  c5 = - pcsi^2*dt/(dx*dx);
  c6 = ptau - 2*c5 - c4;
  a2 = c5 + c6;
//
  aa(1,1) = a1;
  aa(1,2) = c1 - 1;
  aa(1,N+1) = c3;
  aa(N,N-1) = aa(1,2);
  aa(N,N) = aa(1,1);
  aa(N,2*N) = c3;
  aa(N+1,1) = -2*dt;
  aa(N+1,N+1) = a2;
  aa(N+1,N+2) = c5;
```

```
aa(2*N,N) = -2*dt;
aa(2*N,2*N-1) = c5;
aa(2*N,2*N) = a2;
for j=2:N-1
  aa(j,j-1) = c1;
  aa(j,j) = c2;
  aa(j,j+1) = c1;
  aa(j,N+j) = c3;
  aa(N+j,j) = -2*dt;
  aa(N+j,N+j-1) = c5;
  aa(N+j,N+j) = c6;
  aa(N+j,N+j+1) = c5;
end
```

P2. Solving the linear system (4.27)

```
for i = 2:M                    % Loop after the time level i
  im1 = i - 1;
  ip1 = i + 1;
  d = zeros(1,2*N);
  // the right side in (4.27)
  for j = 2:N-1
    d(j) = prho*pc*u(im1,j) + c3*fi(im1,j);
  end
  d(1) = prho*pc*u(im1,1) ...
         +c3*fi(im1,1)+dx*w(1,im1);
  d(N) = prho*pc*u(im1,N) ...
         +c3*fi(im1,N)+dx*w(2,im1);
  for j = 1:N
    jj = N + j;
    d(jj) = ptau*fi(im1,j);
  end
```

```
//
            // New solution in  z
083         z = zeros(1,2*N);
084         [L,U] = lu(aa);                              % see Lemma 5.1
085         z = U\(L\d');
            // Solution on level  i
086          for j = 1:N
087           u(i,j) = z(j);
088           fi(i,j) = z(N+j);
089          end
090      end     % next  i
         // Program End
091      t2 = cputime;
092      timp_CPU = t2 - t1
```

Details about the functioning of the program *P_algfrac1D* are given in the sequel.

The lines 003-069 initialize the work arrays (parameters model, time and space grids, dimensions of arrays, solve the Cauchy problem (4.25), initial conditions, etc).

The lines 070-090 implement the *fractional steps method*:

- the lines 074-082 construct the right side in (4.27);
- the line 085 compute, in vector z, the solution u^i, φ^i, corresponding to the linear system (4.27);

The program *P_algfrac1D* stops working at line 091.

The numerical results obtained by this program are presented in Section 5.3 below.

Lemma 5.1. *If*

$$\tau + \xi^2 \frac{dt}{dx^2} \neq \frac{dt}{2a}, \tag{5.1}$$

then the matrix coefficients in linear system (4.27) *can be factored into the product of a lower-triangular matrix and an upper-triangular matrix* (*LU - factorization*).

Proof. Let the elements of matrix coefficients in linear system (4.27), Chapter 4 denoted by a_{mn}, $m, n = 1, 2, \cdots, 2N$, . Analyzing the main diagonal elements of block matrices A_{11} and A_{22} in (4.20), Chapter 4, we first find that $a_1 = c_1 + c_2 + 1 + dx \cdot h = \rho c + k\frac{dt}{dx^2} + 1 + dx \cdot h \neq 0$ and $c_2 = \rho c - 2c_1 = \rho c + 2k\frac{dt}{dx^2} \neq 0$. Observing now that $c_5 + c_6 \neq 0$ reflects the assumptions expressed in (5.1), as well as $c_6 \neq 0$, we can easily find that $a_{mm} \neq 0\ \forall m = 1, 2, \cdots, 2N$. Therefore, Gaussian elimination can be performed on the system (4.27) without interchanges; consequently A has an LU factorization.

5.3 The comparison between the Newton method - `algnewton1D` and the fractional steps method - `algfrac1D`

The aim of this Section is to present numerical experiments implementing the conceptual algorithms `algnewton1D` and `algfrac1D`. In what follows, all computations were performed taking $T = 2$, $b = 1$, $M = 100$, $N = 40$, while, for the model's parameters (see lines 005-012 and 025-026 in program listing*P_algnewton1D*) we have considered industrial values, which are:

- the casting speed ($c = 12.5$ mm/s),
- physical parameters:
  - the density ($\rho = 7.85$ kg/m^3),
  - the latent heat ($\ell = 65.28\ kcal/kg$),
  - the thermal conductivity ($k = 7.8e - 2$),
  - the length of separating zone ($\xi = .5$),
  - the relaxation time ($\tau = 1.0e + 3 * \xi^2$),
  - the coefficients of heat transfer ($h = 32.012$),
  - $a = \sqrt{\xi}$;

The initial values $\varphi_0(x_j)$, $j = 1, 2, ..., N$, plotted in Fig. (**1**) - left side, were computed via Matlab function `csapi(fi0)` - cubic spline interpolant to the given data `fi0` (see line 023 in program listing *P_algnewton1D*).

In view of the physical nature of the phase-field transition system, it is natural to work with a function $u(t, x)$ verifying together with the solution $\varphi(t, x)$ the stationary equation $\varphi_t = \Delta\varphi = 0$. Accordingly, the initial values $u_j^0 = u_0(x_j)$, $j = 1, 2, ..., N$, plotted in Fig. (**1**) - right side, were computed as solution of the discrete form to the stationary equation $(2a)^{-1}[\varphi_0(x) - \varphi_0^3(x)] + 2u_0(x) = 0$ (see Caginalp [36]), i.e.:

$$\frac{1}{2a}\left[\varphi_0(x_j) - \varphi_0^3(x_j)\right] + 2u_0(x_j) = 0 \quad j = 1, 2, ..., N.$$

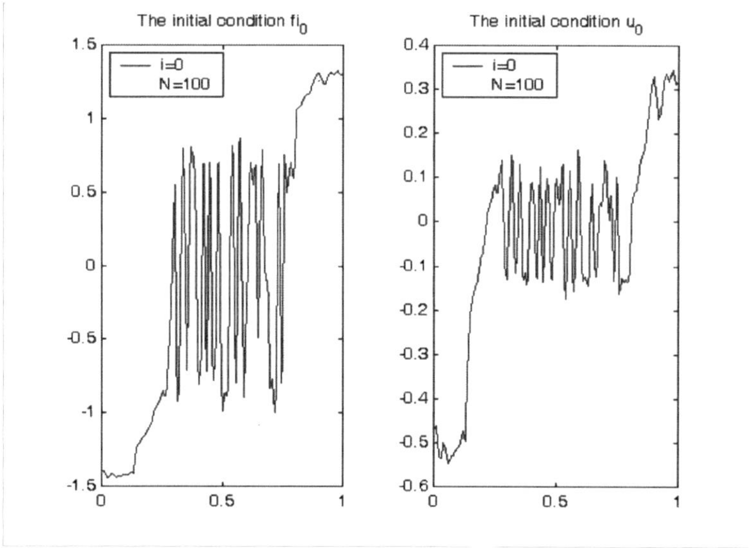

Figure 1 *The initial conditions* φ_0 *and* u_0

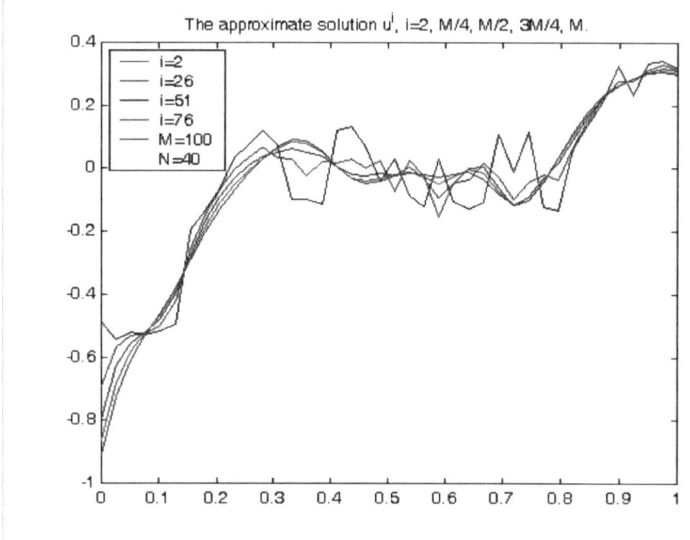

Figure 2 *The approximate solution* $z(\varepsilon, \cdot)$ *of Cauchy problem* (4.10)

So, the starting iterate $v^{(1)}$ for the Newton method corresponding to the time level i (see relation (4.21) and line 054 in program listing *P_algnewton1D*) is $v(t_{i-1})$ which was already computed.

Now we are able to compute (see relations (4.24)-(4.25) in Chapter 4) the vectors: $\varphi^1 = \left(\varphi_j^1\right)_{j=\overline{1,N}}$, $u^1 = \left(u_j^1\right)_{j=\overline{1,N}}$ and $z(\varepsilon,\cdot) = \left(z\left(\varepsilon,\varphi_0(x_j)\right)\right)_{j=\overline{1,N}}$, plotted in Fig. (**2**).

The numerical results, obtained by algorithms `algnewton1D` and `algfrac1D`, were plotted in Figs. (**3-4**) and (**5-6**), respectively.

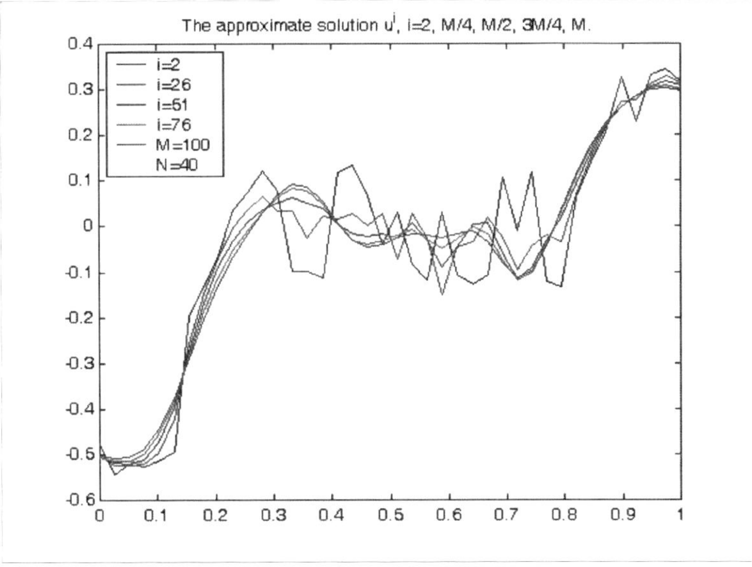

Figure 3 *u^i at different levels of time, obtained by* `algnewton1D`

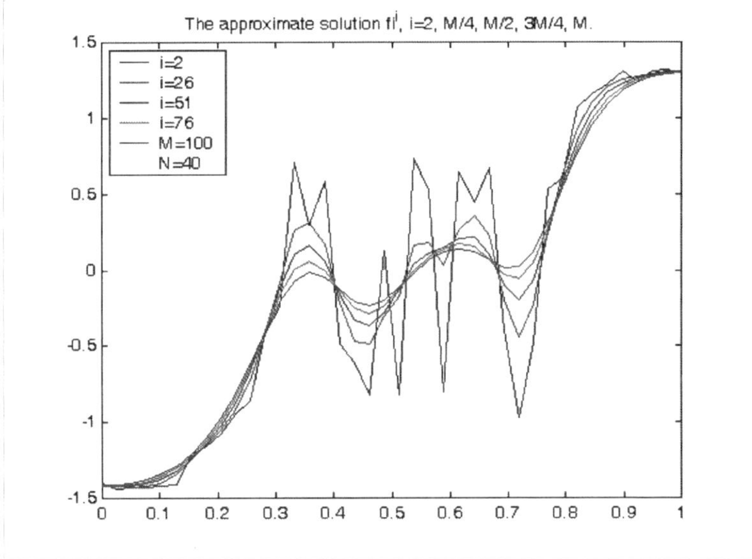

Figure 4 *φ^i at different levels of time, obtained by* `algnewton1D`

For this test we have used the temperature surrounding at $\partial\Omega=\{0,b\}$ given by $w(t_i,0) = -R/2$, $w(t_i,b) = R/4$, $i = 1,2,...,M$ (see lines 033-034 in program listing *P_algnewton1D*).

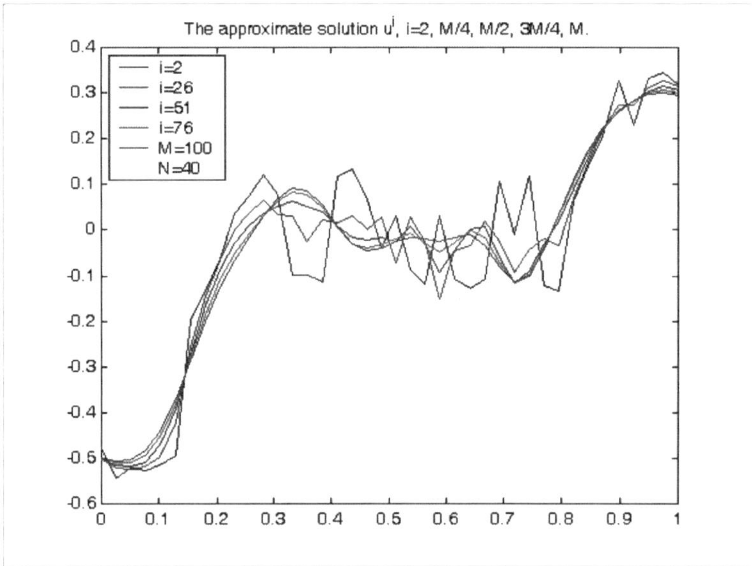

Figure 5 u^i *at different levels of time, obtained by* `algfrac1D`

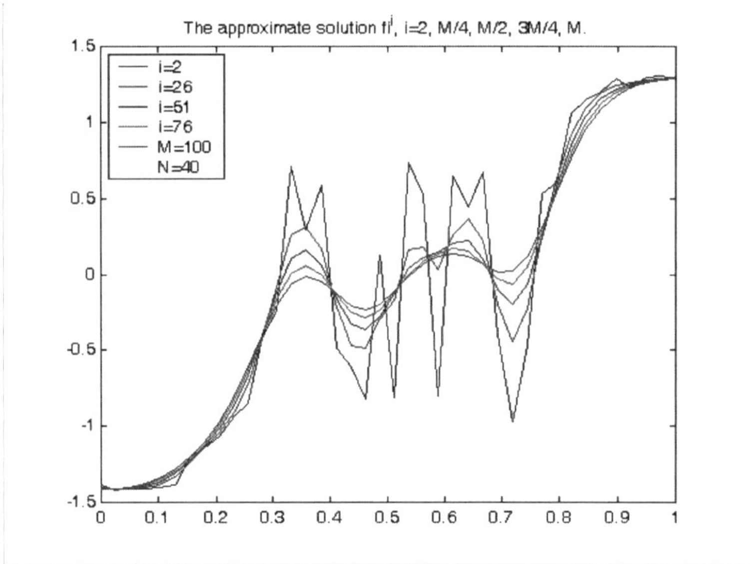

Figure 6 φ^i *at different levels of time, obtained by* `algfrac1D`

Let us remark that *Newton method* is an iterative method, while *fractional steps method* does not require iterations to advance from time level $i-1$ to i (compare `algnewton1D` with `algfrac1D`). Hence, the *fractional steps method* is faster which is showed by the numerical tests. Table **1** below precisely shows the CPU-time (in seconds) taken by algorithms `algnewton1D` and `algfrac1D`.

M	N	$algnewton1D$	$algfrac1D$
100	40	2.969	0.313
200	100	34.063	1.359
300	100	51.140	2.016

Table 1. *CPU-time taken by* `algnewton1D` *and* `algfrac1D`

Numerical stability *via* `algfrac1D`. We continue this Section by presenting numerical experiments regarding the *stability* of equation (4.28) (see Proposition 4.1 in Chapter 4). The shape of the graphs plotted in Fig. (**5**) and (**6**) shows the stability and accuracy of the numerical results obtained by algorithm `algfrac1D`.

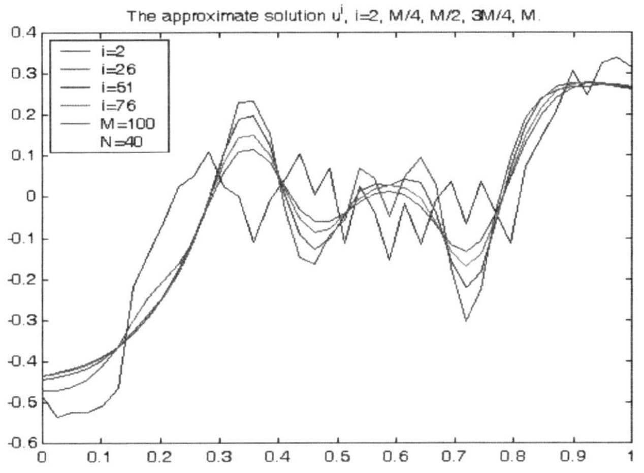

Figure 7 *An example of slight numerical instability*

Taking now k=*.785*, we can verify that $v_{min} - v_{max}$=*-15.2372* which means that the first hypothesis in Proposition 4.1 is not verified. Consequently, the numerical scheme is unstable as shown in Fig. (**7**). Furthermore, if we keep k=*.785* and take τ=*1.0e+2**ξ^2 (in place of

τ=*1.0e+3*$^*\xi^2$), we can also get $v_{min} - v_{max} < 0$, which is also an unstable case. Moreover, analyzing the graph in Fig. (**8**) we found a more pronounced instability. Let us remark that the instability of the solution occurred following a slight change (modification) of only two physical parameters (k and τ in this case). This highlights the strong dependence of approximation scheme regarding physical parameters.

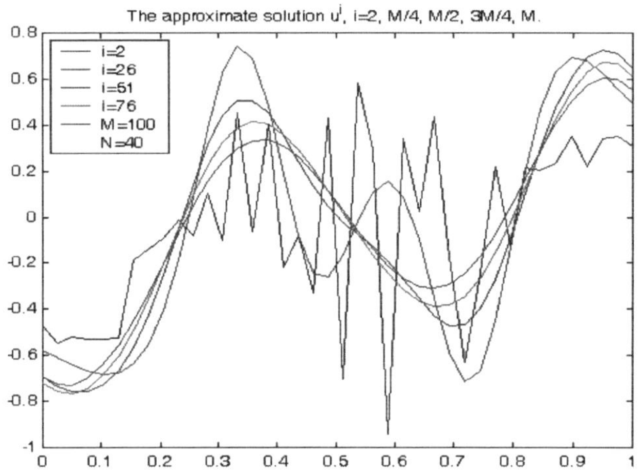

Figure 8 *An example of strong numerical instability*

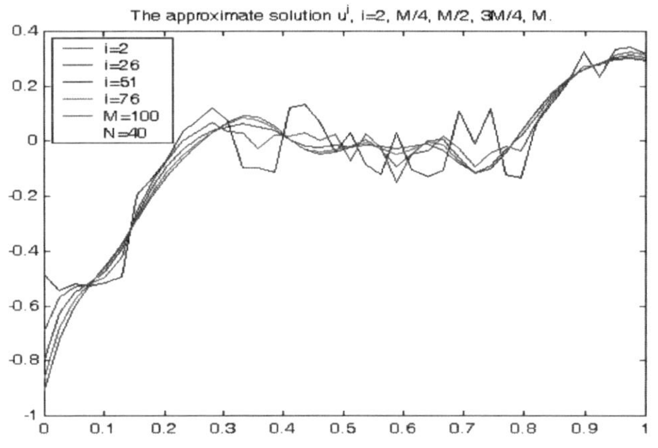

Figure 9 u^i *corresponding to* $w^i(0) = -2R$, *via* `algfrac1D`

We turn to numerical stability conditions and we change the temperature surrounding at $0 \in \partial\Omega$ by setting $w(t_i, 0) = -2 * R$, $i = 1, 2, ..., M$. The numerical results obtained by algorithm `algfrac1D` were plotted in Fig. (**9**). Analyzing the approximations near zero, we observe an instability just for u, due to the nature of boundary conditions that we have considered $(4.5)_1$. In addition we also find a slight difference in the error of approximation.

As a conclusion, let us remark that conditions of stability are sustained by both theory and numerical experiment and that they are significantly dependent on the physical parameters.

Analyzing the numerical results in terms of physical phenomena, we find that *the temperature distribution* tends to become parabolic and *the phase function* distribution i.e., say that the instability of the portion of material will disappear. Moreover, analyzed together (see Fig. (**5**) and (**6**), for example), theoretical meaning assigned to functions u and φ as well as the *zone of separation* between material phases have been highlighted.

The numerical solution obtained through this way can be considered admissible one for the corresponding boundary optimal control problem (from this perspective, compare Fig. (**5**) and (**9**)). Generally, the *fractional steps method* considered here can be used to approximate the solution of a nonlinear parabolic phase-field system containing a general nonlinear part.

5.4 The program implementing the algorithm `algfracfem2D`

In this Section we will describe the program implementing the algorithm `algfracfem2D` stated in Section 4.2.

Before starting, we like to recall the result given by Theorem 2.6 in Chapter 2 which guarantees that the approximate solution of the problem (4.7)-(4.10) can be regarded as approximate solution of the nonlinear phase-field transition system (4.4)-(4.6) in Chapter 4.

The finite element method (**fem**) is a general method for approximating the solution of boundary value problems for partial differential equations. This method is derived from the Ritz (or Gelerkin) method, characteristic for the finite element method being the choice of the finite dimensional space, namely, the *span* of a set of finite element basis functions (see ANNEXE B for more details). The steps in solving a boundary value problem using **fem** are:

P0. (D) The direct formulation of the problem;

P1. (V) A variational (weak) formulation for problem (D);

P2. The construction of a finite element mesh (triangulation);

P3. The construction of the finite dimensional space of test function, **called finite element basis functions;**

P4. (V_{nn}) A discrete analogous of (V);

P5. Assemble the linear system of equations;

P6. Solve the system obtained in P5.

In our concrete case, on the position of (D) we have the problem given by relations (4.7)-(4.8) and, corresponding to (V), the relations (4.56)-(4.57), are stated in Chapter 4. The next steps (**P2**-**P4**) were prepared in Section 4.2 and ending with the discrete equations corresponding to (4.7)-(4.8), i.e.:

$$\begin{cases} Ru_l^i+\frac{\ell}{2}B\varphi_l^i+dt\ khFRu_l^i=B(\rho c\ u_l^{i-1}+\frac{\ell}{2}\varphi_l^{i-1}+dt\ kw^{i-1}), \\ S\varphi_l^i - 2dt\ Bu_l^i = B\tau\varphi_l^{i-1}, \end{cases} \tag{5.2}$$

where u_l^i, φ_l^i, $i=\overline{1,M}$, $l=\overline{1,nn}$, are unknown vectors for time level i.
From the initial conditions (4.9) we have

$$u_l^0 = u_0(N_l), \qquad \varphi_l^0 = \varphi_0(N_l) \qquad l=\overline{1,nn}, \tag{5.3}$$

and therefore the numerical algorithm `algfracfem2D` to compute the approximate solution of the nonlinear system (4.4)-(4.6) by *fractional steps method* can be rewritten as

`Begin` **algfracfem2D**
  $i:=0 \ \rightarrow$ Compute u_l^0, φ_l^0, $l=\overline{1,nn}$ from (5.3)
  `For` $i:=1$ `to` M `do`
    Compute $z_l = z(\cdot, N_l)$, $l=\overline{1,nn}$ from (4.10)-Chapter IV;
    $\varphi^{i-1} := z_l$, $l=\overline{1,nn}$;
    Compute u_l^i, φ_l^i, $l=\overline{1,nn}$, solving the linear system (5.2);
  `End-for`;
`End`.

Below we give the program listing, calling *P_algfracfem2D*, which implement the numerical algorithm `algfracfem2D` (remaining steps **P5-P6**).

Program *P_algfracfem2D*

// Implementation of the conceptual algorithm *algfracfem2D* to

```
// compute the approximate solution of nonlinear phase-field
// transition system by fractional steps method. Case 2D.
//
//   T>0 is an input parameter representing the length of
//         the time interval.
//   M   is an input parameter; it is the dimension of the uniform
//         grid {t_i}_{i=1}^M on [0,T] (t_1 = 0, t_M = T).
//   dt  is an input parameter; the time step size.
//   t   is the vector of length M containing the points t_i, i = 1,M.
//   nkx is an input parameter; it is the number of nodes on Ox.
//   nky is an input parameter; it is the number of nodes on Oy.
//   nkx1=nkx-1.
//   nky1=nky-1.
//   eps is an input parameter, given by fractional steps scheme.
//   prho, pc, pl, pcsi, ptau, ph, pk, pa are input
//         parameters; they are equal to the quantities ρ, c,
//         ℓ, ξ, τ, h, k, a, respectively.
//   nt, nn, SC, Tr, EDGEX1, EDGEX2, EDGEY1, EDGEY2
//         represents the input parameters of the subroutine
//         triangl_cr.
//   h1  is an input parameter; the length of edge on Ox.
//   h2  is an input parameter; the length of edge on Oy.
//   aria  is the value of the area of triangle ẽ.
//   bas[3][3], intval[6], d, s11, s12, s21, s22, e1, e2,
//         e3 represents the input parameters of the subroutines
//         asb(), asg(), asx().
//   u0  is the vector of length nn containing the initial values
//         u_0(x_j), j = 1,nn.
//   fi0 is the vector of length 84 containing the initial values
//         φ_0(x_j), j = 1,84.
```

```
//   z_eps is the vector of length nn; the solution of Cauchy
//          problem (4.10) on [0, ε], see relation (4.25).
//   w is an input parameter; an array of length M*nn that contains
//          the boundary values on Σ.
//   u    is the approximate matrix for unknown u; an array
//          of length M*nn.
//   fi   is the approximate matrix for unknown φ; an array
//          of length M*nn.
//   aa   is the matrix coefficients in (5.2) having the length
//          equal to 2*nn×2*nn
//   dd   is the vector of length 2*nn containing the right-hand side
//          corresponding to the linear system (5.2).
//   z, diag are work array having the length equal to 2*nn.
//   b, g, fr are work array having the length equal to nn*nn.
//   i, ii, j, k, l, itemp, ir, is, i1, ip1, nt2, nx1,
//          ny1, nfipu, doinn, temp1, temp2, x[3], y[3],
//          mt, mtp1, s - are dummy variables.
//   k_uscalar, h_uscalar - are dummy variables.
//
format compact
clear
P0. Initial values
eps =.01;
 // Physical parameters
prho= 7.85;
pc = 12.5;
pl = 65.28;
pcsi = .5;
ptau = 1.e+3*pcsi^2;
ph = 32.012;
```

```
pk = 7.8e-2;
  // Elements of the continuous casting process
hc = 550;                     % the high of the crystallizer
vt = 75/6;                    % the casting speed
  // Construction of the time mesh
T = hc/vt;
M = input('M=');
dt = (T-eps)/(M-1);
t = eps:dt:T;
hci = t.*vt;                  Ω_{t_i}, i = 1, 2, ..., M
  // Construction of the triangulation T_h
slab_width = 1300;
slab_thick = 220;
nkx1 = 1; nky1 = 1;
while rem(nkx1,2) == 1
  disp('Enter a (even) positive integer (10,12,...,66)')
  nkx1 = input('nkx1=');
end
while rem(nky1,2) == 1
  disp('Enter a (even) positive integer (10,12,...,22)')
  nky1 = input('nky1=');
end
h1 = slab_width/nkx1;
h2 = slab_thick/nky1;
nkx = nkx1 + 1;
nky = nky1 + 1;
nn = nkx*nky;
nt = 2*nkx1*nky1;
triangl_cr;
  // local node ordering on ∂Ω   (see Figure B.7 in ANNEXE B)
```

```
skrx = [1 1 2 2];
sksx = [1 2 1 2];
skry = [1 1 3 3];
sksy = [1 3 1 3];
skr23= [2 2 3 3];
sks23= [2 3 2 3];
  // Choose u_0, φ_0
pa = sqrt(pcsi);
pa4 = 1./(4*pa);
f0 = 1.1704;
zz = f0;
u0 = (f0*f0*f0 - f0)*pa4;
ut = 1530;                          % casting temperature (u0*1000)
P1. Compute z(eps,.) from (4.25)
z_eps = zz.*sqrt(pa./(pa + eps.*zz.2̂));
  // store the memory space for:  u, fi
u = zeros(M,nn);
fi = zeros(M,nn);
b = zeros(nn,nn);
g = zeros(nn,nn);
fr= zeros(nn,nn);
  // u^0, fi^0 from  (5.3)
u(1,:)  = u0;
fi(1,:)  = z_eps;
  // Spline interpolation for the walls of crystallizer
spline_temp_crist_pfix;             % generates bcs_fix,
spline_temp_crist_pmobil;           % generates bcs_mobil,
spline_temp_crist_Fright;           % generates bcs_rg,
spline_temp_crist_Fleft;            % generates bcs_lf
    // Assemble the system
```

```
cuadex;
asb;                                  % assemble b_{j,k}, j,k = 1,2,...,nn
asg;                                  % assemble g_{j,k}, j,k = 1,2,...,nn
asfr;                            % assemble fr_{j,k}, j,k = 1,2,...,nn; h2=h1
 // Compute  aa ← the matrix coefficients in (5.2)
aa = zeros(2*nn,2*nn);
for j=1:nn
  for k=1:nn
    aa(j,k) = phro*pc*b(j,k)+dt*pk*g(j,k) ...
                   + dt*pk*ph*fr(j, k);
                                       % contribution of elements r_{kl}
    aa(j,nn+k) = b(j, k)*pl/2;
                                       % contribution of b_{kl} in (5.2)_1
  end
end
for j=1:nn
  l=nn+j;
  for k=1:nn
    aa(l,k) = -2*dt*b(j, k);
                                       % contribution of b_{kl} in (5.2)_2
    aa(l,nn+k)= (tau-dt/(2*pa))*b(j, k) ...
                   + dt*pcsi*pcsi*g(j, k);
                                       % contribution of s_{kl} in (5.2)_2
    end
  end
```

P2. Loop after the time level `ii`

```
for ii=2:M
  // choose w(t,x) from IR,  (t∈ [0,T], x ∈ ∂Ω)
  w = zeros(M,nn);
```

```
      for k=1:nn
        if fix(k/nkx) == k*1./nkx                   % Ox2; Γ2
          w(ii,k) = fnval(bcs_rg,{SC(1,k) ...
                   - slab_width,SC(2,k)});
        end
        if fix((k-1)/nkx) == (k-1)*1./nkx           % Ox2; Γ4
          w(ii,k) = fnval(bcs_lf,{SC(1,k),SC(2,k){);
        end
        if (k>=1 & k<=nkx)                          % Ox1; Γ1
          w(ii,k) = fnval(bcs_mobil,{SC(1,k),SC(2,k){);
        end
        if (k>=nky1*nkx+1 & k<=nn)                  % Ox1; Γ3
          w(ii,k) = fnval(bcs_fix,{SC(1,k),SC(2,k){);
        end
      end
      // Compute  dd ← the right hand side in (5.2)
      right_hand;
//
        // New solution in  z
        z = zeros(1,2*nn);
        z = aa\dd';
        // Solution on level  i
         for j = 1:nn
          u(ii,j) = z(j);
          fi(ii,j) = z(nn+j);
         end
     end     % next  ii
    // Program End
```

Details about the program *P_algfracfem2D* are given in the sequel.

The value of parameter T (line 013) was computed based on the values of two parameters of continuous casting technology: hc - height of crystallizer (mold) and vt - casting speed, while, to generate the triangulation T_ρ (line 035), we consider Ω a cross-section in a slab (thin) of $1300mm \times 220mm$. In Fig. (**10**), the mesh can be seen in the directions of x_1 and x_2 - axis of a rectangular profile.

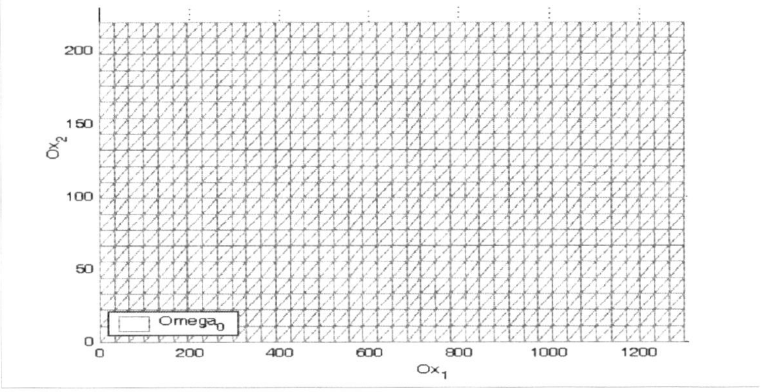

Figure 10 *The triangulation over* Ω =[0,1300]x[0,220]

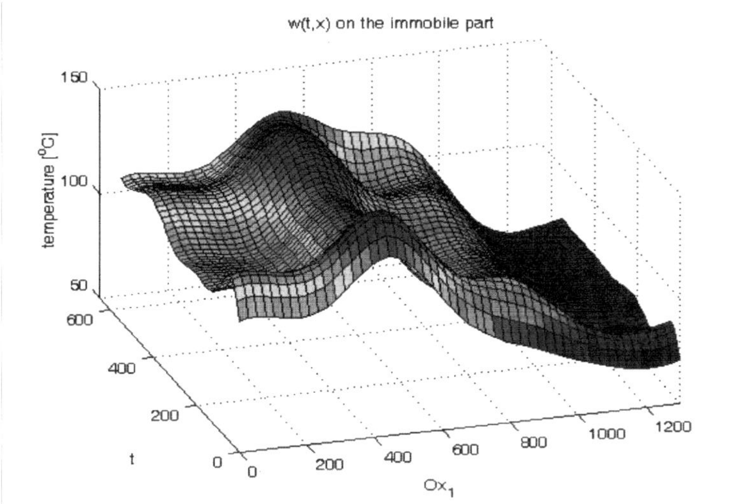

Figure 11 *The values $w(t,x)$ on the immobile part*

The initial solution u^1, φ^1 (see lines 054, 055) was computed as solution of stationary equation $\varphi_t = \Delta\varphi = 0$ (see line 046) and as solution of Cauchy problem - relation (4.25) (see line 048), respectively.

Matlab functions listed below in lines 056-059, give the values of

$w(t,x)$, $(t,x) \in \Sigma$, as a spline interpolation (only the immobile part is illustrated in Fig. (**11**):

```
% spline_temp_crist_pfix
x1=[0 50:200:1250 1300];
x3=[0 40 100 200 300 400 500 550];
z=[113. 114. 114.1 135.8 103.6 85.4 73.4 68.7 70.6 ;
130. 131. 131.3 153. 120.8 109. 86.9 82.3 84.1 ;
117. 118. 118. 139.7 107.5 107.6 74.6 70. 71.8 ;
113. 113. 113.1 134.8 102.6 102.8 78.8 74.2 76.;
112. 113. 113.5 135.2 103. 101.9 76.8 72.2 74. ;
125. 126. 126.8 148.5 116.3 110.6 77.6 72.9 74.8;
119. 120. 120.6 142.3 110.1 110.8 77.8 73.2 75. ;
119. 120. 120.3 142. 129.8 119.5 78.4 73.8 75.6 ];
bcs_fix = csapi({x1,x3},z');
fnplt(bcs_fix)
axis([0 1300 0 650 50 150]);
ylabel(' t');
xlabel(' Ox_1');
zlabel(' temperature [ôC]');
title(' w(t,x) on the immobile part');

% spline_temp_crist_pmobil
x1=[0 50:200:1250 1300];
x3=[0 40 100 200 300 400 500 550];
z=[114. 115. 115.5 127.2 105.8 107.5 62.5 57.9 59.7;
118. 119. 119.3 131. 109.8 109.6 82.6 77.9 79.8 ;
117. 118. 118.1 139.8 107.6 109.5 80.5 75.9 77.7;
121. 122. 122.1 143.8 111.6 109. 82. 77.3 79.2 ;
117. 118. 118.7 140.4 108.2 109. 79. 74.3 76.2 ;
114. 115. 115.6 137.3 105.1 108. 77. 72.3 74.2 ;
115. 115. 115.4 137.1 104.9 100.1 78. 73.4 75.2;
117. 118. 118.5 140.2 108. 101.8 81.8 77.1 79. ];
bcs_mobil=csapi({x1,x3},z');

% spline_temp_crist_Fright
x1=[0 10:100:210 220];
x3=[0 100 200 300 400 500 550];
z=[78. 79. 79.7 79. 78.;
74. 74. 74.2 74. 74.;
73.7 73.7 74. 73.7 73.7;
73.7 73.7 74. 73.8 73.7;
73.5 73.5 73.8 73.5 73.5;
```

```
73.4 73.4 73.6 73.4 73.4;
73.2 73.2 73.4 73.2 73.2];
bcs_rg=csapi({x1,x3},z');

% spline_temp_crist_Fleft
x1=[0 10:100:210 220];
x3=[0 100 200 300 400 500 550];
z=[87. 88. 88.3 88. 87.;
76. 76.8 76.8 76.8 76.;
76. 76.6 76.6 76.6 76.;
76. 76.6 76.6 76.6 76.;
76. 76.4 76.4 76.4 76.;
76. 76. 76.2 76. 76.;
76. 75.2 76. 75.2 76.];
bcs_lf=csapi({x1,x3},z');
```

Lines `061-078` implements the first part in step **P5** - *assemble the linear system.* For details on this topic, the reader may consult ANNEXE B.

Lines `080-093` implement the construction of $w^i(\cdot)$, $i = \overline{1, M}$ and the Matlab function `right_hand` in line `094`, listed below, computes the right hand side in (5.2):

```
% 'right_hand'
dd = zeros(1,doinn);
im1 = ii - 1;
for j=1:nn
  for k=1:nn      dd(j) = dd(j) + b(j, k)*(phro*pc*u(im1,k) + (pl/2)*fi(im1,k));
% Ox1
    if (k>=1 & k<=nkx)     % Gama_1
      dd(j) = dd(j) + dt*pk*w(ii,k);
    end
    if (k>=nky1*nkx+1 & k<=nn)     % Gama_3
      dd(j) = dd(j) - dt*pk*w(ii,k);
    end
```

```
% Ox2
    if fix(k/nkx) == k*1./nkx    % Gama_2
      dd(j) = dd(j) + dt*pk*w(ii,k);
    end
    if fix((k-1)/nkx) == (k-1)*1./nkx    % Gama_4
      dd(j) = dd(j) - dt*pk*w(ii,k);
    end
  end
end
for j=1:nn
  ll = nn + j;
  for k=1:nn
    dd(ll) = dd(ll) + b(j,k)*ptau*fi(ii,k);
  end
end
```

Line 096 implements the last part in step **P5** - *solve the linear system*, after that, it stores the approximate solution on *ii* time level. The program *P_algfracfem2D* ends after the calculation of the approximate solution when time level M is finished.

We shall present now the numerical experiments implementing the conceptual algorithm *algfracfem2D*, (the program *P_algfracfem2D*).

Figs. (**12**), (**14**) represent the approximate solutions u^M ($M = 4$ and $M = 8$, respectively), while Figs. (**13**), (**15**) represent the approximate solutions φ^M ($M = 4$ and $M = 8$, respectively).

The shape of the graphs shows the numerical stability and accuracy of the results obtained by implementing the *fractional steps method* (4.7)-(4.10). The most interesting aspect that we can observe whilw analyzing Figs. (**12-15**) is the presence of *supercooling* and *superheating* phenomena (presence of solid fractions in the liquid, for example).

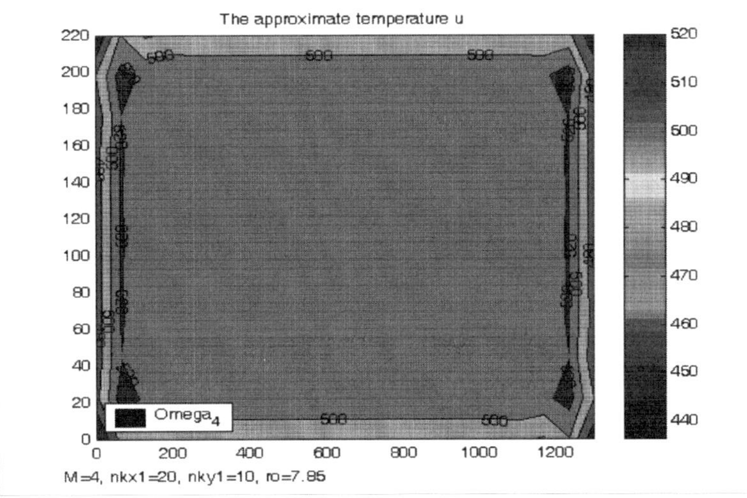

Figure 12 *The approximate temperature* u^4

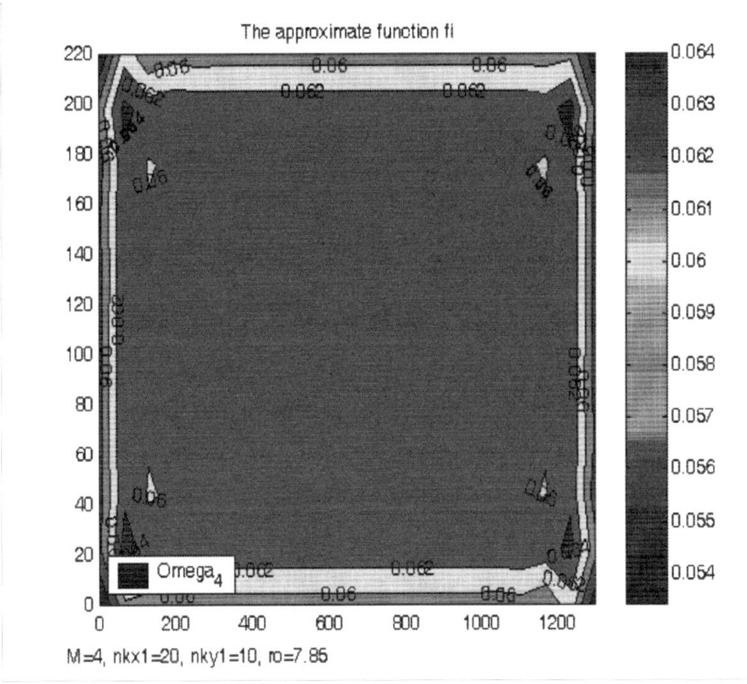

Figure 13 *The approximate solution* φ^4

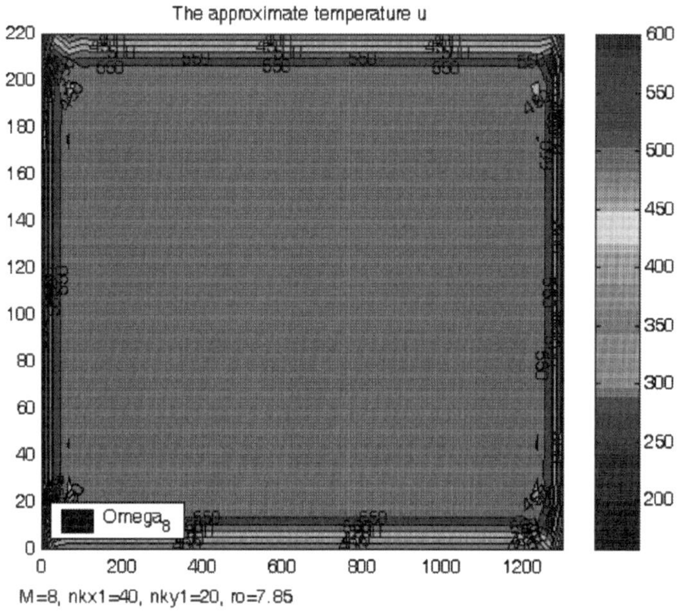

Figure 14 *The approximate temperature* u^8

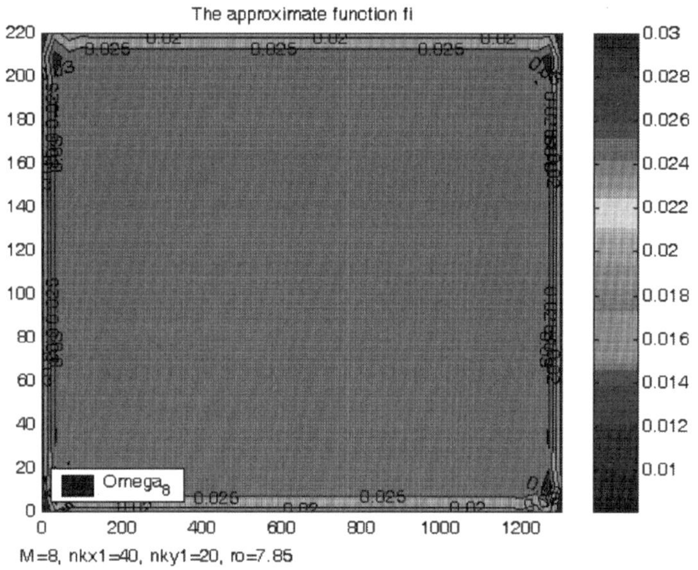

Figure 15 *The approximate solution* φ^8

The numerical solution computed by this way can be considered as an admissible one for the corresponding optimal control problem, formulated in Chapter 3, in order to improve the process optimization of continuous casting.

Analysis and Optimal Control of Phase-Field Transition System: Fractional Steps Methods, 2012, 283-293 283

ANNEXE A

AN AUXILIARY SYSTEM

Costică MOROŞANU

Faculty of Mathematics, "Al.I. Cuza" University of Iasi, Romania

Abstract: The existence and regularity for the solution of one linear system (*auxiliary linear system*), having a structure similar with those that appear in the associated approximating schemes to the nonlinear phase-field transition system, it is proved in this Annexe (**Theorem A.1**). Concerning the methods used in the proof, an essential difference with respect to Chapter 2 consists in that, presently, we make the a priori estimates in $L^2(Q)$ in place of $L^p(Q)$. Estimation technique that is used is also different from those used for the rest of the book.

Consider the linear system

$$\begin{cases} v_t + \dfrac{\ell}{2}\psi_t = k\Delta v + f_1 & \text{in} \quad Q = [0,T] \times \Omega, \\ \dfrac{\partial v}{\partial \nu} + hv = w(t)g(x) & \text{on} \quad \Sigma, \\ v(0,x) = v_0(x) & \text{on} \quad \Omega, \end{cases} \tag{A.1}$$

$$\begin{cases} \tau\psi_t = \xi^2\Delta\psi + c\psi + 2v + f_2 & \text{in} \quad Q, \\ \psi = 0 & \text{on} \quad \Sigma, \\ \psi(0,x) = \psi_0(x) & \text{on} \quad \Omega, \end{cases} \tag{A.2}$$

where $w \in W^{1,\infty}(0,T)$, $w(0) = 0$, and $g \in L^6(\partial\Omega)$ are given functions.

We have

Theorem A.1 *Assume that $f_1, f_2 \in L^p(Q)$, $p > 2$ and $v_0, \psi_0 \in W^{1,\infty}(\Omega)$ satisfying $\frac{\partial}{\partial \nu}v_0 + hv_0 = 0$ $(w(0) = 0)$. Let $c > 0$ be a constant. Then the linear parabolic system* (A.1)-(A.2) *has a unique solution $(v,\psi) \in (W_p^{2,1}(Q) \cap L^\infty(Q))^2$.*

Proof. The proof of this result is known. However we shall outline it because the estimates and the regularity of solution obtained here will be used in the proof of approximation theorems (convergence of the fractional steps schemes). We will apply the Leray-Schauder degree (see [86]-[87]) in the Banach space $B := \{(v,\psi),\ v,\psi \in L^p(Q), p>2\}$. For this we consider the linear operator $T_\lambda : B \to B$, for each $0 \le \lambda \le 1$, which is defined by the unique solution (v,ψ) of the linear problem

$$v_t - k\Delta v = \lambda\Big(-\frac{\ell}{2}\psi_t + f_1\Big) \qquad \text{in} \quad Q, \tag{A.3}$$

$$\tau\psi_t - \xi^2\Delta\psi = c\psi + \lambda(2\eta + f_2) \qquad \text{in} \quad Q, \tag{A.4}$$

$$\frac{\partial v}{\partial \nu} + hv = w(t)g(x) \qquad \text{on} \quad \Sigma, \tag{A.5}$$

$$\psi = 0 \qquad \text{on} \quad \Sigma, \tag{A.6}$$

$$v(0,x) = v_0(x), \quad \psi(0,x) = \psi_0(x), \qquad \text{on} \quad \Omega, \tag{A.7}$$

for all $(\eta,\zeta) \in B$.

Firstly, let us verify that T_λ is well defined (i.e. problem (A.3)-(A.7) has a solution). Using L^p–theory (see O.A. Ladyzhenskaya, B.A. Solonnikov, N.N., Uraltzava [49]), from equation (A.4), with $\eta \in L^p(Q)$, we obtain the existence of a unique solution $\psi \in W_p^{2,1}(Q)$. The embedding theorem of Lions-Peetre (see J.L. Lions [3]) show that $W_q^{2,1}(Q) \subset L^\mu(Q)$, where μ is given by relation (5.3) in Chapter 1.

In our case (i.e., $p > 2$) we have $W_p^{2,1}(Q) \subset L^\infty(Q)$ and then $\psi \in W_p^{2,1}(Q) \cap L^\infty(Q) \subset L^p(Q)$. How $\psi \in W_p^{2,1}(Q)$, then $\psi_t \in L^p(Q)$. Applying now the L^p-theory to equation (A.3) we get the existence of a unique solution v and that $v \in W_p^{2,1}(Q) \cap L^\infty(Q) \subset L^p(Q)$. Therefore the operator

$$T_\lambda(\eta,\zeta) = (v,\psi) \qquad \forall\,(\eta,\zeta) \in B, \tag{A.8}$$

is well defined.

In the sequel by C we denote different positive constants.

Consider now the homotopy $T : B \times [0,1] \to B$, $T(\eta,\zeta,\lambda) = T_\lambda(\eta,\zeta)$, i.e.,

$$(\eta,\zeta,\lambda) \in B \times [0,1] \quad \to \quad (v,\psi) = T_\lambda(\eta,\zeta) \in B. \tag{A.9}$$

Let $(\eta_0, \zeta_0, \lambda_0) \in B \times [0,1]$ be fixed. For all $(\eta, \zeta, \lambda) \in B \times [0,1]$ we have

$$\begin{aligned}\|T(\eta,\zeta,\lambda) - T(\eta_0,\zeta_0,\lambda_0)\|_B \le \|T(\eta,\zeta,\lambda) - T(\eta_0,\zeta_0,\lambda)\|_B \\ + \|T(\eta_0,\zeta_0,\lambda) - T(\eta_0,\zeta_0,\lambda_0)\|_B.\end{aligned}$$

We set $(v_\lambda^{\eta,\zeta}, \psi_\lambda^{\eta,\zeta}) = T_\lambda(\eta,\zeta)$ şi $v^\lambda = v_\lambda^{\eta,\zeta} - v_\lambda^{\eta_0,\zeta_0}$, $\psi^\lambda = \psi_\lambda^{\eta,\zeta} - \psi_\lambda^{\eta_0,\zeta_0}$. Than $T(\eta,\zeta,\lambda) - T(\eta_0,\zeta_0,\lambda)$ is the solution to the system

$$\begin{cases} v_t^\lambda - k\Delta v^\lambda = \lambda\Big(-\dfrac{\ell}{2}\psi_t^\lambda\Big) & \text{in } Q, \\ \tau\psi_t^\lambda - \xi^2\Delta\psi^\lambda - c\psi^\lambda = 2\lambda(\eta - \eta_0) & \text{in } Q, \\ \dfrac{\partial v^\lambda}{\partial \nu} + hv^\lambda = 0 & \text{on } \Sigma, \\ \psi^\lambda = 0 & \text{on } \Sigma, \\ v^\lambda = \psi^\lambda = 0 & \text{on } \Omega. \end{cases} \tag{A.10}$$

The L^p-theory applied to the equation $(A.10)_2$ provides the existence of a solution $\psi^\lambda \in W_p^{2,1}(Q)$ and that the following estimate

$$\|\psi^\lambda\|_{W_p^{2,1}(Q)} \le C\|\eta - \eta_0\|_{L^p(Q)}$$

holds (we have used also that $\lambda \le 1$). The L^p-theory applied now to the equation $(A.10)_1$ guaranty the existence of a unique solution $v^\lambda \in W_p^{2,1}(Q)$ and the following estimate

$$\begin{aligned}\|v^\lambda\|_{W_p^{2,1}(Q)} &\le C\|\psi_t^\lambda\|_{L^p(Q)} \\ &\le C\|\psi^\lambda\|_{W_p^{2,1}(Q)} \le C\|\eta - \eta_0\|_{L^p(Q)}.\end{aligned}$$

But $W_p^{2,1}(Q) \subset L^p(Q)$ and than, from the last inequality we obtain

$$\|T(\eta,\zeta,\lambda) - T(\eta_0,\zeta_0,\lambda)\|_B \le C\|\eta - \eta_0\|_{L^p(Q)}. \tag{A.11}$$

The term $T(\eta_0,\zeta_0,\lambda) - T(\eta_0,\zeta_0,\lambda_0)$ represent the solution of the system

$$
\begin{cases}
\dfrac{\partial}{\partial t}(v_\lambda^{\eta_0,\zeta_0} - v_{\lambda_0}^{\eta_0,\zeta_0}) - k\Delta(v_\lambda^{\eta_0,\zeta_0} - v_{\lambda_0}^{\eta_0,\zeta_0}) & \\
\quad = (\lambda - \lambda_0)f_1 - \dfrac{\ell}{2}\Big(\lambda\dfrac{\partial}{\partial t}\psi_\lambda^{\eta_0,\zeta_0} - \lambda_0\dfrac{\partial}{\partial t}\psi_{\lambda_0}^{\eta_0,\zeta_0}\Big) & \text{in } Q, \\
\tau\dfrac{\partial}{\partial t}(\psi_\lambda^{\eta_0,\zeta_0} - \psi_{\lambda_0}^{\eta_0,\zeta_0}) - \xi^2\Delta(\psi_\lambda^{\eta_0,\zeta_0} - \psi_{\lambda_0}^{\eta_0,\zeta_0}) & \\
\quad -c(\psi_\lambda^{\eta_0,\zeta_0} - \psi_{\lambda_0}^{\eta_0,\zeta_0}) = 2\eta_0(\lambda - \lambda_0) + (\lambda - \lambda_0)f_2 & \text{in } Q, \\
\dfrac{\partial}{\partial \nu}(v_\lambda^{\eta_0,\zeta_0} - v_{\lambda_0}^{\eta_0,\zeta_0}) + h(v_\lambda^{\eta_0,\zeta_0} - v_{\lambda_0}^{\eta_0,\zeta_0}) = 0 & \text{on } \Sigma, \\
\psi_\lambda^{\eta_0,\zeta_0} - \psi_{\lambda_0}^{\eta_0,\zeta_0} = 0 & \text{on } \Sigma, \\
v_\lambda^{\eta_0,\zeta_0} - v_{\lambda_0}^{\eta_0,\zeta_0} = \psi_\lambda^{\eta_0,\zeta_0} - \psi_{\lambda_0}^{\eta_0,\zeta_0} = 0 & \text{on } \Omega.
\end{cases}
\tag{A.12}
$$

The L^p-theory applied to the equation $(A.12)_2$ guaranty the existence of a unique solution $(\psi_\lambda^{\eta_0,\zeta_0} - \psi_{\lambda_0}^{\eta_0,\zeta_0}) \in W_p^{2,1}(Q)$ as well as the estimate

$$
\|\psi_\lambda^{\eta_0,\zeta_0} - \psi_{\lambda_0}^{\eta_0,\zeta_0}\|_{W_p^{2,1}(Q)} \leq C|\lambda - \lambda_0|(\|\eta_0\|_{L^p(Q)} + \|f_2\|_{L^p(Q)})
$$

whence

$$
\left\|\frac{\partial}{\partial t}(\psi_\lambda^{\eta_0,\zeta_0} - \psi_{\lambda_0}^{\eta_0,\zeta_0})\right\|_{L^p(Q)} \leq C|\lambda - \lambda_0|(\|\eta_0\|_{L^p(Q)} + \|f_2\|_{L^p(Q)}).
$$

Applying the same reasoning to the equation $(A.12)_1$ and we deduce the existence of a unique solution $(v_\lambda^{\eta_0,\zeta_0} - v_{\lambda_0}^{\eta_0,\zeta_0}) \in W_p^{2,1}(Q)$ and that the following estimate holds

$$
\begin{aligned}
&\|v_\lambda^{\eta_0,\zeta_0} - v_{\lambda_0}^{\eta_0,\zeta_0}\|_{W_p^{2,1}(Q)} \\
&\quad \leq C\left\|\lambda\frac{\partial}{\partial t}\psi_\lambda^{\eta_0,\zeta_0} - \lambda_0\frac{\partial}{\partial t}\psi_{\lambda_0}^{\eta_0,\zeta_0}\right\|_{L^p(Q)} + C|\lambda - \lambda_0|\|f_1\|_{L^p(Q)} \\
&\quad \leq C|\lambda - \lambda_0|(\|\eta_0\|_{L^p(Q)} + \|f_1\|_{L^p(Q)} + \|f_2\|_{L^p(Q)}).
\end{aligned}
$$

Applying again the embedding theorem, from the last inequality we obtain

$$
\|T(\eta_0, \zeta_0, \lambda) - T(\eta_0, \zeta_0, \lambda_0)\|_B \tag{A.13}
$$

$$\leq C|\lambda - \lambda_0|(\|\eta_0\|_{L^p(Q)} + \|f_1\|_{L^p(Q)} + \|f_2\|_{L^p(Q)}). \qquad (A.14)$$

Thereof one obtains the continuity of the operator T_λ in (w, ζ) and in λ as soon as $(w_n, \zeta_n, \lambda_n) \in B \times [0,1]$ is a sequence such that $(w_n, \zeta_n, \lambda_n) \to (w, \zeta, \lambda)$ as $n \to \infty$. Furthermore, because the imbedding of $W_p^{2,1}(Q)$ into $L^\infty(Q)$ is compact (see Adams [88, Theorem 6.2]), then T_λ is also a compact operator on the Banach space B.

A priori estimates in $L^2(Q)$ for the equation $T_\lambda(v, \psi) = (v, \psi)$, i.e., for the system

$$\begin{cases} v_t - k\Delta v = \lambda\Big(-\dfrac{\ell}{2}\psi_t + f_1\Big) & \text{in } Q, \\ \tau\psi_t - \xi^2\Delta\psi - c\psi = \lambda(2v + f_2) & \text{in } Q, \\ \dfrac{\partial v}{\partial \nu} + hv = w(t)g(x) & \text{on } \Sigma, \\ \psi = 0 & \text{on } \Sigma, \\ v(0,x) = v_0(x), \quad \psi(0,x) = \psi_0(x) & \text{on } \Omega \end{cases} \qquad (A.15)$$

with $0 \leq \lambda \leq 1$.

Multiplying $(\text{A.15})_1$ by $(v + \lambda\frac{\ell}{2}\psi)$, using integration by parts and Green's formula, yields

$$\frac{1}{2}\frac{\partial}{\partial t}\int_\Omega (v + \lambda\frac{\ell}{2}\psi)^2 dx + k\int_\Omega |\nabla v|^2 dx$$
$$+\lambda\frac{k\ell}{2}\int_\Omega \nabla v \nabla\psi dx + hk\int_{\partial\Omega} v^2 dx$$
$$\leq \frac{1}{2}\int_\Omega (v + \lambda\frac{\ell}{2}\psi)^2 dx + \frac{1}{2}\int_\Omega f_1^2 dx + k\int_{\partial\Omega} vw(t)g(x)dx.$$

But

$$k\int_{\partial\Omega} vw(t)g(x)dx \leq \frac{kh}{2}\int_{\partial\Omega} v^2 dx + \frac{k}{2h}\|w\|^2_{L^3(\partial\Omega)}\|g\|^2_{L^6(\partial\Omega)}$$

and than from the above inequality we derive

$$\frac{1}{2}\frac{\partial}{\partial t}\int_\Omega (v + \lambda\frac{\ell}{2}\psi)^2 dx + k\int_\Omega |\nabla v|^2 dx \qquad (A.16)$$

$$+\lambda\frac{k\ell}{2}\int_\Omega \nabla v\nabla\psi dx + hk\int_{\partial\Omega} v^2 dx$$

$$\leq \frac{1}{2}\int_\Omega (v+\lambda\frac{\ell}{2}\psi)^2 dx + C\|f_1\|^2_{L^2(\Omega)} + C\|w\|^2_{L^3(\partial\Omega)}\|g\|^2_{L^6(\partial\Omega)}.$$

We multiply now the equation $(A.15)_2$ with $A\psi$, $A>0$ and we integrate over Ω. Thus we obtain

$$A\frac{\tau}{2}\frac{\partial}{\partial t}\int_\Omega \psi^2 dx - A\xi^2\int_\Omega \Delta\psi\cdot\psi dx$$

$$= c\int_\Omega \psi^2 dx + 2\lambda A\int_\Omega v\psi dx + \lambda A\int_\Omega f_2\psi dx.$$

Green's formula, Cauchy inequality together with the inequality

$$2\lambda A\int_\Omega v\psi dx \leq \lambda A\int_\Omega (v+\lambda\frac{\ell}{2}\psi)^2 dx + \lambda A\int_\Omega \psi^2 dx$$

lead the previous estimate to the following inequality

$$A\frac{\tau}{2}\frac{\partial}{\partial t}\int_\Omega \psi^2 dx + A\xi^2\int_\Omega |\nabla\psi|^2 dx \qquad (A.17)$$

$$\leq C\Big(\int_\Omega (v+\lambda\frac{\ell}{2}\psi)^2 dx + \int_\Omega \psi^2 dx\Big) + C\|f_2\|^2_{L^2(\Omega)}.$$

Adding (A.16), (A.17) and we obtain

$$\frac{1}{2}\frac{\partial}{\partial t}\int_\Omega [(v+\lambda\frac{\ell}{2}\psi)^2 + A\tau\psi^2]dx$$

$$+\int_\Omega [k|\nabla v|^2 + \lambda\frac{k\ell}{2}\nabla v\nabla\psi + A\xi^2|\nabla\psi|^2]dx + \frac{hk}{2}\int_{\partial\Omega} v^2 dx$$

$$\leq C\Big(\int_\Omega (v+\lambda\frac{\ell}{2}\psi)^2 dx + \int_\Omega \psi^2 dx\Big)$$

$$+C\|f_1\|^2_{L^2(\Omega)} + C\|f_2\|^2_{L^2(\Omega)} + C\|w\|^2_{L^3(\partial\Omega)}\|g\|^2_{L^6(\partial\Omega)}.$$

For $A = 1 + \dfrac{\ell^2 k}{4\xi^2}$ we have

$$k\int_\Omega [|\nabla v|^2 + \lambda\frac{\ell}{2}\nabla v\nabla\psi + A\frac{\xi^2}{k}|\nabla\psi|^2]dx$$

$$\geq \frac{3k}{4}\int_\Omega |\nabla v|^2 dx + \xi^2\int_\Omega |\nabla\psi|^2 dx$$

and then the above inequality leads to

$$\frac{1}{2}\frac{\partial}{\partial t}\int_\Omega [(v + \lambda\frac{\ell}{2}\psi)^2 + A\tau\psi^2]dx$$

$$+\frac{3k}{4}\int_\Omega |\nabla v|^2 dx + \xi^2\int_\Omega |\nabla\psi|^2 dx + \frac{hk}{2}\int_{\partial\Omega} v^2 dx$$

$$\leq C\Big(\int_\Omega v^2 dx + \int_\Omega \psi^2 dx\Big)$$

$$+C\|f_1\|^2_{L^2(\Omega)} + C\|f_2\|^2_{L^2(\Omega)} + C\|w\|^2_{L^3(\partial\Omega)}\|g\|^2_{L^6(\partial\Omega)}$$

(we have also used the Cauchy inequality and $\lambda \leq 1$).

We integrate the last inequality on $[0, t]$, $t \in (0, T]$ and we get

$$\frac{1}{2}\Big[\|v(t) + \lambda\frac{\ell}{2}\psi(t)\|^2_{L^2(\Omega)} + A\tau\|\psi(t)\|^2_{L^2(\Omega)}\Big]$$

$$+\frac{3k}{4}\int_0^t\int_\Omega |\nabla v|^2 dxds + \xi^2\int_0^t\int_\Omega |\nabla\psi|^2 dxds + \frac{hk}{2}\int_0^t\int_{\partial\Omega} v^2 dxds$$

$$\leq \frac{1}{2}\Big(\|v_0 + \lambda\frac{\ell}{2}\psi_0\|^2_{L^2(\Omega)} + A\tau\|\psi_0\|^2_{L^2(\Omega)}\Big)$$

$$+C\int_0^t (\|v(s)\|^2_{L^2(\Omega)} + \|\psi(s)\|^2_{L^2(\Omega)})ds$$

$$+C\|f_1\|^2_{L^2(Q)} + C\|f_2\|^2_{L^2(Q)} + C\|w\|^2_{L^3(\Sigma)}\|g\|^2_{L^6(\Sigma)}.$$

For $\tau A > \frac{\lambda^2\ell^2}{4} + \frac{1}{2}$ we have

$$\|v(t) + \lambda\frac{\ell}{2}\psi(t)\|^2_{L^2(\Omega)} + A\tau\|\psi(t)\|^2_{L^2(\Omega)} \geq \frac{1}{2}\|v(t)\|^2_{L^2(\Omega)} + \frac{1}{2}\|\psi(t)\|^2_{L^2(\Omega)}$$

and then the previous estimate becomes

$$\frac{1}{4}\|v(t)\|^2_{L^2(\Omega)} + \frac{1}{4}\|\psi(t)\|^2_{L^2(\Omega)}$$

$$+\frac{3k}{4}\int_0^t\int_\Omega |\nabla v|^2 dxds + \xi^2\int_0^t\int_\Omega |\nabla\psi|^2 dxds + \frac{hk}{2}\int_0^t\int_{\partial\Omega} v^2 dxds$$

$$\leq \widetilde{C} + C\int_0^t (\|v(s)\|^2_{L^2(\Omega)} + \|\psi(s)\|^2_{L^2(\Omega)})ds$$

where $\widetilde{C} > 0$ depends on $\|v_0\|^2_{L^2(\Omega)}$, $\|\psi_0\|^2_{L^2(\Omega)}$, $\|f_1\|^2_{L^2(Q)}$, $\|f_2\|^2_{L^2(Q)}$, $\|w\|^2_{L^3(\Sigma)}$, $\|g\|^2_{L^6(\Sigma)}$). By Gronwall's inequality we derive

$$\|v(t)\|^2_{L^2(\Omega)} + \|\psi(t)\|^2_{L^2(\Omega)} \leq \widetilde{C}e^{CT}, \quad \forall t \in [0, T].$$

Substituting this inequality in the previous, we get

$$\|v(t)\|^2_{L^2(\Omega)} + \|\psi(t)\|^2_{L^2(\Omega)} + k\int_0^t \|\nabla v(s)\|^2_{L^2(\Omega)}ds \qquad (A.18)$$

$$+\xi^2\int_0^t \|\nabla\psi(s)\|^2_{L^2(\Omega)}ds + hk\int_0^t\int_{\partial\Omega} v^2 dxds \leq C, \quad \forall t \in [0, T].$$

Multiplying now $(A.15)_2$ by ψ_t and integrating on Ω. We obtain

$$\tau\int_\Omega \psi_t^2 dx + \xi^2\int_\Omega \psi_t(-\Delta\psi)dx$$

$$= c\int_\Omega \psi\psi_t dx + 2\lambda\int_\Omega v\psi_t dx + \lambda\int_\Omega f_2\psi_t dx.$$

Since $\lambda \leq 1$ and

$$2\lambda\int_\Omega v\psi_t dx \leq \frac{4}{\tau}\|v\|^2_{L^2(\Omega)} + \frac{\tau}{4}\|\psi_t\|^2_{L^2(\Omega)},$$

$$\lambda\int_\Omega f_2\psi_t dx \leq \frac{1}{\tau}\|f_2\|^2_{L^2(\Omega)} + \frac{\tau}{4}\|\psi_t\|^2_{L^2(\Omega)},$$

using Green's formula and relation (A.18), the previous equality becomes

$$\frac{\tau}{2}\int_\Omega \psi_t^2 dx + \frac{\xi^2}{2}\int_\Omega |\nabla\psi|^2 dx \leq \frac{4}{\tau}\|v\|^2_{L^2(\Omega)} + \frac{1}{\tau}\|f_2\|^2_{L^2(\Omega)} + \frac{c}{2}\frac{\partial}{\partial t}\int_\Omega \psi^2 dx$$

$$\leq C + C\|f_2\|^2_{L^2(\Omega)} + \frac{c}{2}\frac{\partial}{\partial t}\int_\Omega \psi^2 dx.$$

Integrating on $[0,t]$, $t \in (0,T]$ and using again (A.18), we obtain

$$\int_0^t \int_\Omega \psi_t^2 dxds + \int_\Omega |\nabla\psi|^2 dx \leq C \quad \forall t \in [0,T]. \qquad (A.19)$$

We multiply $(A.15)_2$ with $\Delta\psi$ and we integrate on Ω. We get

$$\tau\int_\Omega \psi_t \Delta\psi dx - \xi^2\int_\Omega (\Delta\psi)^2 dx$$

$$= c\int_\Omega \psi\Delta\psi dx + 2\lambda\int_\Omega v\Delta\psi dx + \lambda\int_\Omega f_2\Delta\psi dx.$$

Now

$$\tau\int_\Omega \psi_t\Delta\psi dx = -\frac{\tau}{2}\frac{\partial}{\partial t}\int_\Omega |\nabla\psi|^2 dx,$$

$$c\int_\Omega \psi\Delta\psi dx \leq \frac{3c}{2\xi^2}\|\psi\|^2_{L^2(\Omega)} + \frac{\xi^2}{6}\|\Delta\psi\|^2_{L^2(\Omega)},$$

$$2\lambda\int_\Omega v\Delta\psi dx \leq \frac{6}{\xi^2}\|v\|^2_{L^2(\Omega)} + \frac{\xi^2}{6}\|\Delta\psi\|^2_{L^2(\Omega)},$$

$$\lambda\int_\Omega f_2\Delta\psi dx \leq \frac{3}{2\xi^2}\|f_2\|^2_{L^2(\Omega)} + \frac{\xi^2}{6}\|\Delta\psi\|^2_{L^2(\Omega)},$$

the above relation becomes

$$\frac{\tau}{2}\frac{\partial}{\partial t}\int_\Omega |\nabla\psi|^2 dx + \frac{\xi^2}{2}\int_\Omega (\Delta\psi)^2 dx$$

$$\leq C(\|v\|^2_{L^2(\Omega)} + \|\psi\|^2_{L^2(\Omega)} + \|f_2\|^2_{L^2(\Omega)}).$$

Integrating this on $[0, t]$ and using (A.18), we get the estimate

$$\int_\Omega |\nabla\psi(t,x)|^2 dx + \int_0^t \int_\Omega (\Delta\psi)^2 dx \le C \quad \forall t \in [0,T]. \tag{A.20}$$

The possible different constants C from (A.18)-(A.20) depends on $|\Omega|$ (measure of Ω), T, $\|v_0\|_{L^2(\Omega)}$, $\|\psi_0\|_{L^2(\Omega)}$, $\|\nabla\psi_0\|_{L^2(\Omega)}$, $\|f_1\|_{L^2(Q)}$, $\|f_2\|_{L^2(Q)}$, $\|w\|_{L^3(\Sigma)}$, $\|g\|_{L^6(\Sigma)}$ as well as on the physical parameters τ, ξ, a, h, k.

From (A.19) and (A.20) we obtain the estimates

$$\|\psi_t\|_{L^2(Q)} \le C, \tag{A.21}$$

$$\|\Delta\psi\|_{L^2(Q)} \le C, \tag{A.22}$$

for each (v, ψ) solution of (A.15). By elliptic boundary regularity this implies that $\|\psi\|_{L^2(0,T;H^2(\Omega))} \le C$ (see V. Barbu [15]) and so (A.21) and (A.22) lead to the estimate

$$\|v\|_{L^p(Q)} + \|\psi\|_{L^p(Q)} \le C(\|v\|_{W_2^{2,1}(Q)} + \|\psi\|_{W_2^{2,1}(Q)}) \le C \tag{A.23}$$

(we have used also the L^p estimates for (v, ψ) and the inequality given of imbedding theorem).

By (A.23) it is clearly now that we may choose a number $r > 0$ such that

$$\|(v, \psi)\|_B \le C \le r \tag{A.24}$$

whenever (v, ψ) is a solutio of problem (A.15).

Let $r > 0$ be large enough. Then the operator $T_\lambda : B \longrightarrow B$, $\lambda \in [0, 1]$, have the property

$$T_\lambda(v, \psi) \ne (v, \psi), \quad \forall (v, \psi) \in \partial B_r, \quad \forall \lambda \in [0, 1], \tag{A.25}$$

where $B_r := \{(v, \psi) \in B,\ \|(v, \psi)\|_B < r\}$ (see and (A.24)), i.e. the equation $T_\lambda(v, \psi) = (v, \psi)$ has not the solution on ∂B_r. Thus, using (A.25) and the compactness of homotopy T_λ we may define the topological degree (the Leray-Schauder degree) of $I_B - T_\lambda$ in 0 relative to B_r, namely

$$\deg\,(I_B - T_\lambda, B_r, 0), \quad \forall\ 0 \le \lambda \le 1, \tag{A.26}$$

where I_B represents the identity of the space B (see Schwartz [87]). Particullary, there exists deg $(I_B - T_\lambda, B_r, 0)$ for $\lambda = 0$ and $\lambda = 1$.

Because T_0 is a constant map, it is easy to show that $\deg(I_B - T_0, B_r, 0) = \deg(I_B) = 1$. From this and from homotopy invariance of the Leray-Schauder degree we may conclude that the equation $(v, \psi) = T_1(v, \psi)$ has a solution $(v, \psi) \in B_r \subset B$. From (A.8) and (A.15) follows than that the problem (A.1)-(A.2) has a solution $(v, \psi) \in B = L^p(Q) \times L^p(Q)$, $p > 2$. The L^p-theory of parabolic linear equations and the Lions-Peetre imbedding theorem, already used several times, give us the regularity of the solution stated in our theorem, i.e., $(v, \psi) \in (W_p^{2,1}(Q) \cap L^\infty(Q))^2$.

Note. In the end I want to add that references [89]-[115] represents a small part of other works consulted by the author, works warmly recommended to readers who want to look and to other aspects (both theoretical and numerical) on the phase-field transition system.

ANNEXE B

IMPLEMENTATION OF FEM - FINITE ELEMENT METHOD

Costică MOROŞANU

Faculty of Mathematics, "Al.I. Cuza" University of Iasi, Romania

Abstract: The finite element method (**fem** for short) is a general method for approximating the solution of boundary value problems for partial differential equations. This method derives from the Ritz (or Gelerkin) method, characteristic for the finite element method being the choice of the finite dimensional space, namely, in the case of **fem** the finite dimensional space, corresponding to the original space of functions, is the *span* of a set of finite element basis functions, as we will see in the sequel.

The steps in solving a boundary value problem using **fem** are

P0. (D) The direct formulation of the problem;

P1. (V) A variational (weak) formulation for problem (D);

P2. The construction of a finite element mesh (triangulation);

P3. The construction of the finite dimensional space of test function (called *finite element basis functions*);

P4. (V_{nn}) A discrete analogous of (V);

P5. Assemble the system;

P6. Solve the system in P5.

We shall show by an example how the **fem** can be implemented.

P0. The direct formulation of the problem. Consider the boundary value problem

$$\begin{cases} -\Delta u + u = 0 & \text{in } \Omega, \\ \frac{\partial}{\partial \nu} u = g & \text{on } \Gamma = \partial\Omega, \end{cases} \tag{D}$$

where $\Omega \subset I\!R^2$, g is a given function and,

$$u \in \mathcal{U} = \Big\{ u \big| u : \Omega \to I\!R, u \in C(\Omega)$$

$$\frac{\partial}{\partial x}u, \frac{\partial}{\partial y}u \text{ are piecewise continuous on } \Omega,$$

$$\int_\Omega |\nabla u|^2 dxdy < \infty, \frac{\partial}{\partial \nu}u = g \text{ on } \Gamma \Big\}.$$

For simplicity we shall assume that Ω is a polygonal domain, $\Omega = [0, a] \times [0, b]$, $a, b > 0$. Denote by $\Gamma = \Gamma_1 \cup \Gamma_2 \cup \Gamma_3 \cup \Gamma_4$ (see Fig. (**1**).

P1. A weak formulation for problem (D). Making use of Green's formula, the problem (D) has the following variational formulation

$$\text{find } u \in \mathcal{U}, \quad \int_\Omega \nabla u \nabla v dxdy + \int_\Omega uvdxdy = \int_\Gamma gvdxdy \quad \forall v \in \mathcal{U}. \quad (V)$$

P2. The construction of a finite element mesh. Let us now realize a mesh (triangulation) of domain Ω. For this, we divide Ω in a set of "nt" nonoverlaping triangles (see Fig. **1**) whose vertices are the nodes of the triangulation

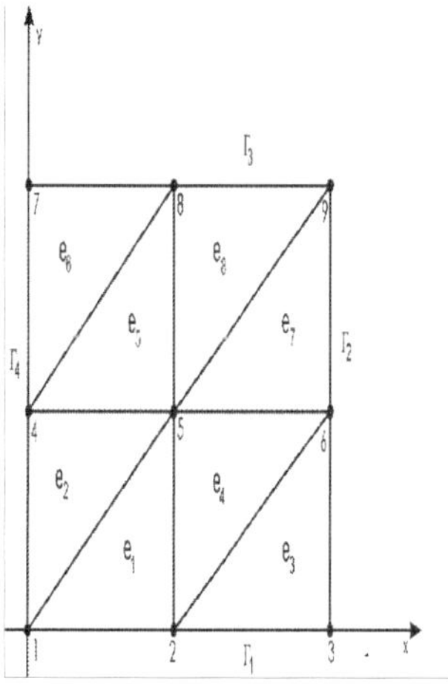

Figure 1 (nt=8)

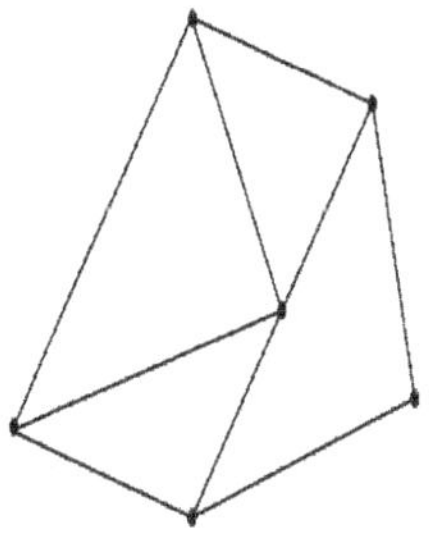

Figure 2

$$T_h = \{K_i \big| K_i \subset \Omega, K_i - \text{triangle }, i = 1, 2, ..., nt\}$$

i.e.,

$$\text{for } i \neq j, i, j = \overline{1, nt}, \quad K_i \cap K_j = \begin{cases} \text{a node, or} \\ \text{an edge, or} \\ \text{an empty set,} \end{cases}$$

$$\Omega = \bigcup_{K \in T_h} K = K_1 \cup K_2 \cup ... \cup K_{nt},$$

and no vertex (node) of one triangle lies on the edge of another triangle (see Fig. **2**)

A vertex can belong to many elements and an edge to at most two elements. The choice of elements and nodes defines a triangulation or *"a finite element mesh"*. We suppose that all elements in a mesh are the same type. For the sake of easy reference, we collect here the most important notation that will be used in this section

nt	- number of triangles in a finite element mesh,
nn	- number of nodes in a finite element mesh,
$e_\ell, \ell = \overline{1, nt}$	- the elements,
$N_j = (x_j, y_j), j = \overline{1, nn}$	- the nodes,
t	- number of nodes in a single element,
$i_r^{(\ell)}, r = \overline{1, t}$	- node numbers of nodes in e_ℓ,
$\phi_i(x, y), i = \overline{1, nn}$	- global basis functions (defined on Ω),
$\phi_r^{(\ell)}(x, y), r = \overline{1, t}$	- local basis functions (defined on e_ℓ),
$\tilde{e}$	- the standard element,
$\tilde{\phi}_r(\tilde{x}, \tilde{y}), r = \overline{1, t}$	- standard local basis functions (defined on $\tilde{e}$),

P3. The construction of the finite element basis functions. We associate with each node N_i a function $\phi_i(x, y)$, defined on $\bar{\Omega}$, with the following properties:

(a) The restriction of ϕ_i to any element e_ℓ is a linear function

$$\phi_i(x, y) = a_{1i}^{(\ell)} x + a_{2i}^{(\ell)} y + a_{3i}^{(\ell)}, \quad (x, y) \in e_\ell; \qquad (B.1)$$

(b) $$\phi_i(N_j) = \delta_{ij} = \begin{cases} 1, & i = j \\ 0, & i \neq j, \end{cases} \qquad i, j = \overline{1, nn};$$

(c) ϕ_i is uniquely determined on every "element edge" by its value at the nodes belonging to that edge.

Let

$$V_{nn} = \text{SPAN}\{\phi_1, \phi_2, \cdots, \phi_{nn}\}.$$

We call V_{nn} a *"finite element space"* and $\phi_1, \phi_2, \cdots, \phi_{nn}$, *"finite element basis functions"* or *"global basis functions"*. Property (b) implies that $\phi_1, \phi_2, \cdots, \phi_{nn}$ are linearly independent and V_{nn} has dimension nn. The coefficients in (B.1) are determined from (b) (will see in the sequel). Because the matrix of linear system of equations obtained by (b) is nonsingular then, ϕ_i are well defined on e_ℓ. Property (c) ensures that ϕ_i is continuous at element edges. Thus $\phi_i \in C(\bar{\Omega})$. Suppose node N_i is not in e_ℓ. Then it follows from (B.1) and (b) that ϕ_i vanishes everywhere in e_ℓ. Hence

"ϕ_i assumes nonzero values only in those elements to which N_i belongs" (see Fig. **3**).

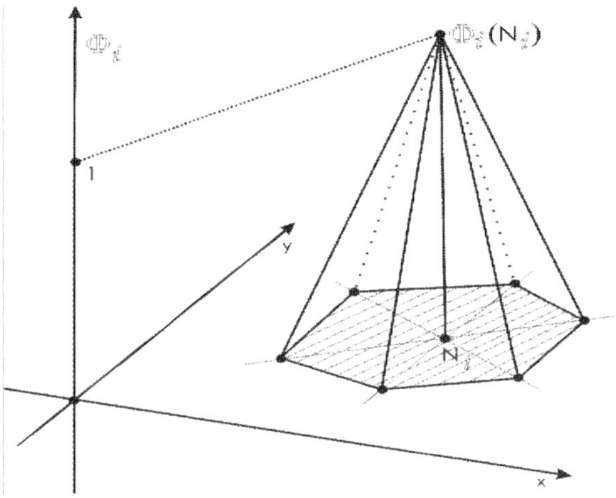

Figure 3 *The support of basis function ϕ_i*

We call the union of elements to which N_i belongs the *"support"* of ϕ_i. By definition, a function $U \in V_{nn}$ if and only if it can be expressed as

$$U(x,y) = \sum_{\ell=1}^{nn} U_\ell \phi_\ell(x,y) \quad (x,y) \in \bar{\Omega}, \qquad (B.2)$$

where $U_\ell = U(N_\ell), \ell = \overline{1, nn}$. Every such a function is a continuous, piecewise linear over $\bar{\Omega}$.

For any $u \in C(\bar{\Omega})$, let $u_\ell \in V_{nn}$ be defined by

$$u_\ell(x,y) = \sum_{\ell=1}^{nn} u(N_\ell)\phi_\ell(x,y) \qquad (x,y) \in \bar{\Omega}.$$

Applying (b) we find that

$$u(N_\ell) = u_\ell(N_\ell) \qquad \ell = \overline{1,nn}. \tag{B.3}$$

That is, u_ℓ "interpolates" u at the nodes. Obviously, $u_\ell(x,y) = u(x,y)$, $\forall (x,y) \in \bar{\Omega} \Leftrightarrow u \in V_{nn}$.

P4. A discrete analogous of (V). We look now for a discrete approximation of the solution of (V) in the space (V_{nn}). For this we denote by $u_{nn} \in V_{nn}$ the V_{nn}-interpolant of u, i.e.,

$$u_{nn}(x,y) = \sum_{\ell=1}^{nn} u_\ell \phi_\ell(x,y) \qquad (x,y) \in \bar{\Omega},$$

where $u_\ell = u(N_\ell) = u_{nn}(N_\ell)$, $\ell = \overline{1,nn}$ (see (B.3)). We will find out the unknown values of u_{nn} in the nodes of triangulation T_h. This writes as follows

$$\text{find} \quad u_{nn} \in V_{nn}, \tag{V_{nn}}$$

$$\int_\Omega \nabla u_{nn} \nabla U dxdy + \int_\Omega u_{nn} U dxdy = \int_\Gamma gU dxdy \quad \forall U \in V_{nn}.$$

Let us denote by

$$\bar{g}(x,y) = \begin{cases} g(x,y), & (x,y) \in \Gamma \\ 0, & (x,y) \in \Omega, \end{cases}$$

and by $\bar{g}_\ell$ the V_{nn}-interpolant of $\bar{g}$. Then

$$\bar{g}(x,y) = \sum_{\ell=1}^{nn} \bar{g}_\ell \phi_\ell(x,y) \qquad (x,y) \in \bar{\Omega},$$

where $\bar{g}_\ell = \bar{g}(N_\ell)$, $\ell = \overline{1,nn}$. Then, the problem (V_{nn}) can be rewritten as

$$\sum_{k=1}^{nn} U_k \Big[\sum_{\ell=1}^{nn} \Big(\int_\Omega \nabla\phi_k \nabla\phi_\ell dxdy\Big) u_\ell\Big] + \sum_{k=1}^{nn} U_k \Big[\sum_{\ell=1}^{nn} \Big(\int_\Omega \phi_k \phi_\ell dxdy\Big) u_\ell\Big]$$

$$= \sum_{k=1}^{nn} U_k \Big[\sum_{\ell=1}^{nn} \Big(\int_{\bar{\Omega}} \phi_k \phi_\ell dxdy \Big) \bar{g}_\ell \Big]$$

or

$$\sum_{\ell=1}^{nn} \Big(\int_\Omega \nabla\phi_k \nabla\phi_\ell dxdy \Big) u_\ell + \sum_{\ell=1}^{nn} \Big(\int_\Omega \phi_k \phi_\ell dxdy \Big) u_\ell$$

$$= \sum_{\ell=1}^{nn} \Big(\int_{\bar{\Omega}} \phi_k \phi_\ell dxdy \Big) \bar{g}_\ell \qquad k = \overline{1, nn}.$$

Setting

$$b_{k\ell} = \int_\Omega \phi_k \phi_\ell dxdy, \quad \bar{b}_{k\ell} = \int_{\bar{\Omega}} \phi_k \phi_\ell dxdy,$$

$$c_{k\ell} = \int_\Omega \nabla\phi_k \nabla\phi_\ell dxdy,$$

the above linear system can be writhen in matrix form as

$$A\bar{u} = d \qquad (B.4)$$

where

$$\begin{aligned}
&A = B + C, \\
&B = \big(b_{k\ell}\big)_{k,\ell=\overline{1,nn}}, \\
&C = \big(c_{k\ell}\big)_{k,\ell=\overline{1,nn}}, \\
&d[k] = \bar{B}\bar{g}, \quad k = \overline{1, nn}, \\
&\bar{u} = (u_1, u_2, \cdots, u_{nn})^T \quad \text{(the vector of unknowns)}, \\
&\bar{B} = \big(\bar{b}_{k\ell}\big)_{k,\ell=\overline{1,nn}}, \\
&\bar{g} = (\bar{g}_1, \bar{g}_2, \cdots, \bar{g}_{nn})^T.
\end{aligned}$$

P5. Assemble the system. First of all, we need a representation of the triangulation T_h. We observe that the assignment of integers i from 1 until nn, can be done in $nn!$ different ways. So the enumerations of the nodes defines a "*global node ordering*". It is convenient to have also a "*local node ordering*" of the nodes in each element, as illustrated in Fig. (**4 b**).

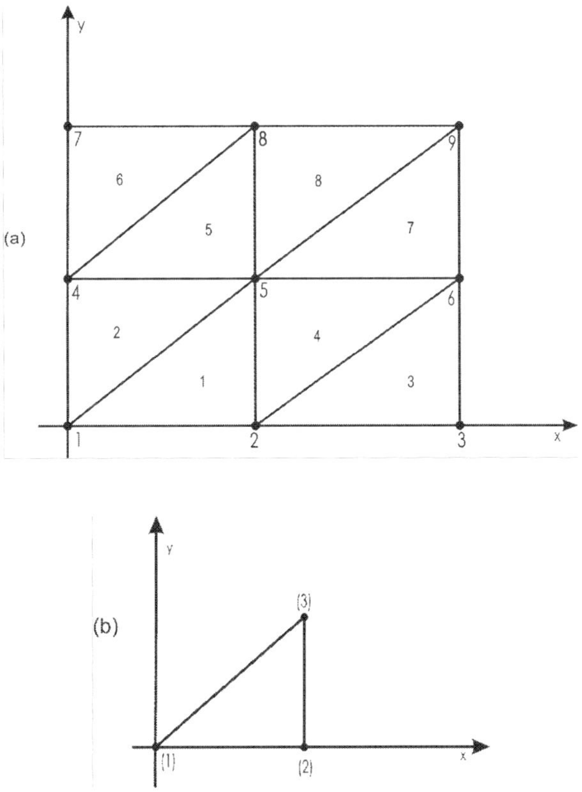

Figure 4 a) *Global node numbers.* $nt = 8$, $nn = 9$, $t = 3$
b) *Local node numbers* **(writhen in brackets)**

The task of generating a suitable finite element mesh over a given domain Ω is a quite complicated one. Let N_i, $i = 1, 2, ..., nn$ and K_ℓ, $\ell = 1, 2, ..., nt$ be enumerations of the nodes and triangles of T_h, respectively (see Fig. **4 a**) and Fig. (**1**). Then, the triangulation T_h may be specified using two matrices (arrays): SC (spatial coordinates) of dimension $2 \times nn$, containing in each SC_{1j} and SC_{2j} the Cartesian coordinates of node N_j, $j = \overline{1, nn}$ and T (global node numbers of every element) of dimension $3 \times nt$, containing in each column the numerical labels of the nodes in the triangle K_ℓ, $\ell = \overline{1, nt}$. Therefore, in our case (see Fig. **1**), if we consider $a = b = 0.2$, we have

$$SC: \begin{pmatrix} 0.0 & 0.1 & 0.2 & 0.0 & 0.1 & 0.2 & 0.0 & 0.1 & 0.2 \\ 0.0 & 0.0 & 0.0 & 0.1 & 0.1 & 0.1 & 0.2 & 0.2 & 0.2 \end{pmatrix},$$

$$T: \begin{pmatrix} 1 & 1 & 2 & 2 & 4 & 4 & 5 & 5 \\ 2 & 5 & 3 & 6 & 5 & 8 & 6 & 9 \\ 5 & 4 & 6 & 5 & 8 & 7 & 9 & 8 \end{pmatrix}.$$

We define

nkx - number of nodes on Ox,
nky - number of nodes on Oy,
$nkx1 = nkx - 1$,
$nky1 = nky - 1$.

Then

$nn = nkx * nky$,
$nt = 2 * nkx1 * nky1$ (nt is a even number).

In connection with boundary conditions, we need a representation of all "element edges" on Γ. Let

EDGEX1[nkx1], EDGEX2[nkx1],
EDGEY1[nky1], EDGEY2[nky1]

the vectors of numbers of all elements with a edge on Γ_1, Γ_3, Γ_4 and Γ_2, respectively. In our case (see Fig. **1**) we have

EDGEX1[2]=(1,3),
EDGEX2[2]=(6,8),
EDGEY1[2]=(2,6),
EDGEY2[2]=(3,7).

The following BORLAND C++ implements the arrays

SC, T, EDGEX1, EDGEX2, EDGEY1 and EDGEY2.

```
//
// Subroutine "triangl"
//
#include<stdio.h>
#include<math.h>
void triangl();
```

```
{
  extern int nkx,nky,nkx1,nky1,nt,nn;
  extern int T[3][nt],EDGEX1[nkx1], EDGEX2[nkx1],
             EDGEY1[nky1], EDGEY2[nky1];
  extern float SC[2][nn],h;
  int i,j,nt2,itemp,ℓ,k;
  float temp1,temp2;
//
//  nkx   is an input parameter; it is the number of nodes on Ox.
//  nky   is an input parameter; it is the number of nodes on Oy.
//  nkx1=nkx-1.
//  nky1=nky-1.
//  nt    is an input parameter; the number of triangles in a mesh.
//  nn    is an input parameter; the number of nodes in a mesh.
//  h     is an input parameter; it is the length of a edge.
//  SC   is an output parameter; is the matrix of length
//          2*nn containing in each SC_1j and SC_2j the
//          Cartesian coordinates of node N_j, j = 1,nn.
//  T    is an output parameter; is the matrix of length
//          3*nt containing in each column the numerical labels
//          of the nodes of the triangle K_ℓ, ℓ = 1,nt.
//  EDGEX1  is an output parameter; is the vector of length
//          nkx1 containing the numbers of all elements with
//          a edge on Γ1.
//  EDGEX2  is an output parameter; is the vector of length
//          nkx1 containing the numbers of all elements with
//          a edge on Γ3.
//  EDGEY1  is an output parameter; is the vector of length
//          nky1 containing the numbers of all elements with
//          a edge on Γ4.
//  EDGEY2  is an output parameter; is the vector of length
//          nky1 containing the numbers of all elements with
```

```
//        a edge on Γ2.
//   i, j, k, ℓ, nt2, itemp, temp1,temp2 are dummy variables.
//
for(i=1; i<=nkx; i++)
   {temp1 = h*(i-1);
    for(j=1; j<=nky; j++)
      { temp2 = h*(j-1);
        k = (j-1)*nkx+i;
        SC[0][k] = temp1;
        SC[1][k] = temp2;
       }
    }
nt2 = nt/2;
for(k=1; k<=nt2; k++)
   {j = (k-0.1)/nkx1;
    i = k-j*nkx1;
    itemp = i+ j*nkx;
    ℓ = 2*k-1;
     T[0][ℓ] = itemp;
     T[1][ℓ] = itemp+1;
     T[2][ℓ] = itemp+nkx+1;
    ℓ = 2*k;
     T[0][ℓ] = itemp;
     T[1][ℓ] = itemp+nkx+1;
     T[2][ℓ] = itemp+nkx;
   }
for(k=1; k<=nkx1; k++)
    EDGEX1[k] = 2*k-1;
EDGEY1[1] = 2;
for(k=2; k<=nky1; k++)
    EDGEY1[k] = EDGEY1[k-1] + 2*nky1;
for(k=1; k<=nkx1; k++)
```

```
        EDGEX2[k] = EDGEY1[nky1] + 2*nkx1*(k-1);
    for(k=1; k<=nky1; k++)
        EDGEY2[k] = EDGEX1[nkx1] + 2*(k-1);
}     // end of subroutine "triangl".
```

Local basis functions. Standard element. Let us defined by

$$\phi_r^{(\ell)}(x,y) = \phi_{i_r^{(\ell)}}(x,y) \qquad (x,y) \in e_\ell, \quad r = \overline{1,t},$$

the *"local basis functions"* over e_ℓ. A "local basis function" is simply the restriction of some "global basis function" to e_ℓ.

We assume that the "global basis functions" $\phi_1, \phi_2, ..., \phi_{nn}$, have the property that:

It is possible to choose a *"standard element"* $\tilde{e}$ in the $\tilde{x} - \tilde{y}$ plane, with local basis functions $\tilde{\phi}_1(\tilde{x},\tilde{y}),...,\tilde{\phi}_t(\tilde{x},\tilde{y})$, of type (B.1), and find for every element e_ℓ, $\ell = \overline{1,nt}$, an invertible affine variable transformation

$$\begin{cases} x = x(\tilde{x},\tilde{y}) = s_{11}\tilde{x} + s_{12}\tilde{y} + b_1 \\ y = y(\tilde{x},\tilde{y}) = s_{21}\tilde{x} + s_{22}\tilde{y} + b_2 \end{cases} \quad (\tilde{x},\tilde{y}) \in \tilde{e} \qquad (B.5)$$

dependent on ℓ, such that (B.5) maps $\tilde{e}$ onto e_ℓ and

$$\tilde{\phi}_r(\tilde{x},\tilde{y}) = \phi_r^{(\ell)}(x(\tilde{x},\tilde{y}), y(\tilde{x},\tilde{y})) \quad (\tilde{x},\tilde{y}) \in \tilde{e}. \qquad (B.6)$$

Fig. (**5**) shows our choice of the standard element $\tilde{e}$.

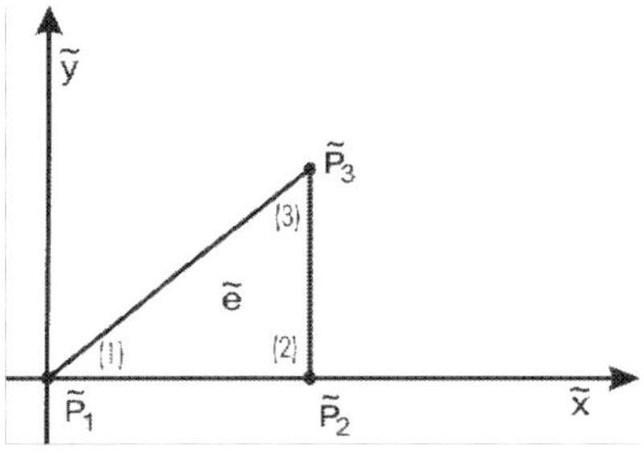

Figure 5 *The standard element*

Denoting the inverse transformation in (B.5) by

$$\begin{cases} \tilde{x} = \tilde{x}(x, y) \\ \tilde{y} = \tilde{y}(x, y) \end{cases} \quad (x, y) \in e_\ell,$$

we can rewrite (B.5) as

$$\phi_r^{(\ell)}(x, y) = \tilde{\phi}_r(\tilde{x}(x, y), \tilde{y}(x, y)).$$

Relation (B.5) implies that the "local basis functions" $\phi_r^{(\ell)}$, $\ell = \overline{1, nt}$, can be reduced by variable transformations to the single function $\tilde{\phi}_r$. Because the "global basis functions" ϕ_1, ..., ϕ_{nn} are composites of the "local basis functions", this means that all operations on the "global basis functions" can be reduced to operations on $\tilde{\phi}_1$,...,$\tilde{\phi}_t$. We call $\tilde{\phi}_1$,...,$\tilde{\phi}_t$ the *"standard local basis functions"* (s.l.b.f. for short). By (B.1), the s.l.b.f. $\tilde{\phi}_r$ has the form

$$\tilde{\phi}_r(\tilde{x}, \tilde{y}) = a_{1r}\tilde{x} + a_{2r}\tilde{y} + a_{3r} \quad (\tilde{x}, \tilde{y}) \in \tilde{e}, \quad r = 1, 2, 3. \qquad (B.7)$$

Using (b) in **P3**, we will determine the coefficients a_{1r}, a_{2r}, a_{3r} so that $\tilde{\phi}_r$ has the value 1 (unity) at node r and 0 (zero) at the other two nodes. So, simple calculations establish that

$$\tilde{\phi}_1(\tilde{x}, \tilde{y}) = -\tilde{x} + 1, \quad \tilde{\phi}_2(\tilde{x}, \tilde{y}) = \tilde{x} - \tilde{y}, \quad \tilde{\phi}_3(\tilde{x}, \tilde{y}) = \tilde{y}. \qquad (B.8)$$

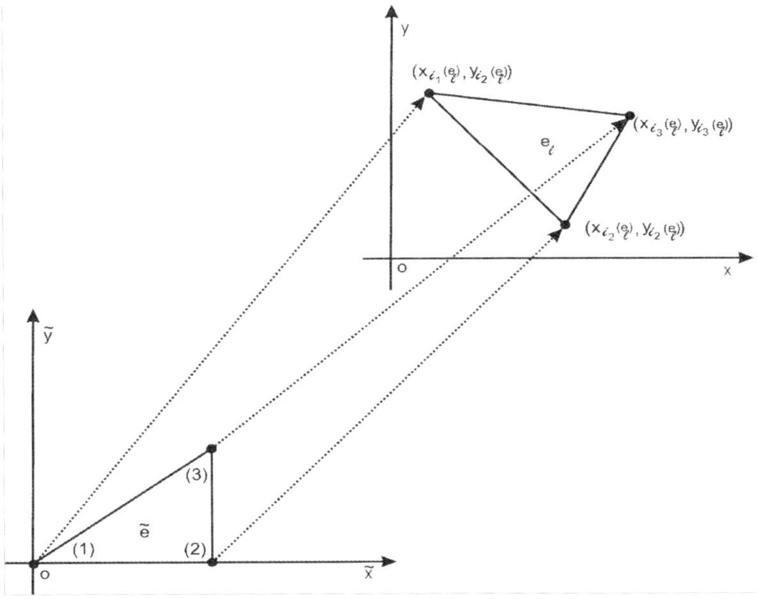

Figure 6 *The mapping of* $\tilde{e}$ *onto* e_ℓ.

Let us find now the affine transformation in (B.5) which maps the vertices of $\tilde{e}$ onto those of e_ℓ as is indicated in Fig. (**6**).
(1) $\to (x_{i_1^{(\ell)}}, y_{i_1^{(\ell)}})$ is equivalent with

$$\begin{cases} x_{i_1^{(\ell)}} = s_{11} \cdot 0 + s_{12} \cdot 0 + b_1 \\ y_{i_1^{(\ell)}} = s_{21} \cdot 0 + s_{22} \cdot 0 + b_2 \end{cases}$$

which leads to $b_1 = x_{i_1^{(\ell)}}$, $b_2 = y_{i_1^{(\ell)}}$.
(2) $\to (x_{i_2^{(\ell)}}, y_{i_2^{(\ell)}})$ is equivalent with

$$\begin{cases} x_{i_2^{(\ell)}} = s_{11} \cdot 1 + s_{12} \cdot 0 + b_1 \\ y_{i_2^{(\ell)}} = s_{21} \cdot 1 + s_{22} \cdot 0 + b_2 \end{cases}$$

which leads to $s_{11} = x_{i_2^{(\ell)}} - x_{i_1^{(\ell)}}$, $s_{21} = y_{i_2^{(\ell)}} - y_{i_1^{(\ell)}}$.
(3) $\to (x_{i_3^{(\ell)}}, y_{i_3^{(\ell)}})$ is equivalent with

$$\begin{cases} x_{i_3^{(\ell)}} = s_{11} \cdot 1 + s_{12} \cdot 1 + b_1 \\ y_{i_3^{(\ell)}} = s_{21} \cdot 1 + s_{22} \cdot 1 + b_2 \end{cases}$$

which yields $s_{12} = x_{i_3^{(\ell)}} - x_{i_2^{(\ell)}}$, $s_{22} = y_{i_3^{(\ell)}} - y_{i_2^{(\ell)}}$. Thus, the affine transformation (B.5) and its inverse can be expressed in the matrix-vector form as

$$\begin{pmatrix} x \\ y \end{pmatrix} = C\begin{pmatrix} \tilde{x} \\ \tilde{y} \end{pmatrix} + \begin{pmatrix} b_1 \\ b_2 \end{pmatrix}, \quad \begin{pmatrix} \tilde{x} \\ \tilde{y} \end{pmatrix} = C^{-1}\begin{pmatrix} x \\ y \end{pmatrix} + C^{-1}\begin{pmatrix} b_1 \\ b_2 \end{pmatrix}, \tag{5.5}$$

where

$$C = \begin{pmatrix} x_{i_2^{(\ell)}} - x_{i_1^{(\ell)}} & x_{i_3^{(\ell)}} - x_{i_2^{(\ell)}} \\ y_{i_2^{(\ell)}} - y_{i_1^{(\ell)}} & y_{i_3^{(\ell)}} - y_{i_2^{(\ell)}} \end{pmatrix}, \quad \begin{pmatrix} b_1 \\ b_2 \end{pmatrix} = \begin{pmatrix} x_{i_1^{(\ell)}} \\ y_{i_1^{(\ell)}} \end{pmatrix}.$$

Since every affine transformation maps straight lines onto straight lines, it follows that (5.5) maps $\tilde{e}$ onto e_ℓ.

We shall now verify that (B.6) is satisfied. For $r = 1$, we consider that (see relation (B.6))

$$\varphi_1^{(\ell)}(\tilde{x}, \tilde{y}) = \phi_1^{(\ell)}(x(\tilde{x}, \tilde{y}), y(\tilde{x}, \tilde{y})) \quad (\tilde{x}, \tilde{y}) \in \tilde{e}.$$

We must to obtain that $\varphi_1^{(\ell)}(\tilde{x}, \tilde{y}) = \tilde{\phi}_1(\tilde{x}, \tilde{y})$ (see 5.4). By definition (see **P3**), we know that $\phi_1^{(\ell)}(x, y)$ is a linear function of x and y and that

$$\phi_1^{(\ell)}(N_{i_s^{(\ell)}}) = \delta_{1s}, \quad s = 1, 2, 3,$$

i.e.,

$$\begin{aligned} \varphi_1^{(\ell)}(0,0) &= \phi_1^{(\ell)}(x(0,0), y(0,0)) = \phi_1^{(\ell)}(N_{i_1^{(\ell)}}) = 1, \\ \varphi_1^{(\ell)}(1,0) &= \phi_1^{(\ell)}(x(1,0), y(1,0)) = \phi_1^{(\ell)}(N_{i_2^{(\ell)}}) = 0, \\ \varphi_1^{(\ell)}(1,1) &= \phi_1^{(\ell)}(x(1,1), y(1,1)) = \phi_1^{(\ell)}(N_{i_3^{(\ell)}}) = 0. \end{aligned}$$

But from (B.1),

$$\phi_1^{(\ell)}(x, y) = a_{11}^{(\ell)} x + a_{21}^{(\ell)} y + a_{31}^{(\ell)} \quad (x, y) \in e_\ell, \tag{5.6}$$

and then the above equalities can be rewrites as

$$\begin{cases} \phi_1^{(\ell)}(N_{i_1^{(\ell)}}) = 1 \\ \phi_1^{(\ell)}(N_{i_2^{(\ell)}}) = 0 \\ \phi_1^{(\ell)}(N_{i_3^{(\ell)}}) = 0 \end{cases} \Leftrightarrow \begin{cases} a_{11}^{(\ell)} x_{i_1^{(\ell)}} + a_{21}^{(\ell)} y_{i_1^{(\ell)}} + a_{31}^{(\ell)} = 1 \\ a_{11}^{(\ell)} x_{i_2^{(\ell)}} + a_{21}^{(\ell)} y_{i_2^{(\ell)}} + a_{31}^{(\ell)} = 0 \\ a_{11}^{(\ell)} x_{i_3^{(\ell)}} + a_{21}^{(\ell)} y_{i_3^{(\ell)}} + a_{31}^{(\ell)} = 0 \end{cases}$$

$$\Leftrightarrow \quad \begin{cases} a_{11}^{(\ell)}(x_{i_2^{(\ell)}} - x_{i_1^{(\ell)}}) + a_{21}^{(\ell)}(y_{i_2^{(\ell)}} - y_{i_1^{(\ell)}}) = -1 \\ a_{11}^{(\ell)}(x_{i_3^{(\ell)}} - x_{i_2^{(\ell)}}) + a_{21}^{(\ell)}(y_{i_3^{(\ell)}} - y_{i_2^{(\ell)}}) = 0 \end{cases}$$

$$\Leftrightarrow \quad \begin{cases} a_{11}^{(\ell)} s_{11} + a_{21}^{(\ell)} s_{21} = -1 \\ a_{11}^{(\ell)} s_{12} + a_{21}^{(\ell)} s_{22} = 0. \end{cases}$$

Solving this linear system we find the solution

$$a_{11}^{(\ell)} = -\frac{s_{22}}{D}, \quad a_{21}^{(\ell)} = \frac{s_{12}}{D}, \quad a_{31}^{(\ell)} = 1 + \frac{s_{22}}{D} x_{i_1^{(\ell)}} - \frac{s_{12}}{D} y_{i_1^{(\ell)}},$$

where $D = s_{11}s_{22} - s_{21}s_{12}$. So (5.6) becomes

$$\phi_1^{(\ell)}(x,y) = -\frac{s_{22}}{D}x + \frac{s_{12}}{D}y + 1 + \frac{s_{22}}{D}x_{i_1^{(\ell)}} - \frac{s_{12}}{D}y_{i_1^{(\ell)}}. \tag{5.7}$$

Now, making use of (B.1) (see also (5.5)) and (5.7), we get

$$\begin{aligned}
\phi_1^{(\ell)}(x,y) &= \phi_1^{(\ell)}(s_{11}\tilde{x} + s_{12}\tilde{y} + b_1, s_{21}\tilde{x} + s_{22}\tilde{y} + b_2) \\
&= \phi_1^{(\ell)}(s_{11}\tilde{x} + s_{12}\tilde{y} + x_{i_1^{(\ell)}}, s_{21}\tilde{x} + s_{22}\tilde{y} + y_{i_1^{(\ell)}}) \\
&= -\frac{s_{22}}{D}(s_{11}\tilde{x} + s_{12}\tilde{y} + x_{i_1^{(\ell)}}) + \frac{s_{12}}{D}(s_{21}\tilde{x} + s_{22}\tilde{y} + y_{i_1^{(\ell)}}) \\
&\quad + 1 + \frac{s_{22}}{D}x_{i_1^{(\ell)}} - \frac{s_{12}}{D}y_{i_1^{(\ell)}} \\
&= -\frac{1}{D}(s_{11}s_{22} - s_{12}s_{21})\tilde{x} + \frac{1}{D}(s_{12}s_{22} - s_{22}s_{12})\tilde{y} + 1 \\
&= -\tilde{x} + 1.
\end{aligned}$$

So $\varphi_1^{(\ell)}(\tilde{x},\tilde{y}) = \tilde{\phi}_1(\tilde{x},\tilde{y})$. Similar calculations establish that $\tilde{\phi}_2(\tilde{x},\tilde{y}) = \phi_2^{(\ell)}(x(\tilde{x},\tilde{y}), y(\tilde{x},\tilde{y}))$ and $\tilde{\phi}_3(\tilde{x},\tilde{y}) = \phi_3^{(\ell)}(x(\tilde{x},\tilde{y}), y(\tilde{x},\tilde{y}))$, and so (B.6) is established.

The calculation of the data in (B.4). The calculation of the element matrix A and of the right hand side vector d in (B.4) can be performed in the following major steps:

(i) $b_{kj} = \int\limits_{\Omega} \phi_k(x,y)\phi_j(x,y)dxdy \quad k,j = \overline{1,nn}$,

(ii) $c_{kj} = \int\limits_{\Omega} \nabla\phi_k(x,y)\nabla\phi_j(x,y)dxdy \quad k,j = \overline{1,nn}$,

(iii) $d[k] \quad k = \overline{1, nn}$.

We remark that the first two steps are independent of the boundary conditions. We shall now describe in detail the implementation of each step

(i) According to mesh of Ω (see **P1**), the term b_{kj} is decomposed as

$$b_{kj} = \sum_{\ell=1}^{nt} b_{kj}^{(\ell)},$$

where

$$b_{kj}^{(\ell)} = \int_{e_\ell} \phi_k^{(\ell)}(x,y)\phi_j^{(\ell)}(x,y)dxdy \qquad k,j = \overline{1, nn}.$$

But $b_{kj}^{(\ell)}$ is nonzero only if N_k and N_j belong to e_ℓ. Thus, if $b_{kj}^{(\ell)}$ is nonzero, then clearly:

• among the node numbers $i_1^{(\ell)}$, $i_2^{(\ell)}$ $i_3^{(\ell)}$ are two, $i_r^{(\ell)}$ and $i_s^{(\ell)}$, such that $k = i_r^{(\ell)}$, $j = i_s^{(\ell)}$;

• $b_{kj}^{(\ell)} = \displaystyle\int_{e_\ell} \phi_r^{(\ell)}(x,y)\phi_s^{(\ell)}(x,y)dxdy.$

We shall exploit here the variable transformation (B.5) and its inverse (5.5) to express $b_{kj}^{(\ell)}$ as an integral in the $\tilde{x}$, $\tilde{y}$ variables over the "standard element" $\tilde{e}$. The change of variables requires $dxdy$ to be replaced by $|\tilde{D}|d\tilde{x}\tilde{y}$, where $\tilde{D}$ is the Jacobian of (B.5), i.e.,

$$\tilde{D} = \begin{vmatrix} \dfrac{\partial x}{\partial \tilde{x}} & \dfrac{\partial x}{\partial \tilde{y}} \\ \dfrac{\partial y}{\partial \tilde{x}} & \dfrac{\partial y}{\partial \tilde{y}} \end{vmatrix} = \begin{vmatrix} s_{11} & s_{12} \\ s_{21} & s_{22} \end{vmatrix} = s_{11}s_{22} - s_{21}s_{12} \tag{5.8}$$

$$= (x_{i_2^{(\ell)}} - x_{i_1^{(\ell)}})(y_{i_3^{(\ell)}} - y_{i_2^{(\ell)}}) - (x_{i_3^{(\ell)}} - x_{i_2^{(\ell)}})(y_{i_2^{(\ell)}} - y_{i_1^{(\ell)}}).$$

So

$$b_{kj}^{(\ell)} = |\tilde{D}| \int_{\tilde{e}} \tilde{\phi}_r(\tilde{x},\tilde{y})\tilde{\phi}_s(\tilde{x},\tilde{y})d\tilde{x}d\tilde{y} \qquad k,j = \overline{1, nn}. \tag{5.9}$$

Let us compute the integral in (5.9). We consider two linear functions

$\phi_i(x,y) = a_{1i}x + a_{2i}y + a_{3i}$ and $\phi_j(x,y) = a_{1j}x + a_{2j}y + a_{3j}$. We have

$$\begin{aligned}\phi_i(x,y)\phi_j(x,y) = a_{1i}a_{1j}x^2+\\ a_{2i}a_{2j}y^2+\\ (a_{1i}a_{2j} + a_{2i}a_{1j})xy+\\ (a_{1i}a_{3j} + a_{3i}a_{1j})x+\\ (a_{2i}a_{3j} + a_{3i}a_{2j})y+\\ a_{3i}a_{3j}.\end{aligned}$$

Using the formula $\int_{\tilde{e}} f(x,y)dxdy = \int_0^1 \left(\int_0^x f(x,y)dy\right)dx = \int_0^1 \left(\int_1^y f(x,y)dx\right)dy$, we have

$$\int_{\tilde{e}} \tilde{x}^2 d\tilde{x}d\tilde{y} = \frac{1}{4}, \quad \int_{\tilde{e}} \tilde{y}^2 d\tilde{x}d\tilde{y} = \frac{1}{12}, \quad \int_{\tilde{e}} \tilde{x}\tilde{y}d\tilde{x}d\tilde{y} = \frac{1}{8},$$
$$\int_{\tilde{e}} \tilde{x}d\tilde{x}d\tilde{y} = \frac{1}{3}, \quad \int_{\tilde{e}} \tilde{y}d\tilde{x}d\tilde{y} = \frac{1}{6}, \quad \int_{\tilde{e}} d\tilde{x}d\tilde{y} = \frac{1}{2}.$$

We may compute the coefficients a_{1i}, a_{2i}, a_{3i}, $i = 1,2,3$, from the relations

$$\phi_i(\tilde{x},\tilde{y})\phi_j(\tilde{x},\tilde{y}) = \tilde{\phi}_i(\tilde{x},\tilde{y})\tilde{\phi}_j(\tilde{x},\tilde{y}) \quad \forall i,j = 1,2,3. \tag{5.10}$$

We will obtain

$$\begin{array}{lll} a_{11} = -1 & a_{21} = 0 & a_{31} = 1 \\ a_{12} = 1 & a_{22} = -1 & a_{32} = 0 \\ a_{13} = 0 & a_{23} = 1 & a_{33} = 0. \end{array}$$

The following BORLAND C++ code implements the computation of the integral in (5.9):

```
//
// Subroutine "cuadex"
//
01   void cuadex();
```

```
{
  extern float bas[3][3],intval[6];
//
//  bas is an output parameter; is the matrix of length
//          3*3 containing the coefficients of $\tilde{\phi}_i * \tilde{\phi}_j$.
//  intval is an output parameter; is the vector of length 6
//          containing the integral values on $\tilde{e}$ from $\tilde{x}^2$, ...
//
  bas[0][0] = -1;      bas[2][1] = 0;
  bas[1][0] = 0;       bas[0][2] = 0;
  bas[2][0] = 1;       bas[1][2] = 1;
  bas[0][1] = 1;       bas[2][2] = 0;
  bas[1][1] = -1;
  intval[0] = 0.25;
  intval[1] = 1./12;
  intval[2] = 0.125;
  intval[3] = 1./3;
  intval[4] = 1./6;
  intval[5] = 0.5;
}
```

We can assemble A from the contribution of elements b_{kj} by following subroutine:

```
//
// Subroutine "asb"
//
#include<stdio.h>
#include<math.h>
  void asb();
{
  extern int i,j,k,ℓ,itemp,ir,is;
  extern int nn,nt,T[3][nt];
```

```
extern float d,temp1,temp2,x[3],y[3];
extern float SC[2][nn],bas[3][3],intval[6];
extern float a[nn][nn];
//
// ir,is the number of nodes in "standard element".
// d     the value of D̃.
// a     is an output parameter; is the matrix of length
//         nn*nn containing the values b_kj.
//
for(ℓ=1; ℓ<=nt; ℓ++)
   {for(i=0; i<3; i++)
      { itemp = T[i][ℓ];
        x[i] = SC[0][itemp];
        y[i] = SC[1][itemp];
      }
     d = (x[1]-x[0])*(y[2]-y[1])-
                   (x[2]-x[1])*(y[1]-y[0]);
     for(ir=0; ir<3; ir++)
      {k = T[ir][ℓ];
       for(is=0; is<3; is++)
         {temp2 = bas[0][ir]*bas[0][is]*intval[0]
 +bas[1][ir]*bas[1][is]*intval[1]
 +(bas[0][ir]*bas[1][is]+bas[1][ir]*bas[0][is])*intval[2]
 +(bas[0][ir]*bas[2][is]+bas[2][ir]*bas[0][is])*intval[3]
 +(bas[1][ir]*bas[2][is]+bas[2][ir]*bas[1][is])*intval[4]
 +(bas[2][ir]*bas[2][is]*intval[5];
          temp1 = fabs(d)*temp2;
          j = T[is][ℓ];
          a[k][j] = a[k][j] + temp1;
         }
      }
   }
```

25 }

(ii) As we see in step (i), the term c_{kj} may be decomposed as

$$c_{kj} = \sum_{\ell=1}^{nt} c_{kj}^{(\ell)},$$

where

$$c_{kj}^{(\ell)} = \int_{e_\ell} \nabla\phi_r^{(\ell)}(x,y)\nabla\phi_s^{(\ell)}(x,y)dxdy \qquad k,j=\overline{1,nn},$$

with $k = i_r^{(\ell)}$ and $j = i_s^{(\ell)}$, $r,s = 1,2,...,t$. We must to pass at the integral onto $\tilde{e}$. Firstly, we have

$$\nabla\phi_r^{(\ell)}(x,y)\nabla\phi_s^{(\ell)}(x,y)$$

$$= \frac{\partial\phi_r^{(\ell)}(x,y)}{\partial x}\frac{\partial\phi_s^{(\ell)}(x,y)}{\partial x} + \frac{\partial\phi_r^{(\ell)}(x,y)}{\partial y}\frac{\partial\phi_s^{(\ell)}(x,y)}{\partial y}.$$

Making use of (5.3), we get

$$\frac{\partial\phi_r^{(\ell)}(x,y)}{\partial x} = \frac{\partial\tilde{\phi}_r}{\partial\tilde{x}}\frac{\partial\tilde{x}}{\partial x} + \frac{\partial\tilde{\phi}_r}{\partial\tilde{y}}\frac{\partial\tilde{y}}{\partial x},$$

$$\frac{\partial\phi_r^{(\ell)}(x,y)}{\partial y} = \frac{\partial\tilde{\phi}_r}{\partial\tilde{x}}\frac{\partial\tilde{x}}{\partial y} + \frac{\partial\tilde{\phi}_r}{\partial\tilde{y}}\frac{\partial\tilde{y}}{\partial y}.$$

So (see (5.8))

$$c_{kj}^{(\ell)} = |\tilde{D}| \int_{\tilde{e}} \Big[E_1 \frac{\partial\tilde{\phi}_r}{\partial\tilde{x}}\frac{\partial\tilde{\phi}_s}{\partial\tilde{x}} \tag{5.11}$$

$$+E_2\left(\frac{\partial\tilde{\phi}_r}{\partial\tilde{x}}\frac{\partial\tilde{\phi}_s}{\partial\tilde{y}} + \frac{\partial\tilde{\phi}_r}{\partial\tilde{y}}\frac{\partial\tilde{\phi}_s}{\partial\tilde{x}}\right) + E_3\frac{\partial\tilde{\phi}_r}{\partial\tilde{y}}\frac{\partial\tilde{\phi}_s}{\partial\tilde{y}}\Big]d\tilde{x}d\tilde{y},$$

where

$$E_1 = \left(\frac{\partial\tilde{x}}{\partial x}\right)^2 + \left(\frac{\partial\tilde{x}}{\partial y}\right)^2, \qquad E_2 = \left(\frac{\partial\tilde{y}}{\partial x}\right)^2 + \left(\frac{\partial\tilde{y}}{\partial y}\right)^2,$$

$$E_3 = \frac{\partial \tilde{x}}{\partial x}\frac{\partial \tilde{y}}{\partial x} + \frac{\partial \tilde{x}}{\partial y}\frac{\partial \tilde{y}}{\partial y}.$$

We recall (see (5.5))

$$\begin{cases} \tilde{x} = \dfrac{1}{\tilde{D}}(s_{22}x - s_{12}y) + C^{-1}b_1, \\ \tilde{y} = \dfrac{1}{\tilde{D}}(-s_{21}x + s_{11}y) + C^{-1}b_2 \end{cases}$$

and then,

$$\begin{array}{ll} \dfrac{\partial \tilde{x}}{\partial x} = \dfrac{s_{22}}{\tilde{D}} & \dfrac{\partial \tilde{x}}{\partial y} = -\dfrac{s_{12}}{\tilde{D}} \\ \dfrac{\partial \tilde{y}}{\partial x} = -\dfrac{s_{21}}{\tilde{D}} & \dfrac{\partial \tilde{y}}{\partial y} = \dfrac{s_{11}}{\tilde{D}} \end{array}$$

and so,

$$E_1 = \frac{s_{12}^2 + s_{22}^2}{\tilde{D}^2}, \quad E_3 = \frac{s_{11}^2 + s_{21}^2}{\tilde{D}^2},$$
$$E_2 = -\frac{s_{11}s_{12} + s_{21}s_{22}}{\tilde{D}^2}.$$

Because (see, for example, (5.3′), (5.10) and subroutine "cuadex")

$$\tilde{\phi}_r(\tilde{x}, \tilde{y}) = bas[1][r]\tilde{x} + bas[2][r]\tilde{y} + bas[3][r], \quad r = 1, 2, 3,$$

we have

$$\frac{\partial \tilde{\phi}_r(\tilde{x}, \tilde{y})}{\partial \tilde{x}} = bas[1][r], \quad \frac{\partial \tilde{\phi}_r(\tilde{x}, \tilde{y})}{\partial \tilde{y}} = bas[2][r], \quad r = 1, 2, 3,$$

and then (5.11) becomes

$$c_{kj}^{(\ell)} = \frac{1}{|\tilde{D}|} \int_{\tilde{e}} \Big[(s_{12}^2 + s_{22}^2) bas[1][r] bas[1][s] \tag{5.12}$$

$$-(s_{11}s_{12} + s_{21}s_{22})(bas[1][r]bas[2][s] + bas[2][r]bas[1][s])$$
$$+(s_{11}^2 + s_{21}^2) bas[2][r] bas[2][s]\Big] \, d\tilde{x} d\tilde{y}$$

$$= \frac{\text{area of } \tilde{e}}{|\tilde{D}|} \left[(s_{12}^2 + s_{22}^2) bas[1][r]bas[1][s] \right.$$

$$-(s_{11}s_{12} + s_{21}s_{22})(bas[1][r]bas[2][s] + bas[2][r]bas[1][s])$$

$$\left. +(s_{11}^2 + s_{21}^2) bas[2][r]bas[2][s] \right].$$

We can assemble A from the contribution of elements c_{kj} by following subroutine:

```
//
// Subroutine "asc"
//
#include<stdio.h>
#include<math.h>
    void asc();
  {
    extern int i,j,k,ℓ,itemp,ir,is;
    extern int nn,nt,T[3][nt];
    extern float aria,d,temp1,temp2,s11,s12,s21,s22,
                                  e1,e2,e3,x[3],y[3];
    extern float SC[2][nn],bas[3][3];
    extern float a[nn][nn];
//
//  ir,is the number of nodes in "standard element".
//  d     the value of D̃.
//  a     is an output parameter; is the matrix of length
//            nn*nn containing the values b_kj.
//  aria  the value of the area of triangle ẽ.
//
    aria = 0.5;
    for(ℓ=1; ℓ<=nt; ℓ++)
       {for(i=0; i<3; i++)
          { itemp = T[i][ℓ];
            x[i] = SC[0][itemp];
```

```
          y[i] = SC[1][itemp];
        }
      s11 = x[1]-x[0];     s12 = x[2]-x[1];
      s21 = y[1]-y[0];     s22 = y[2]-y[1];
      d = s11*s22-s12*s21;
      temp1 = fabs(d);
      e1 = s12*s12+s22*s22;
      e2 = s11*s12+s21*s22;
      e3 = s11*s11+s21*s21;
      for(ir=0; ir<3; ir++)
        {k = T[ir][ℓ];
         for(is=0; is<3; is++)
           {temp2 = bas[0][ir]*bas[0][is]*e1
 -(bas[0][ir]*bas[1][is]+bas[1][ir]*bas[0][is])*e2
 +bas[1][ir]*bas[1][is]*e3;
            temp2 = temp2*aria;
            j = T[is][ℓ];
            a[k][j] = a[k][j] + temp2/temp1;
           }
        }
      }
 }
```

Remark. To assemble the global matrix $A = B + C$ in (B.4), we examine by subroutines "asb()" and "asc()" all elements e_ℓ, $\ell = \overline{1, nt}$, of triangulation T_h, stored in the array $T[3][nt]$ and, successively, add in the contributions from different elements. We also have used the array $SC[2][nn]$ to find the node coordinates.

(iii) To compute $d[k]$, $k = \overline{1, nn}$, we recall firstly that $\bar{g}(x, y) = 0$ for $(x, y) \in \Omega$ and then

$$\begin{aligned}\int_{\bar{\Omega}} \bar{g}(x, y)U(x, y)dxdy &= \int_{\Gamma} g(x, y)U(x, y)dxdy \\ &= U_k g_j \int_{\Gamma} \phi_k \phi_j dxdy,\end{aligned}$$

where $g_j = g(N_j)$, $N_j \in \Gamma$. But $\int_\Gamma \phi_k \phi_j dxdy$ may be nonzero only if N_k, N_j belong to the same edge on Γ. Then we have to compute a edge integral of the form

$$b_{kj}^{(\ell)} = \int_{N_k}^{N_j} \phi_r^{(\ell)}(x,y)\phi_s^{(\ell)}(x,y)dxdy \quad k,j = \overline{1,nn}, \tag{5.13}$$

where $\phi_r^{(\ell)}$, $\phi_s^{(\ell)}$, as usual, denote local basis functions. If the "edge element" on Γ belong to Γ_1 then (5.13) means

$$b_{kj}^{(\ell)} = \int_{N_k}^{N_j} \phi_r^{(\ell)}(x,0)\phi_s^{(\ell)}(x,0)dx \tag{5.14}$$

and, if the "edge element" on Γ belong to Γ_2, then (5.13) means

$$b_{kj}^{(\ell)} = \int_{N_k}^{N_j} \phi_r^{(\ell)}(0,y)\phi_s^{(\ell)}(0,y)dy. \tag{5.15}$$

To compute the integrals in (5.14)-(5.15) we apply again the variable transformation (B.5), expressing these as an integral along to the edge of the "standard element" $\tilde{e}$ with vertex nodes $r, s \in \{1,2,3\}$. Let us denote by h_1- the length of edge on Ox and by h_2- the length of edge on Oy. By (5.1) we have $dx = h_1 d\tilde{x}$, $dy = h_2 d\tilde{y}$ and so (5.14)-(5.15) becomes

$$\begin{aligned} b_{kj}^{(\ell)} &= h_1 \int_{N_k}^{N_j} \tilde{\phi}_r(\tilde{x},0)\tilde{\phi}_s(\tilde{x},0)d\tilde{x} \\ &= h_1 \Big[a_{1r}a_{1s}\int_0^1 \tilde{x}^2 d\tilde{x} + (a_{1r}a_{3s} + a_{3r}a_{1s})\int_0^1 \tilde{x}d\tilde{x} + a_{3r}a_{3s}\Big], \end{aligned} \tag{5.16}$$

$$\begin{aligned} b_{kj}^{(\ell)} &= h_2 \int_{N_k}^{N_j} \tilde{\phi}_r(0,\tilde{y})\tilde{\phi}_s(0,\tilde{y})d\tilde{y} \\ &= h_2 \Big[a_{2r}a_{2s}\int_0^1 \tilde{y}^2 d\tilde{y} + (a_{2r}a_{3s} + a_{3r}a_{2s})\int_0^1 \tilde{y}d\tilde{y} + a_{3r}a_{3s}\Big]. \end{aligned} \tag{5.17}$$

To know what are the "standard element basis functions" $\tilde{\phi}_r$, $\tilde{\phi}_s$ that can appear in (5.16)-(5.17), we need to know the vertices corresponding to the "local node ordering" for each "element edge" on Γ. These can be specified using the following vectors (see Fig. **7**).

$$skrx[4] = \{1,1,2,2\}, \quad sksx[4] = \{1,2,1,2\}, \quad \text{for } \Gamma_1;$$
$$skry[4] = \{1,1,3,3\}, \quad sksy[4] = \{1,3,1,3\}, \quad \text{for } \Gamma_4;$$

and

$$skr23[4] = \{2,2,3,3\}, \quad sks23[4] = \{2,3,2,3\}, \quad \text{for } \Gamma_2, \Gamma_3.$$

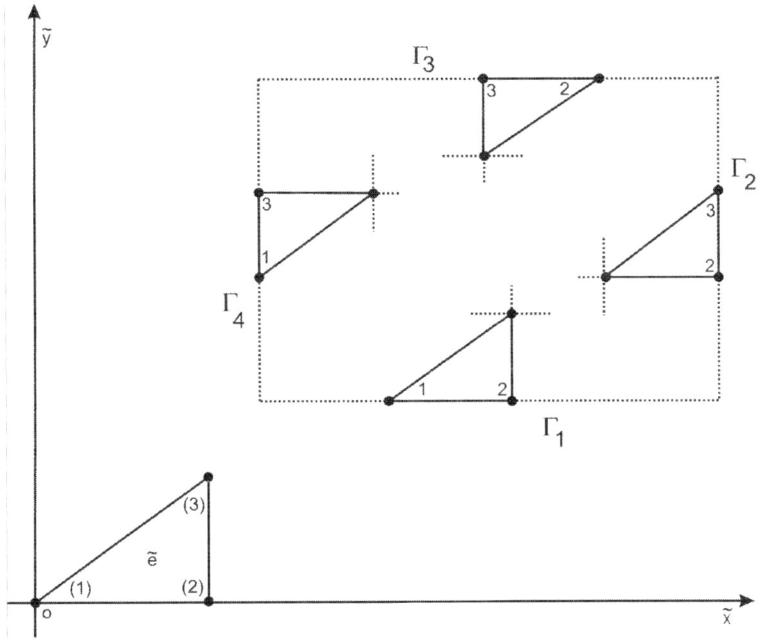

Figure 7 *The position of* "standard element" *relative to* Γ

For example, a pair $(skrx[], sksx[])$ saying who are the functions $\tilde{\phi}_r$, $\tilde{\phi}_s$ from relation (5.16).

$d[k]$, $k = \overline{1, nn}$, can thus be assembled as follows:

```
//
// Subroutine "asd"
//
#include<stdio.h>
#include<math.h>
01   void asd();
```

```
{
  extern int i,j,k,ℓ,ir,is,i1,nt,nn;
  extern int nkx1,nky1,T[3][nt],skrx[4],sksx[4]
                 skry[4],sksy[4],skr23[4],sks23[4];
  extern int EDGEX1[nkx1],EDGEX2[nkx1],
                 EDGY1[nky1],EDGEY2[nky1];
  extern float temp1,h1,h2,bas[3][3],d[nn];
//
//  ir,is the number of nodes in "standard element".
//  h1  is an input parameter; the length of edge on Ox.
//  h2  is an input parameter; the length of edge on Oy.
//  skrx[4], sksx[4] is an input parameter; "local node
//        ordering" for an "element edge" on Γ_1.
//  skry[4], sksy[4] is an input parameter; "local node
//        ordering" for an "element edge" on Γ_4.
//  skr23[4], sks23[4] is an input parameter; "local node
//        ordering" for an "element edge" on Γ_2, Γ_3.
//  EDGEX1[nkx1] is an input parameter; the numbers of all
//        elements with a edge on Γ_1.
//  EDGEX2[nkx1] is an input parameter; the numbers of all
//        elements with a edge on Γ_3.
//  EDGEY1[nky1] is an input parameter; the numbers of all
//        elements with a edge on Γ_4.
//  EDGEY2[nkx1] is an input parameter; the numbers of all
//        elements with a edge on Γ_2.
//
// Assemble on Γ_1:
  for(i=1; i<=nkx1; i++)
     {ℓ=EDGEX1[i];
      for(i1=1; i1<=4; i1++)
        { ir = skrx[i1];
          is = sksx[i1];
```

```
            temp1 = bas[0][ir]*bas[0][is]/3.
      +(bas[0][ir]*bas[2][is]+bas[2][ir]*bas[0][is])*0.5;
      +bas[2][ir]*bas[2][is];
            temp1 = temp1*h1;
            k = T[ir][ℓ];
            d[k] = d[k] + temp1;
          }
        }
// Assemble on Γ4:
    for(i=1; i<=nky1; i++)
      {ℓ=EDGEY1[i];
       for(i1=1; i1<=4; i1++)
         { ir = skry[i1];
           is = sksy[i1];
           temp1 = bas[1][ir]*bas[1][is]/3.
      +(bas[1][ir]*bas[2][is]+bas[2][ir]*bas[1][is])*0.5;
      +bas[2][ir]*bas[2][is];
           temp1 = temp1*h2;
           k = T[ir][ℓ];
           d[k] = d[k] + temp1;
         }
       }
// Assemble on Γ3:
    for(i=1; i<=nkx1; i++)
      {ℓ=EDGEX2[i];
       for(i1=1; i1<=4; i1++)
         { ir = skr23[i1];
           is = sks23[i1];
           temp1 = bas[0][ir]*bas[0][is]/3.
      +(bas[0][ir]*bas[2][is]+bas[2][ir]*bas[0][is])*0.5;
      +bas[2][ir]*bas[2][is];
           temp1 = temp1*h1;
```

```
            k = T[ir][ℓ];
            d[k] = d[k] + temp1;
          }
        }
// Assemble on Γ2:
    for(i=1; i<=nky1; i++)
      {ℓ=EDGEY2[i];
       for(i1=1; i1<=4; i1++)
         { ir = skr23[i1];
           is = sks23[i1];
           temp1 = bas[1][ir]*bas[1][is]/3.
   +(bas[1][ir]*bas[2][is]+bas[2][ir]*bas[1][is])*0.5;
   +bas[2][ir]*bas[2][is];
           temp1 = temp1*h2;
           k = T[ir][ℓ];
           d[k] = d[k] + temp1;
         }
      }
  }
```

The following arrays shown the matrix A and the vector d after assembly:

$$A = \begin{pmatrix} 2a_{11} & a_{12} & 0 & a_{14} & 2a_{15} & 0 & 0 & 0 & 0 \\ a_{21} & 3a_{22} & a_{23} & 0 & 2a_{25} & 2a_{26} & 0 & 0 & 0 \\ 0 & a_{32} & a_{33} & 0 & 0 & a_{36} & 0 & 0 & 0 \\ a_{41} & 0 & 0 & 3a_{44} & 2a_{45} & 0 & a_{47} & 2a_{48} & 0 \\ 2a_{51} & 2a_{52} & 0 & 2a_{54} & 6a_{55} & 2a_{56} & 0 & 2a_{58} & 2a_{59} \\ 0 & 2a_{62} & a_{63} & 0 & 2a_{65} & 3a_{66} & 0 & 0 & a_{69} \\ 0 & 0 & 0 & a_{74} & 0 & 0 & a_{77} & a_{78} & 0 \\ 0 & 0 & 0 & 2a_{84} & 2a_{85} & 0 & a_{87} & 3a_{88} & a_{89} \\ 0 & 0 & 0 & 0 & 2a_{95} & a_{96} & 0 & a_{98} & 2a_{99} \end{pmatrix}$$

$$d = \begin{pmatrix} (2b_{11} + b_{12} + b_{14})g_1 \\ (b_{21} + 2b_{22} + b_{23})g_2 \\ (b_{32} + 2b_{33} + b_{36})g_3 \\ (b_{41} + 2b_{44} + b_{47})g_4 \\ 0 \\ (b_{63} + 2b_{66} + b_{69})g_6 \\ (b_{74} + 2b_{77} + b_{78})g_7 \\ (b_{87} + 2b_{88} + b_{89})g_8 \\ (b_{96} + b_{98} + 2b_{99})g_9 \end{pmatrix}.$$

REFERENCES

[1] V. Barbu, Optimal Control of Variational Inequalities. Research Notes in Mathematics 100, Pitman, London, Boston, Melbourne, 1984.

[2] V. Barbu, and A.N. Barron, "Bang-bang controllers for an optimal cooling problems," *Control & Cybernetics,* vol. 16, pp. 91-102, 1987.

[3] J.L. Lions, Control of distributed singular systems, Gauthier- Villars, Paris, 1985.

[4] C. Saguez, Contrôle optimal de systèmes à frontiére libre. Thése d'Etat de l'Université de Technologie de Compiégne, 1980.

[5] V. Arnăutu, and P. Neittaanmäki, Optimal Control from Theory to Computer Programs. Kluwer Academic Publishers, Dordrecht, Boston, London, 2003.

[6] D. Tiba, "Optimality conditions for distributed control problems with nonlinear state equations", *SIAM J. Control and Optimiz.*, vol. 23, pp. 85-110, 1985.

[7] G. Caginalp, "An analysis of a phase field model of a free boundary," *Arch. Rat. Mech. Anal.*, vol. 92, pp. 205-245, 1986.

[8] V. Arnăutu, and C. Moroşanu, "Numerical approximation for the phase-field transition system," *Intern. J. Com. Math.,* vol. 62, no. 3-4, pp. 209-221, 1996.

[9] T. Benincasa, and C. Moroşanu, "Fractional steps scheme to approximate the phase-field transition system with non-homogeneous Cauchy-Neumann boundary conditions," *Numer. Funct. Anal. & Optimiz.,* vol. 30, no. 3-4, pp. 199-213, 2009.

[10] T. Benincasa, and C. Moroşanu, "A product formula approach to an inverse problem governed by nonlinear phase-field transition system. Case 1D," *ROMAI J.*, vol. 6, no. 2, pp. 15-26, 2010.

[11] T. Benincasa, and C. Moroşanu, "A product formula approach to an inverse problem governed by nonlinear phase-field transition system. Numerical implementation in 1D case," In XVIIIth Conference on Applied and Industrial Mathematics - CAIM2010, Oct. 14-17, 2010, Iaşi, Romania, pp. 14, 2010.

[12] T. Benincasa, A. Favini, and C. Moroşanu, "A Product Formula Approach to a Non-homogeneous Boundary Optimal Control

Costică Moroşanu

Problem Governed by Nonlinear Phase-field Transition System. PART I: A Phase-field Model," *J. Optim. Theory and Appl.*, vol. 148, no. 1, pp. 14-30, 2011.

[13] T. Benincasa, A. Favini, and C. Moroşanu, "A Product Formula Approach to a Non-homogeneous Boundary Optimal Control Problem Governed by Nonlinear Phase-field Transition System. PART II: Lie-Trotter Product Formula," *J. Optim. Theory and Appl.*, vol. 148, no. 1, pp. 31-45, 2011.

[14] C. Moroşanu, "Numerical approach of an inverse problem in the phase field equations", *An. Şt. Univ. "Al.I.Cuza" Iaşi*, T XXXIX, s. I-a, f.4, pp.419-436, 1993.

[15] C. Moroşanu, "Approximation and numerical results for phase field system by a fractional step scheme", *Revue d'Analyse Numérique et de Théorie de l'Approximation,* Tome 25, no. 1-2, pp. 137-151, 1996.

[16] C. Moroşanu, "Approximation of the phase-field transition system via fractional steps method", *Numer. Funct. Anal. & Optimiz.*, vol. 18(5& 6), pp. 623-648, 1997.

[17] C. Moroşanu, Analiza şi controlul sistemului tranziţiei de fază. Metode de tip paşi fracţionari, Editura Univ. "Al.I. Cuza" Iaşi, ISBN 973-8243-18-1 (in romanian), 2001.

[18] C. Moroşanu, "On the numerical stability of the cubic splines approximation to solution of phase-field transition system", *PanAmerican Math. J.*, vol. 12, no. 2, pp. 31-46, 2002.

[19] C. Moroşanu, "Boundary optimal control problem for the phase-field transition system using fractional steps method", *Control & Cybernetics*, vol. 32, no. 1, pp. 05-32, 2003.

[20] C. Moroşanu, "Analysis and optimal control of phase-field transition system", *Non. Funct. Anal. & Appl.*, vol. 8, no. 3, pp. 433-460, 2003.

[21] C. Moroşanu, and V. Melnig, "Optimal Control Model via the Phase Field System", 2nd *International Conference on "Biomaterials & Medical Devices"*, BiomMedD'2006, Conference Proceeding, Iasi, Romania, pp. 214-214, 2006.

[22] C. Moroşanu, "Fractional steps method for approximation the solid region via phase field transition system", 6th *International Conference APLIMAT 2007*, PART II, Bratislava, Slovak Republic, February 06-09, pp. 235-240, 2007.

[23] C. Moroşanu, "Approximation of solid region in the continuous casting process of steel via phase field transition system", XVth *Conference on Applied and Industrial Mathematics*, CAIM2007, Oct. 12-14, Mioveni, Romania, pp. 42-43, 2007.

[24] C. Moroşanu, "Approximation of the solid region in the continuous casting process of steel via phase-field transition system", 6th *European Conference of Continuous Casting 2008*, Riccione, Italy, 3-6 June, 2008.

[25] C. Moroşanu, Gh. Iorga, and S. C. Cocindău, "Numerical simulation of the solid region via phase field transition system", *Metalurgia International*, vol. XIII, no. 12, pp. 91-95, 2008.

[26] C. Moroşanu, "An inverse problem in the phase field transition system. The 2D case", 11th *International Conference on Aluminium Alloys -Simulation and modeling*, Aachen, Germany, 22-26 september, 2008.

[27] C. Moroşanu, I. Crudu, Gh. Iorga, and I. Ivan, "Research Concerning the Evolution of Solidification Front via Phase-Field Transition System", CEx05-D11-Prog.no. **84**/28.10.2005, IFA Bucharest, 2008.

[28] C. Moroşanu, Gh. Iorga, and I. Tofan, "Numerical simulation of the thickness accretions in the secondary cooling zone of a continuous casting machine", *Metalurgia International*, vol. XIV, no. 1, pp. 72-75, 2009.

[29] C. Moroşanu, "An inverse problem in the phase-field transition system. The 2D case", *ROMAI J.*, vol. 6, no. 1, pp. 135-143, 2010.

[30] C. Moroşanu, and Ana-Maria Moşneagu, "On the numerical approximation of the phase-field system with non-homogeneous Cauchy-Neumann boundary conditions. Case 1D", (to appear).

[31] C. Moroşanu, D. Motreanu, "An extension of the nonlinear parabolic equation in Caginalp's model, *Comm. in Appl. Anal.*, **2**, no. 2, pp. 159-168, 1998.

[32] C. Moroşanu, and D. Motreanu, "A generalized phase field system", *J. Math. Anal. Appl.*, vol. 237, pp. 515-540, 1999.

[33] C. Moroşanu, and D. Motreanu, "Uniqueness and Approximation for the Nonlinear Parabolic Equation in Caginalp's Model", *Intern. J. of Appl. Math.*, vol. 2, no. 1, pp. 113-129, 2000.

[34] C. Moroşanu, and D. Motreanu, "The phase field system with a general nonlinearity", *Intern. J. of Diff. Equs. and Appl.*, vol. 1, no. 2, pp. 187-204, 2000.

[35] C. Moroşanu, and D. Motreanu, "An extension of the Lie-Trotter product formula", *Nonl. Funct. Anal. & Appl.*, vol. 7, no. 4, pp. 517-530, 2002.

[36] C. Moroşanu, and C. Trenchea, "Identification of nonlinear periodic wave equation", *Appl. Math. Optim.*, vol. 44, pp. 87-104, 2001.

[37] C. Moroşanu, and G. Wang, "State constraint optimal control for the phase field transition system", *Numer. Funct. Anal. & Optimiz.*, vol. 28(3-4), pp. 379-403, 2007.

[38] V. Barbu, and M. Iannelli, "Approximating some non-linear equations by a Fractional step scheme," *Diff. and Integral Eqs.*, vol. 1, pp. 15-26, 1993.

[39] V. Barbu, Nonlinear Semigroups and Differential Equations in Banach Spaces. Nordhoff, Leyden, 1976.

[40] H. Brézis, Operateurs maximaux monotones et semigroupes de contractions dans les espaces de Hilbert, Math. Studies, no. 5, North-Holland, 1973.

[41] Z. Chen, and K.-H. Hoffmann, "Numerical solutions of the optimal control problem governed by a phase field model," *Optim. Control of Partial Diff. Eqs.,* W. Desch, F. Kappel, and K. Kunisch, eds., Basel, Boston, Berlin, Birkhauser Verlag, 1991.

[42] M. Heinkenschloss, and E.W. Sachs, "Numerical solution of a constrained control problem for a phase field model", In: *Optim. Control of Partial Diff. Eqs.,* W. Desch, F. Kappel, and K. Kunisch, eds., Int. Ser. of Numer. Math., vol. 118, Basel, Boston, Berlin, Birkhauser Verlag, pp. 171-187, 1994.

[43] M. Heinkenschloss, and F. Tröltzsch, "Analysis of the Lagrange-SQP-Newton Method for the Control of a Phase Field Equation", Tech. Report 95-03-01, Interdisciplinary Center for Applied Mathematics, Virginia Polytechnic Institute and State University, Blacksburg, VA 24061, 1995.

[44] K.-H. Hoffmann, and L. Jiang, "Optimal control problem of a phase field model for solidification", *Numer. Funct. Anal. & Optimiz.*, vol. 13, pp. 11-27, 1992.

[45] A. Friedman, "Optimal control for parabolic variational inequalities", *SIAM J. Control and Optimiz.*, vol. 27, 1987.

[46] V. Barbu, "A product formula approach to nonlinear optimal control problems," *SIAM J. Control and Optimiz.,* vol. 26, pp. 496-520, 1988.

[47] V. Arnăutu, "On approximation of the inverse one - phase Stefan problem," *Intern. Ser. of Numer. Math.,* vol. 99, Birkhäuser Verlag Basel, pp. 69-81, 1991.

[48] S.J. Ruuth, "Implicit-explicit methods for reaction-diffusion problems in pattern formation", *J. Math. Biol.*, vol. 34, pp. 148-176, 1995.

[49] O.A. Ladyzhenskaya, B.A. Solonnikov, and N.N., Uraltzava, Linear and quasi-linear equations of parabolic type, Prov. Amer. Math. Soc., 1968.

[50] V. Barbu, and T. Precupanu, Convexity and Optimization in Banach Spaces. 2rd ed., Editura Academiei Bucureşti and D. Reidel Publ. Co., Dordrecht, Boston, Lancester, 1986.

[51] H. Brézis, and A. Pazy, "Semigroups of nonlinear contractions on convex sets," *J. Funct. Anal.*, vol. 6, pp. 367-383, 1970.

[52] D. Motreanu, and N. Pavel, Tangency, flow invariance for differential equations, and optimization problems, Marcel Dekker, Inc., New York, Basel, 1999.

[53] H. Brézis, Analyse Fonctionnelle. Théorie et Applications, Masson, Paris, 1983.

[54] H. Brézis, and A. Pazy, "Convergence and approximation of semigroups of nonlinear operators in Banach spaces," *J. Funct. Anal.*, vol. 9, pp. 63-74, 1972.

[55] A. Pazy, "The Lyapunov method for semigroups of nonlinear contractions in Banach spaces", *J. Anal. Math.*, vol. 40, pp. 239-262, 1982.

[56] Y. Kobayashi, "Product formula for nonlinear semigroups in Hilbert spaces", *Proc. Japan Acad.*, vol. 58, Ser. A, pp. 425-428, 1982.

[57] O. Axelson, and V. Barker, Finite element solution of boundary value problems. Academic Press, 1984.

[58] P.G. Ciarlet, The Finite Element Method for Elliptic Problems, North - Holland, 1978.

[59] P.M. Prenter, Splines and variational methods, J. Wiley, New York, London, 1979.

[60] L.I. Rubinstein, The Stefan problem, Transl. Math. Monographs, vol. 27, American Mathematical Society, Providence, Rhode Island, 1971.

[61] O.A. Oleinik, "A method of solution of the general Stefan problem", *Sov. Math. Dokl.*, vol. 1, pp. 1350-1354, 1960.

[62] G. Caginalp, and X. Chen, "Phase field equations in the singular limit of sharp interface problems," In: "IMA," vol. 43, pp. 1-28, 1990-91.

[63] G. Caginalp, X. Chen, and C. Eck, "A rapidly converging phase field model," *Discrete Contin. Dyn. Syst.*, vol. 15, pp. 1017-1034, 2006.

[64] G. Caginalp, X. Chen, and C. Eck, "Numerical tests of a phase field model with second order accuracy," *SIAM J. Appl. Math.*, vol. 68 (6), pp. 1518-1534, 2008.

[65] G. Caginalp, and J.T. Lin, "A numerical analysis of an anisotropic phase field model," *IMA J. Appl. Math.*, vol. 39, pp. 51-66, 1987.

[66] J.W. Cahn, and J.E. Hilliard, "Free energy of a nonuniform system I. Interfacial free energy," *J. Chem. Phys.*, vol. 28, pp. 258-267, 1957.

[67] I. Fonseca, and W. Gangbo, Degree Theory in Analysis and Applications, Clarendon, Oxford, 1995.

[68] P.W. Bates, and Zheng Songmu, "Inertial manifolds and inertial sets for the phase-field equations," *J. Dyn. Diff. Eqs.*, vol. 4, pp. 375-398, 1992.

[69] O. Penrose, and P. C. Fife, "Thermodynamically consistent models of phase-field type for kinetics of phase transitions", *Phys. D.*, vol. 43, pp. 44-62, 1990.

[70] P. Colli, F. Luterotti, G. Schimperna, and U. Stefanelli, "Global existence for a class of generalized systems for irreversible phase changes", *NoDEA Nonl. Diff. Eqs. Appl.*, vol. 9 (3), pp. 255-276, 2002.

[71] A. Miranvile, "Generalized Chan-Hilliard equations based on a microforce balance", *J. Appl. Math.*, vol. 4, pp. 165-185, 2003.

[72] U. Stefanelli, "Analysis of a variable time-step discretization for the Penrose-Fife phase relaxation problem", *Nonl. Anal.*, vol. 45(2), pp. 213-240, 2001.

[73] Liu Xiyuan, "Determination of a parameter $h(t)$ for a phase field model", *J. Partial Diff. Eqs.*, vol. 10, pp. 136-148, 1997.

[74] S. Agmon, A. Douglis, and L. Nirenberg, "Estimates near the boundary for solutions of elliptic partial differential equations satisfying general boundary conditions," *Comm. Pure Appl. Math.*, vol. 12, pp. 623-727, 1959.

[75] V. Barbu, Analysis and Control of Nonlinear Infinite Dimensional Systems. Academic Press, vol. 190, 1993.

[76] V. Barbu, Partial Differential Equations and Boundary Value Problems. Kluwer Academic Publishers, Dordrecht, Boston, London, vol. 441, 1998.

[77] V. Barbu, "Local controllability of the phase field system," *Nonl. Anal. Ser. A: Theory Meth.*, vol. 50, no. 3, pp. 363-372, 2002.

[78] V. Barbu, and G. Wang, "State Constrained Optimal Control Problems Governed by Semilinear Equations," *Numer. Funct. Anal. & Optim.*, vol. 1, pp. 368-385, 2000.

[79] G. Wang, "Optimal control of parabolic differential equations with two point boundary state constraints", *SIAM J. Control and Optimiz.*, vol. 38, pp. 1639-1654, 2000.

[80] P. Aubin, and A. Cellina, *Differential Inclusions.* Springer-Verlag, Berlin, New-York, Heidelberg, 1984.

[81] X. Li, and J. Yong, "Optimal Control Theory for Infinite-Dimensional Systems", In: *Systems Control Found. Appl.*, Birkhäuser Boston, Boston, 1995.

[82] P.C. Fife, and G.S. Gill, "A phase-field description of mushy zones", *Phys. D*, vol. 35, pp. 267-275, 1989.

[83] G. Fix, "Numerical simulation of free boundary problems using phase field models," In: *The Mathematics of Finite Elements and Applications IV*, MAFELAP 1981, J.R. Whiteman, ed., Academic Press, London, New York, pp. 265-279, 1982.

[84] G.D. Smith, Numerical Solution of Partial Differential Equations: Finite Difference Methods, 3rd Edition, Clarendon Press, Oxford, 1985.

[85] V. Barbu, Mathematical Methods in Optimization of Differential Systems. Kluwer Academic Publishers, Dordrecht, Boston, London, vol. 310, 1994.

[86] D. Pascali, and S. Sburlan, Nonlinear Mappings of Monotone type, Sijthoff & Nordhoff, Leiden, 1978.

[87] J.T. Schwartz, Nonlinear functional analysis, Gordon and Breach eds., New York, 1969.

[88] R.A. Adams, Sobolev spaces. Academic Press, Orlando, San Diego, New-York, 1975.

[89] P.W. Bates, and P. C. Fife, "Spectral comparison principles for the Cahn-Hilliard and phase-field equations, and time scales for coarsening," *Phys. D.*, vol. 43, pp. 335-348, 1990.

[90] V. Berti, M. Fabrizio, and C. Giorgi, "A three-dimensional phase transition model in ferromagnetism: Existence and uniqueness," *J. Math. Anal. Appl.*, vol. 355, pp. 661-674, 2009.

[91] J.F. Blowey, and C.M. Elliott, "A phase-field model with a double obstacle potential," In: *Motion by mean curvature and related topics*, Trento - 1992, 1-22, de Gruyter, Berlin, 1994.

[92] J.L. Boldrini, B.M.C. Caretta, and E.Fernández-Cara, "Analysis of a two-phase field model for the solidification of an alloy," *J. Math. Anal. Appl., Vol.*, vol. 357, pp. 25-44, 2009.

[93] G. Caginalp, and Y. Nishiura, "The existence of traveling waves for phases field equations and convergence to sharp interface models in the singular limit," *Quarterly Appl. Math.*, vol. XLIX, 1, pp. 147-162, 1991.

[94] J.W. Cahn, "Theory of crystal growth and interface motion in crystalline materials," *Acta Metallurgica*, vol. 8, 554-562, 1960.

[95] J. Céa, Lectures on Optimization. Theory and Algorithms, Tata Institute of Fundamental Research, Springer-Verlag, Bombay,1978.

[96] B. Démidovitch, and I. Maron, Élements de calcul numérique, Editions Mir, Moscou, 1973.

[97] C.M. Elliott, and Zheng Songmu, "Global existence and stability of solutions to the phase field equations," In: *Free Boundary Problems*, K-H. Hoffman and J.Sprekels, eds., Int. Ser. of Numer. Math. vol. 95, Birkhäuser Verlag, Basel 1990.

[98] K. Glashoff, and E. Sachs, "On theoretical and numerical aspects of the bang-bang principle," *Number. Math.*, vol. 29, pp. 93-113, 1977.

[99] R. Glowinski, Numerical Methods for Non-Linear Variational Problems, Tata Institute of Fundamental Research, Springer-Verlag, Bombay, 1980.

[100] W.A. Gruver, and E Sachs, Algorithmic Methods in Optimal Control, Research Notes in Mathematics, vol. 47, Pitman, London, 1981.

[101] K.-H. Hoffman, and M. Niezgodka, Control of parabolic systems involving free boundaries. Free boundary problems, theory and applications, vol. II, Fasanc and Primicerio eds. Pitman, pp. 431-462, 1983.

[102] N. Kenmochi, and M. Niezgódka, "Evolution systems of nonlinear variational inequalities arising from phase change problems", *Non. Anal.*, vol. 22, pp. 1163-1180, 1994.

[103] J.L. Lions, and E. Magenes, Nonhomogeneous Boundary Value Problems and Applications, Springer-Verlag, New York, 1972.

[104] P. Neittaanmäki, and D. Tiba, Optimal Control of Nonlinear Parabolic Systems. Theory, Algorithms, and Applications, Marcel Dekker, Inc., New York, Basel, Hong Kong, 1994.

[105] J.T. Oden, and J.N. Reddy, An Introduction to the Mathematical Theory of Finite Elements, Wiley, New York, 1976.

[106] N.H. Pavel, Differential Equations, flow invariance and applications, Research Notes in Mathematics, 113, Pitman Advanced Publishing Program, 1984.

[107] E. Polak, Computational Methods in Optimization. A Unified Approach, Academic Press, New York, 1983.

[108] C. Popa, "Trotter product formulae for Hamilton-Jacobi equations in infinite dimensions", *Diff. and Integral Eqs.*, vol. 4, pp. 1251-1268, 1991.

[109] C. Popa, "Feedback laws for nonlinear distributed control problems via Trotter - type product formulae", *SIAM J. Control and Optimiz.*, vol. 33, 1995.

[110] J. Sprekels, and S. Zheng, "Global existence and asymptotic behaviour for a nonlocal phase-field model for non-isothermal phase transitions", *J. Math. Anal. Appl.*, vol. 279(1), pp. 97-110, 2003.

[111] B. Stinner, "Weak solutions to a multi-phase field system of parabolic equations related to alloy solidification", *Adv. Math. Sci. Appl.*, vol. 17(2), pp. 589-638, 2007.

[112] A. Visintin, Models of phase transitions, Preprint of Dipartimento di Matematica dell'Università di Trento, 1996.

[113] I. Vrabie, Compactness Methods for Nonlinear Evolutions, Longman Scientific and Technical, London 1987.

[114] K. Yosida, Functional Analysis, Springer Verlag, Berlin, New York, 1977.

[115] S. Zheng, "Global existence for a thermodynamically consistent model of phase field type", *Diff. Integral Eqs.*, vol. 5, pp. 241-253, 1992.

332 *Analysis and Optimal Control of Phase-Field Transition System: Fractional Steps Methods*, 2012, 332-336

Index

A

B

C

D

Costică Moroşanu

E

F

G

H

I

J

L

M

N

O

P

R

S

T

V

W

Y

Made in the USA
Monee, IL
03 May 2026

49395996R00212